NITROALKENES

NITROALKENES
CONJUGATED
NITRO COMPOUNDS

V. V. Perekalin
Russian State Pedagogical University, St Petersburg, Russia

E. S. Lipina
Russian State Pedagogical University, St Petersburg, Russia

V. M. Berestovitskaya
Russian State Pedagogical University, St Petersburg, Russia

D. A. Efremov
Staffordshire University, Stoke-on-Trent, UK
St Petersburg Institute of Cinema and Television, St Petersburg, Russia

JOHN WILEY & SONS
Chichester · New York · Brisbane · Toronto · Singapore

CHEM

Other Wiley Editorial Offices

John Wiley & Sons, Inc., 605 Third Avenue,
New York, NY 10158-0012, USA

Jacaranda Wiley Ltd, 33 Park Road, Milton,
Queensland 4064, Australia

John Wiley & Sons (Canada) Ltd, 22 Worcester Road,
Rexdale, Ontario M9W 1L1, Canada

John Wiley & Sons (SEA) Pte Ltd, 37 Jalan Pemimpin #05-04,
Block B, Union Industrial Building, Singapore 2057

Library of Congress Cataloging-in-Publication Data

Nitroalkenes : conjugated nitrocompounds / V. V. Perekalin ...
[*et al.*].
        p.   cm.
    Includes bibliographical references and index.
    ISBN 0 471 94318 5
    1. Nitroalkenes. I. Perekalin, V. V. (Vsevolod Vasil'evich)
QD305.N8N59   1994
574'.041—dc20   93-39727
                                                     CIP

British Library Cataloguing in Publication Data

A catalogue record for this book is available from the British Library

ISBN 0 471 94318 5

Typeset in 10/12 pt Times by Thomson Press (India) Ltd., New Delhi
Printed and bound in Great Britain by Biddles Ltd, Guildford, Surrey

CONTENTS

PREFACE

The present monograph *Nitroalkenes* describes many synthetic methods and chemical transformations for conjugated nitroalkenes.

Owing to the limited volume of this monograph, it was not possible to discuss fully the mechanisms of nitroalkene reactions, the problems of nitroalkene stereochemistry or the transformations through alkylation, arylation, rearrangement and photochemical reactions. For the same reason it was impossible to detail many physicochemical methods of structural investigation or quantum-chemical calculations.

Unsaturated nitrosugars and fluoroorganic nitro compounds are thoroughly discussed in *Nitro Compounds: Recent Advances in Synthesis and Chemistry*, edited by H. Feuer and A. T. Nielsen (VCH, New York, 1990), and therefore these items are not included here. Similarly, 1,3-dipolar cycloadditions are not covered in detail here as they are reviewed by A. Baranskii and V. I. Kelarev in *Chemistry of Heterocyclic Compounds* (1990).

The relevant literature up to 1993 has been reviewed, drawing particular attention to the publications of the last decade.

The authors would appreciate very much any comments or remarks.

ACKNOWLEDGEMENTS

It is a great pleasure to thank our colleagues whose help was so valuable in preparing this manuscript, namely N. I. Aboskalova, I. E. Efremova, E. V. Trukhin and T. Yu. Nikolaenko of the Russian State Pedagogical University, St Petersburg. We also extend our appreciation to Professor B. I. Ionin of St Petersburg Technological University for his very useful advice in preparing the final version of the text.

INTRODUCTION

Chemical transformations of conjugated nitroalkenes that involve molecular rearrangement provide invaluable synthetic approaches to numerous new groups of heterocyclic, carbocyclic and elementoorganic compounds.

The specific structure of the unsaturated nitro compound restricts the regiodirected and stereodirected processes where there is a fixed orientation of substituents. The significance of such substances is emphasized by their application as convenient synthones for drugs, pesticides and synthetic block construction, i.e. as the fragments of complicated natural compounds (antibiotics, alkaloids, pyrethroids, prostaglandins and others).

It would not be an overestimation to say that unsaturated nitro compounds are very fruitful starting points for all main classes of organic compound.

1 SYNTHESIS OF UNSATURATED NITRO COMPOUNDS

1.1 MONONITROALKENES

1.1.1 NITRATION OF UNSATURATED HYDROCARBONS

The very first investigation of the nitration of unsaturated hydrocarbons by HNO_3 (specific gravity 1.52) dates back to the middle of the Nineteenth Century [1]. In 1893 Konovalov [2] proved the formation of nitro compounds in the reaction of alkanes with dilute HNO_3. Nitration of saturated hydrocarbons in the gaseous phase was of great significance [3]. The reactions of alkenes and functionally substituted alkenes with concentrated HNO_3 result in the formation of conjugated nitroalkenes [4–9]. Sometimes such processes involve decomposition. The study of alkenecarbonic acid nitration exposed a certain correlation between the nature and the yield of the major products, namely the relationship between HNO_3 concentration and the amounts of α-nitrocarboxylic esters (1) and furoxanecarboxylic esters (2) formed (equation 1). The maximum yield (53.5%) of (1) is achieved with 80% HNO_3 and a 31–66% yield of (2) is obtained for 99.5% HNO_3 [5, 6]. Nitration by 75% HNO_3 leads to a mixture of (1) and (2).

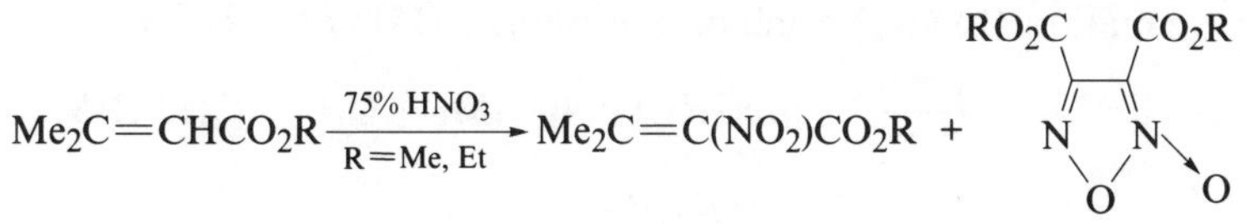

$$Me_2C{=}CHCO_2R \xrightarrow[R=Me, Et]{75\% HNO_3} Me_2C{=}C(NO_2)CO_2R \ + \ \underset{(2)}{\overset{RO_2C\quad CO_2R}{\underset{O\qquad O}{N\diagup\diagdown N}}} \qquad (1)$$

$$\underset{(1)}{}$$

Treatment of alkenecarboxy esters (3) with fuming HNO_3 (equation 2) leads to unsaturated α-nitroalkenecarboxylic esters (4) in 32–39% yield and α-hydroxy-

β-nitroalkanecarboxylic esters (5) in 18–39% yield [7–9].

$$R^1CH{=}CHCO_2R^2 \xrightarrow[<5\,^\circ C]{HNO_3}$$

(3)

$$R^1CH{=}C(NO_2)CO_2R^2 + R^1CH(NO_2)CH(OH)CO_2R^2 \qquad (2)$$

(4) 32–39%   (5) 18–39%

$R^1 =$ Me, Et, Pr; $R^2 =$ Me, Et

2-Nitroalkenecarboxylic esters are produced under the action of a nitrating mixture [9, 10].

Interesting results have been obtained in the reaction of fuming HNO_3 with unsaturated fatty acid esters [11]. The treatment of ester (6) with HNO_3 in the presence of $NaNO_2$ and glacial AcOH at room temperature (equation 3) gave two nitroalkenes (8) and (9) in small quantity and the major product was methyl 11-hydroxy-10-nitroundecanoate (7).

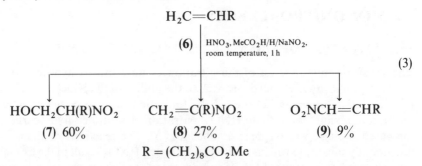

$$H_2C{=}CHR$$

(6) | HNO_3, $MeCO_2H/H/NaNO_2$,
      room temperature, 1 h

(3)

HOCH$_2$CH(R)NO$_2$ CH$_2{=}$C(R)NO$_2$ O$_2$NCH${=}$CHR

(7) 60%   (8) 27%   (9) 9%

$$R = (CH_2)_8CO_2Me$$

β-Nitrostyrene and its homologues can be synthesized by the direct nitration of styrenes [12–15]. 2-(4-Nitrophenyl)-1-nitroethene was obtained by introducing p-nitrostyrene into a mixture of 98% HNO_3 and concentrated H_2SO_4 at 0 °C [16].

1,1-Disubstituted alkenes can be transformed to the corresponding nitroalkenes in low yield under the action of alkyl nitrites. The other products of the process are vicinal nitroalcohols and dinitro compounds [17].

The reaction of 2-butene and acetyl nitrate (equations 4a, 4b) leads to 3-nitro-2-butyl acetate (10) and 3-nitro-1-butene 11. The treatment of (10) with Et_3N in Et_2O (equation 4b) gives a mixture of isomers (11) and (12) [18].

$$MeCH{=}CHMe \xrightarrow[-20\,^\circ C\ to\ 11\,^\circ C]{AcONO_2} MeCHCH(NO_2)Me + CH_2{=}CHCH(Me)NO_2$$

$$\underset{\displaystyle OCOMe}{|}$$

(10) 38%   (11) 27%   (4a)

$$(10) \xrightarrow[Et_2O]{Et_3N} MeCH{=}C(Me_3)NO_2 + (11) \qquad (4b)$$

(12) 84%

(E)-Stilbene is transformed into *threo*-1-acetoxy-1,2-diphenyl-2-nitroethane (70% yield), and the latter gives (Z)-nitrostilbene in quantitative yield under the action of Et_3N in Et_2O [19]. The nitration of other stilbenes and styrenes also leads to β-nitroacetates. The (Z) isomers react slower than the (E) isomers. The bidentate structure of cerium (IV) ammonium nitrate (CAN) promotes the nitration of methoxy-substituted cinnamic acids [20]. Along with the products of aromatic ring nitration are obtained the products of uncommon *ipso* subtitution of the CO_2H group, i.e. nitrostyrenes (14). The best results were achieved for (E)-3,4-methoxycinnamic acid and 2,4-dimethoxycinnamic acids. The nitration of monomethoxycinnamic acid three leads to the formation of the products of substitutional nitration of 14 in 3–39% yield along with nitrocinnamic acids 15 isolated in the form of methyl esters (20–68%) and products of oxidation (equation 5). CAN nitrates aliphatic alkenes to give α-nitroalkenes and β-nitroalcohols as the prevailing products, the ratio of which depends upon the nature of the solvent used and reaction conditions [20].

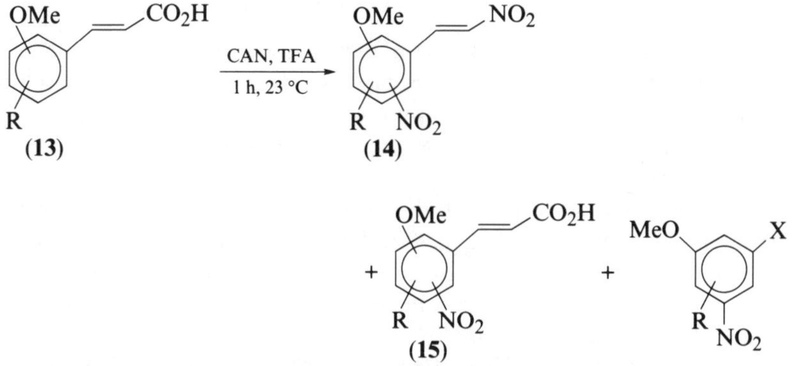

$$R = H, OMe; \quad X = CHO, CO_2H$$
$$TFA = \text{trifluoroacetic acid}$$

Direct nitration is observed for unsaturated heterocycles of the furan and thiophene series [21–23]. A mixture of fuming HNO_3 and H_2SO_4 (equation 6) transforms (16) to (17) at 5 °C [21].

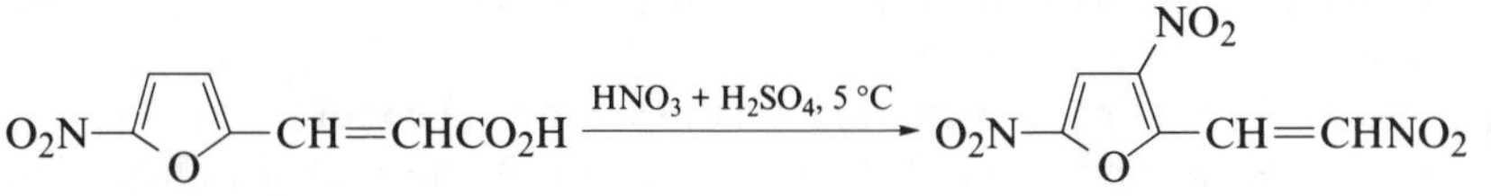

β-Nitroarylethenes can be prepared by the reaction of tetranitromethane (TNM) with the corresponding conjugated arylethenes in high yield [24–27]. For example, isosafrol (18) is isolated in 72.5% yield (equation 7).

$$\text{(18)} \tag{7}$$

The interaction between TNM and π,π-electron and p,π-electron systems such as arylalkenes and alkadienes is characterized by various pathways and results in the formation of tetranitroalkanes, dinitroisoxazolidines, nitroketones and nitroalkenes [26].

The substitutional nitration of tin(IV) organic compounds by TNM [28] is illustrated (equation 8) by the following transformation of (19) to (20).

$$\text{(19)} \qquad \xrightarrow[\substack{\text{DMSO} \\ 18\,^\circ\text{C, 2.5 h}}]{C(NO_2)_4} \qquad \text{(20)}\ 87\% \tag{8}$$

Nitration by dinitrogen tetroxide is a widely used synthetic approach to nitroalkenes. The addition of this nitrating agent to alkenes gives rise to a mixture of nitrated products, such as dinitro compounds and nitronitrites (nitronitrates), which under the action of base undergo elimination of the nitrous acid elements to give nitroalkenes.

Levi *et al.* [29] have extensively studied this type of reaction. Subsequently, large numbers of alkenes, arylalkenes and cyclenes have been investigated [30–33], making it possible to improve the synthetic methods for the resulting products without isolation of the intermediates. For example, 1-octadiene and cyclooctene were transformed directly into 1-nitro-1-octadecene and 1-nitro-cyclooctene in 77% and 95% yields, respectively. The process included addition of the alkenes to an ether solution of N_2O_4 saturated with oxygen and subsequent treatment of the reaction mixture with triethylamine (TEA) [30, 31].

Nitration of styrenes and stilbenes leads to nitroalkenes (22) in low yield, nitroalcohols and various products of oxidation. This effect can be explained by the radical addition of two $\cdot NO_2$ molecules and intermediate formation of the nitronitrite [34]. Further evidence for the radical pathway of the process was obtained on the basis of Hammett constants [34] for a number of *para*-substituted styrenes (equation 9).

The reaction with nitrogen oxides under experimental conditions is a radical process and its starting point is the formation of the nitroalkyl radical. This has been proven by electron spin resonance (ESR) experiments for a series of 1-alkenes [35].

The ^{15}N chemical nuclear polarization method was applied to the N_2O_4 nitration products of 2-ethyl-1-butene, i.e. the corresponding dinitro and nitro-nitrite derivatives, in an attempt to define the process pathway (equation 10).

$$PhCH{=}CHR \xrightarrow[\substack{CCl_4,\ 25\,°C \\ 10\,000\ ppm}]{·NO_2} \underset{\substack{ONO\quad NO_2 \\ (21)}}{PhCH{-}CHR} \xrightarrow[\substack{-HNO_2 \\ R = H}]{} PhCH{=}C\overset{NO_2}{\underset{R}{\diagdown}} \qquad (9)$$

(22) 12% (E)-Ph, 16%
(Z)-Ph, 33%

R = H

$$\underset{\substack{HO\quad NO_2 \\ (23)\quad 24\%}}{PhCHCH_2} \qquad \underset{\substack{O\ NO_2 \\ 24\%}}{PhCCH_2} \qquad \underset{19\%}{PhCHO} \quad \underset{1\%}{PhCO_2H} \qquad \underset{\substack{O \\ 1\%}}{PhCCN}$$

The primary, reversible addition of the $·NO_2$ molecule with the formation of the most stable of the two possible radicals is followed by transformation of the latter into a radical pair under the action of the second molecule of $·NO_2$ [36].

$$Et_2C{=}CH_2 \underset{·NO_2}{\rightleftarrows} Et_2\dot{C}CH_2NO_2$$
$$\xrightarrow{·NO_2} Et_2\dot{C}(NO_2)CH_2·NO_2 \longrightarrow Et_2C(NO_2)CH_2NO_2 \quad (10)$$

Denitration and dehydration of the products of dinitrogen tetroxide addition to alkenes make it possible to synthesize the functional derivatives of the conjugated nitroalkenes [32]. For example, the interaction between (24) and N_2O_4 (equation 11) leads to the corresponding nitro ester (25) [37].

$$\underset{(24)}{ArCH{=}CHCOOR} \xrightarrow[\substack{Et_2O,\,CHCl_3 \\ -10\,°C}]{N_2O_4} \underset{(25)\ \ 29{-}49\%}{ArCH{=}C(NO_2)CO_2R} \qquad (11)$$

$$Ar = Ph,\ C_6H_4NO_2\text{-}p$$

Series of furylnitroethenes and thienylnitroethenes have been prepared under similar conditions by Sitkin *et al.* [22, 23]. It was found that nitrohalogenation of alkenes followed by dehydrohalogenation of the β-nitrohalo derivatives (26; equation 12) can be used as the preparative approach to numerous nitroalkenes (27) [7, 32, 38].

$$RCH{=}CH_2 \xrightarrow[\substack{or\ N_2O_4 + Hal_2}]{NO_2Hal} \underset{(26)}{RCH(Hal)CH_2NO_2} \xrightarrow[-HHal]{} \underset{(27)}{RCH{=}CHNO_2} \quad (12)$$

$$Hal = Cl,\ I$$

Nitrile iodide, which was prepared for the first time in 1932 by the interaction of silver nitrite and iodine [39], is a good, selective reagent for the production of iodonitroalkanes in the reaction with unsaturated hydrocarbons [8, 40–43a]. Recently, Sy and By [43b] introduced NO_2I (equation 13a) into the reaction with substituted styrenes (28) and the second-step dehydrohalogenation of the intermediate (29) allows the preparation of nitroalkenes (30).

$$ArCH{=}CHR \xrightarrow{NO_2I} ArCH(I)CH(NO_2)R \xrightarrow[-HI]{base} ArCH{=}C(NO_2)R \quad (13a)$$

$$\textbf{(28)} \qquad\qquad\qquad \textbf{(29)} \qquad\qquad\qquad \textbf{(30)} \; 45\text{--}82\%$$

$$R = H, \, Me; \quad Ar = Ph, p\text{-}MeOC_6H_4, \, 2,3,5,\text{-}(MeO)_3C_6H_2$$

Various unsaturated nitro compounds have been synthesized in a similar way [43a].

The application of ethylene glycol (equation 13b) as a solvent has made it possible to prepare unsaturated nitro compounds without the elimination step [43c].

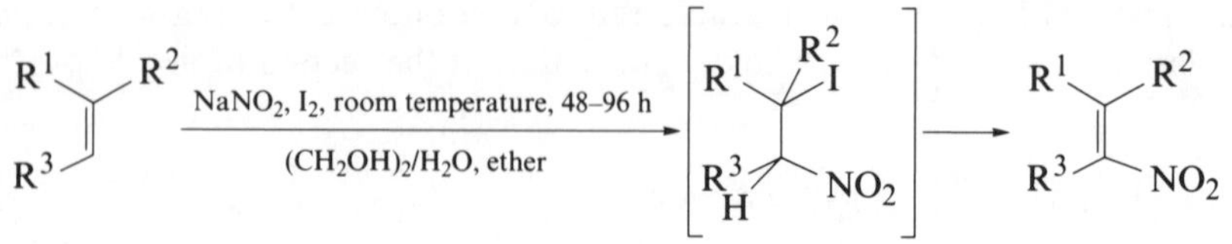

R^1	R^2	R^3	Yield (%)
Ph	H	H	81
(CH$_2$)$_4$		H	72
(CH$_2$)$_4$		H	49
CO$_2$Bπ	H	H	80

(13b)

1,2-Dihydro-2-nitronaphthalene (82%) can be obtained by means of a similar method.

In another case, 2-iodo-1-nitroalkanes (28) were formed readily (60–90%) in the reaction of alkanes with a dinitrogen tetroxide/iodine 1:1 mixture [44]. This method has also been used in the synthesis of nitrocamphene, methyl 3-nitroacrylate and (Z)-α-nitrostilbene.

It has been reported [45] that nitroalkenes are formed in 41–63% yield in the electrolytic oxidation of aliphatic alkenes in an aqueous solution of $NaNO_2/NaNO_3$. The stereospecific nitromercuration of alkenes and subsequent mercury elimination, can be said to be a general synthetic method for the preparation of cyclic, conjugated nitroalkenes and their derivatives [46]. Synthesis of 1-nitrocyclohexene (32; equation 14) with high purity is a good example of this method because the approaches that had been developed earlier had some preparative disadvantages.

This method can be used for the synthesis of 3-nitrocyclo-2-alkenones from cycloalkenones. The latter are introduced into the reaction in the form of ketals. The elimination from the product of nitromercuration gave 33 that was hydrolyzed in 34 (equation 15) [47].

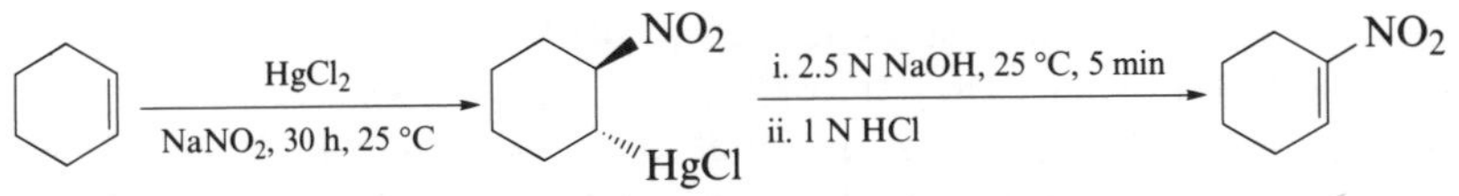

$$\text{(14)}$$

(31) 80% (32) 98%

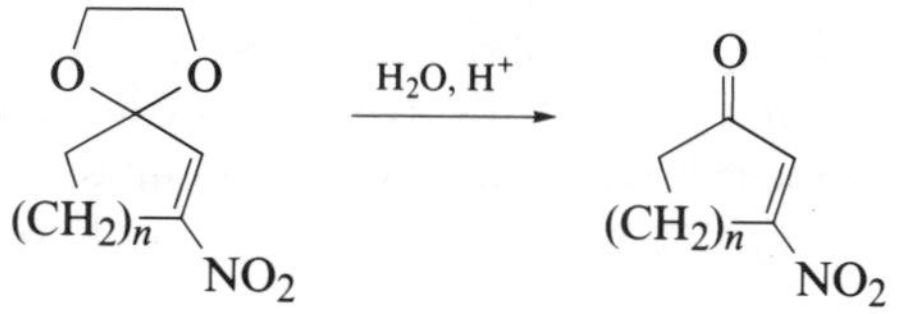

$$\text{(15)}$$

(33) (34) $n = 1, 2$

The interaction between phenylselenyl bromide, silver nitrate and alkenes (35; equation 16) (nitroselenation) results in the preparation of (36) along with (37). The compounds (37) can be transformed into (38) [48, 49]. The formation of products (36) may be depressed pronouncedly by introducing $HgCl_2$ into the reaction mixture [49].

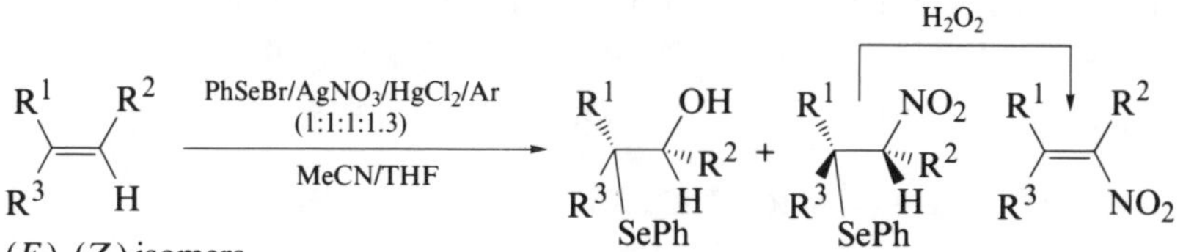

$$\text{(16)}$$

(35) (36) (37) 58–83% (38)

This is the synthetic approach to aliphatic and cyclic nitroalkenes (38–58% yield) [49] as well as to 2-nitro-1-alkenylsilanes [50]. The latter are synthesized

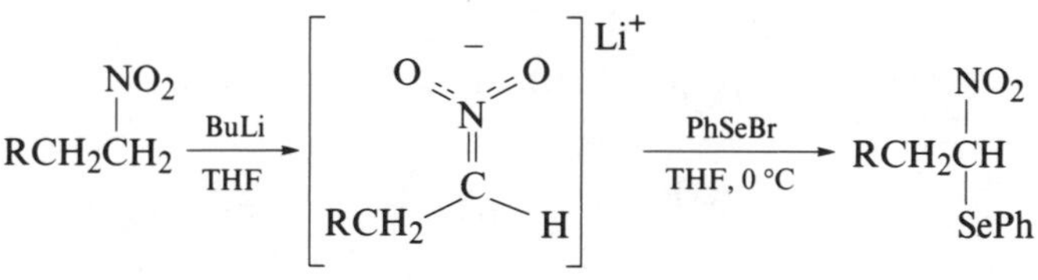

(39) (40) (41)

$$\xrightarrow{35\% H_2O_2} RCH{=}CHNO_2$$

(42)

R	Yield (%)
C_6H_{13}	80
C_7H_{15}	83

$$\text{(17)}$$

by nitroselenation of vinylsilanes (37–66% yield) followed by deselenation (89–96% yield). Both steps are stereospecific.

The 1-nitro-1-phenylselenoalkanes that are accumulated in nitroalkane selenation may also undergo oxidative deselenation. Oxidative transformation of nitroalkanes (39; equation 17) into nitroalkenes (42) proceeds via selenation [51] of nitronates (40) with phenylselenyl bromide, the formation of (41) and its oxidation into (42).

This method has been used in the synthesis of nitrocyclohexenes (55–61% yield). The last step proceeds regioselectively under mild conditions without base, making it possible to synthesize nitrocyclenes [51] of type (43)

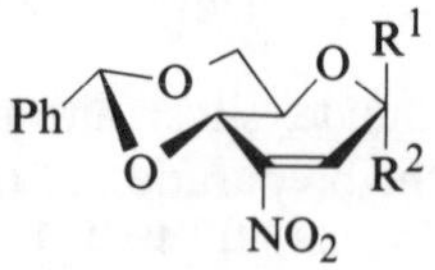

R^1	R^2	Yield (%)
OPh .	H	80
OMe	H	75
H	OMe	67
H	H	71

(43)

1-Nitro-1-selenoalkanes (44; equation 18) can be useful for the synthesis of functionalized nitroalkene derivatives [52], for example compound (46).

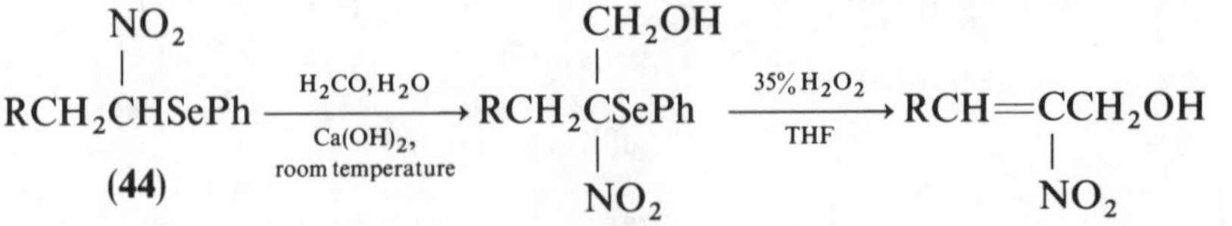

$$RCH_2CHSePh \xrightarrow[\substack{Ca(OH)_2, \\ room\ temperature}]{H_2CO, H_2O} RCH_2CSePh \xrightarrow[THF]{35\% H_2O_2} RCH=CCH_2OH \quad (18)$$

(44) with NO_2 above, CH_2OH and NO_2 on (45) 71–81%, NO_2 on (46) R = C_6H_{13} 81%

R = Me, Et, C_6H_{13}, C_7H_{15}

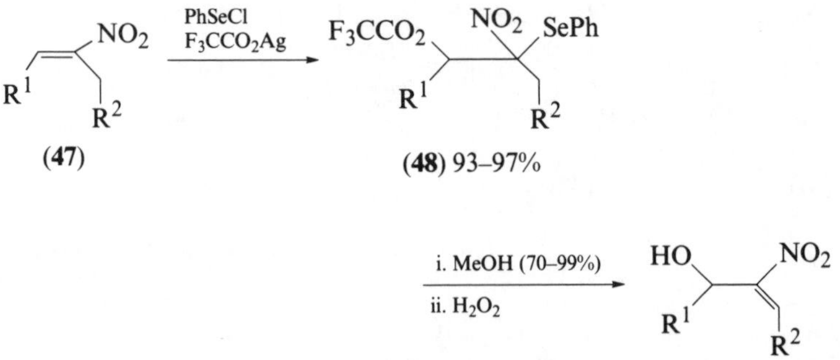

$$\text{(47)} \xrightarrow[F_3CCO_2Ag]{PhSeCl} \text{(48) 93–97\%}$$

(47) R^1, R^2, NO$_2$

(48) F$_3$CCO$_2$, NO$_2$, SePh, R^1, R^2

$$\xrightarrow[\text{ii. H}_2O_2]{\text{i. MeOH (70–99\%)}} \quad (19)$$

(49) HO, NO$_2$, R^1, R^2

(49) 60–85%

R^1 = H, Me, Et, Bu, H, C_5H_{11}; R^2 = H, H, H, H, Bu, H; R^1 = $(CH_2)_2$, R^2 = CH_2

Selenation–deselenation of 2-nitrocyclohexanone acetal under similar conditions results in the corresponding nitroalkene [53]. The same method was used for the intermediate step in the synthesis of the antibiotic thienamycin [54]. Recently, the oxoselenation [55] of nitroalkenes (47) with phenylselenyl chloride and the silver trifluoroacetate was performed, resulting in phenylselenyl trifluoroacetates (48). Hydrolysis and regioselective elimination of the phenyl-selenyl group led to the formation of substituted nitroallylic alcohols (49; equation 19). Note the C=C bond shift in comparison with the starting material.

Seebach has proposed that alcohol pivalates containing 2-nitroallylic fragments may be used as multiple coupling reagents; consequently, many novel nitroalkenes have been synthesized [55].

1.1.2 ALKENYLATION OF NITRO COMPOUNDS

One synthetic approach to unsaturated nitro compounds is the Henry method (1894). It involves the interaction between nitroalkanes and aldehydes or sometimes ketones. The first stage of the process is the condensation of an aliphatic aldehyde with a nitroalkane, resulting in the formation of a β-nitroalcohol. Such nitroalcohols are frequently stabilized by acylation. Further elimination of water or acetic acid leads to the nitroalkene (equation 20).

$$O_2N-\overset{R^1}{\diagup} + \underset{O}{\overset{R^2\diagdown C \diagup R^3}{\|}} \quad \Longleftrightarrow \quad O_2N \overset{R^1}{\diagup} \underset{R^3}{\overset{OH}{\diagdown}} R^2 \quad \longrightarrow \quad O_2N \overset{R^1}{\diagup} \overset{R^2}{=} \underset{R^3}{\diagdown} \quad (20)$$

One-step synthetic approaches to nitroalkenes through the reaction of aliphatic aldehydes with nitroalkanes are rather scarce. The nitroalcohols that arise in the condensation of aromatic or heterocyclic aldehydes with nitroalkanes undergo spontaneous dehydration to give unsaturated nitro compounds [32, 56–60].

The majority of works devoted to the synthesis of alkylnitroalkenes, arylnitro-alkenes and heterylnitroalkenes include studies of condensing agents. The most widely used condensing agents are alcohol solutions of alkalis and alkaline earth metal alkoxides, as well as amines. Application of equivalent quantities of such agents has made it possible to synthesize a large number of unsaturated nitro compounds. One typical example is the aldehyde condensation with nitromethane in methanol (or ethanol) in the cold with the addition of an equivalent quantity of sodium or potassium hydroxide or alkoxide, resulting in the formation of nitroalcohol salts that may be easily transformed into the corresponding nitroalkenes (50; equation 21).

$$RCHO + MeNO_2 \xrightarrow{\text{B}^-} [RCH(OH)CH\!=\!NOO^-]Na^+ \xrightarrow[-NaCl]{\text{HCl}}$$

$$[RCH(OH)CH_2NO_2] \xrightarrow[-H_2O]{} RCH\!=\!CHNO_2 \quad (21)$$

$$\textbf{(50)}$$

R = Alk, Ar, Het

This method is the best available for the reaction of aromatic aldehydes with nitromethane. Other primary nitroalkanes such as 1-nitropropane and phenyl-nitromethane do not undergo chemical transformations under such conditions [32]. This is a synthetic method for a wide range of fatty, aromatic and heterocyclic nitroalkenes such as furylnitroalkenes [61, 62] and thienylnitroalkenes [61, 63]. These compounds are useful in the synthesis of pharmacologically active substances (equations 22 and 23).

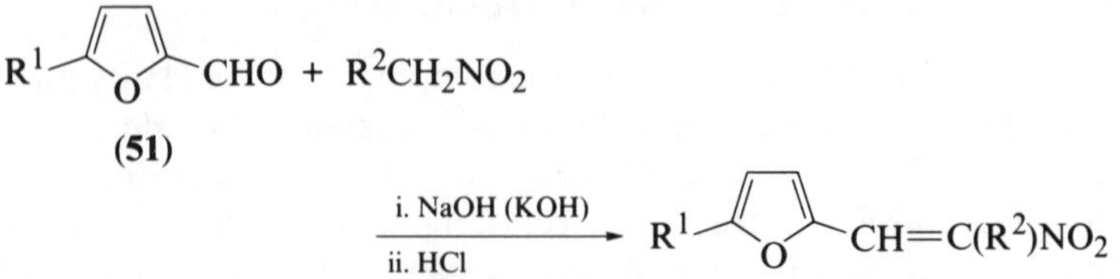

R^1⟨furan⟩CHO + R^{2}CH$_2$NO$_2$

(51)

$$\xrightarrow[\text{ii. HCl}]{\text{i. NaOH (KOH)}} R^1\text{⟨furan⟩}CH\!=\!C(R^2)NO_2 \quad (22)$$

(52)

R^1 = H, Me, Cl, Br, I, NO$_2$, MeO$_2$C, C$_6$H$_4$Cl-m, HOOCS, $\begin{smallmatrix}Me\\Ph\end{smallmatrix}$N, O⟨N⟩;

R^2 = H, Me

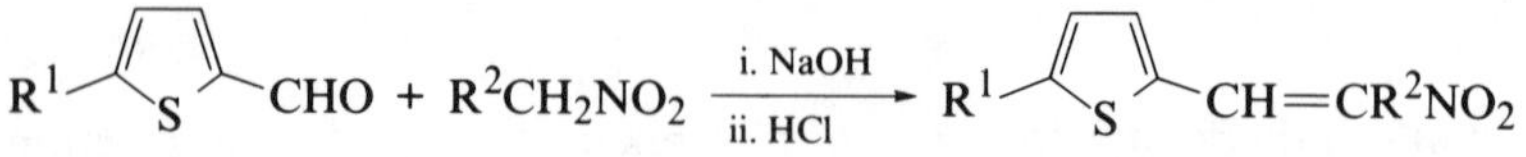

R^1⟨thiophene⟩CHO + R^{2}CH$_2$NO$_2$ $\xrightarrow[\text{ii. HCl}]{\text{i. NaOH}}$ R^1⟨thiophene⟩CH$=$CR^{2}NO$_2$ (23)

(53)   **(54)**

R^1 = H, Me, Et, Pr, Br, Cl; R^2 = H, Me

Interaction between nitro-containing heterocyclic aldehydes that have one [64] or two [65] heteroatoms and nitroalkanes under the action of alkali gives rise to the corresponding heterylnitroalkenes. In some cases, the introduction of aromatic or heterocyclic aldehydes into the process makes it possible to

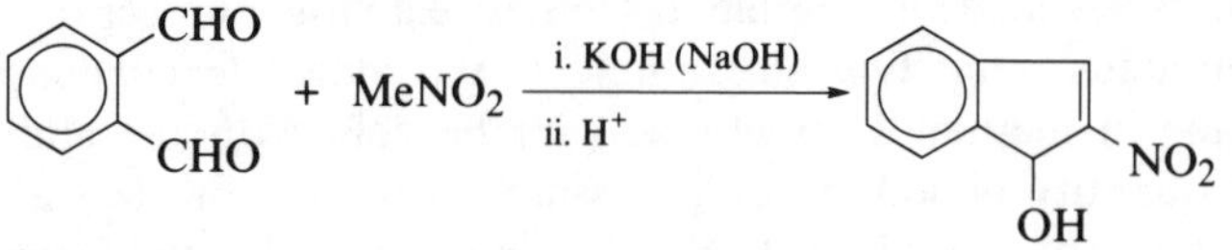

⟨structure with CHO, CHO⟩ + MeNO$_2$ $\xrightarrow[\text{ii. H}^+]{\text{i. KOH (NaOH)}}$ ⟨indene structure with NO$_2$, OH⟩ (24)

(55)   **(56)**

produce β-nitroalcohols that can be further transformed into nitroalkenes [66]. The participation of (55; equation 24) in such a reaction results in the formation of the 1-oxy-2-nitroindene (56) [67].

In 1968, Shiga *et al.* [68] delivered the very first synthesis of 2-nitro-1- • ferrocenylethenes (58) from (57), catalysed by MeONa (equation 25).

$$FcCHO + RCH_2NO_2 \xrightarrow{\text{i. MeONa; ii. HCl}} FcCH{=}CRNO_2 \qquad (25)$$

$$(57) \qquad\qquad\qquad\qquad (58)$$

$$R = H, Me; \quad Fc = Ferrocenyl$$

Monodehydration of nitrodiols (equation 26) proceeds via the intermediates (59) and (60) to give unsaturated nitroalcohol (61) [69].

$$\underset{\substack{\| \\ NOO^- Na^+}}{HOCH_2CCH_2OH} \xrightarrow[\substack{-20\,°C\ to\ -30\,°C}]{40\%\ HNO_3} \underset{\substack{\| \\ NOOH}}{[HOCH_2CCH_2OH]} \xrightarrow[-H_2O]{} \underset{\substack{| \\ NO_2}}{HOCH_2C{=}CH_2}$$

$$(59) \qquad\qquad\qquad (60) \qquad\qquad (61)\ 52\% \quad (26)$$

In an earlier publication [70] the compound (61) had been synthesized by deacetylation of the diacetoxy derivative of 2-nitro-1,3-propanediol. Primary and secondary amines direct the Henry reaction towards spontaneous dehydration. Owing to this effect it is possible to obtain arylnitroalkenes directly from aromatic aldehydes [71–74]. The preparation of (63) from arylnitroalkane (62) takes place only under specific conditions (equation 27) [74].

$$\underset{(62)}{\overset{O_2N}{\underset{CH_2-CH_2}{\diagdown}}\underset{Ar^1}{\diagdown}} + Ar^2CHO \xrightarrow[\substack{CH(OMe)_3}]{\substack{MeNH_2\cdot HCl \\ MeCO_2R}} \underset{(63)}{\overset{O_2N}{\underset{Ar^1H_2C}{\diagup}}{=}\overset{H}{\underset{Ar^2}{\diagdown}}} \qquad (27)$$

Interaction between nitromethane and the substituted derivatives of salicylic aldehyde in the presence of a catalytic quantity of the salt mixture $Me_2NH\cdot HCl/KF$ results in the formation of β-(o-oxyphenyl)-α-nitroalkenes in high yield [75].

Ammonium acetate in acetic acid is widely used in the synthesis of arylnitroalkenes [71,72,76–80]. This catalyst is very efficient in the reactions of nitromethane homologues [71] and fatty aromatic nitroalkanes [32,74].

Application of NH_4OAc makes the process time shorter and depresses completely the polymerization of nitroalkenes even with continuous heating, especially in the condensations with oxyaldehydes and alkoxyaldehydes [81a]. For example, condensation of (64) with nitromethane (equation 28) under the action of this catalyst leads to the product (65) [81b]. Reflux of a mixture of

2,5-dimethoxybenzaldehyde or 3,5-dimethoxybenzaldehyde, nitroethane and ammonium acetate yields 98% of the nitroalkene [82, 83].

$$R^1O\!\!-\!\!\text{CHO} \xrightarrow[\text{NH}_4\text{OAc}]{\text{MeNO}_2} R^1O\!\!-\!\!\text{CH}\!\!=\!\!\text{CHNO}_2 \qquad (28)$$

(64)   (65)

$$R^1 = H, Me; \quad R^2 = OCH_2Ph, H$$

The process with the latter aldehyde under milder conditions, i.e. at 20 °C with the catalyst KF/18-crown-6, gives the nitroalcohol which is isolated in the form of a diastereoisomeric mixture in 87% yield [83].

Such conditions are used in the reactions of silicon-containing aromatic aldehyde (66; equation 29). It is worth mentioning that the trimethylsilyl group activates β-aminoarylethanes in the controlled regioselective synthesis of isoquinoline alkaloids [76].

$$\text{CHO} \atop \text{SiMe}_3 \quad \xrightarrow[\text{reflux}]{\text{MeNO}_2 \atop \text{MeCONH}_4,} \quad \text{NO}_2 \atop \text{SiMe}_3 \qquad (29)$$

OMe   OMe
(66) 85%   (67) 97%

It has been shown [73] that high-pressure (10–15 kbar; 1 bar $= 10^5$ Pa) condensation of aromatic aldehydes containing electron-donating substituents results in nitrostyrenes with much higher yields than under normal conditions.

Ammonium acetate is also an effective catalyst in the synthesis of various heterocyclic nitroalkenes. Its participation activates the condensation of such species as thiophenylaldehydes [77], 3-indolylaldehydes [84, 85] and 3-formyl-7-azaindole [86] with AlkNO$_2$ (Alk = Me, Et, Pr).

The condensation of nitroalkanes with ketones leads to the formation of complex mixtures of products in low yield [56, 57]. Only C_5–C_8 cycloalkanones react with nitromethane in the presence of secondary amines to give nitroallylic isomers, i.e. 1-nitromethyl-1-cycloalkenes.

The problem of the low yield of condensation products derived from nitromethane and ketones was solved in the synthesis of steroid nitroalkenes (70; equation 30) by application of the difunctional ethylenediamine as a catalyst, which probably helps to shorten the distance between the reaction centres and decreases the steric hindrance. The base also makes the proton transfer easier and then initiates ethylenediamine β-elimination [87].

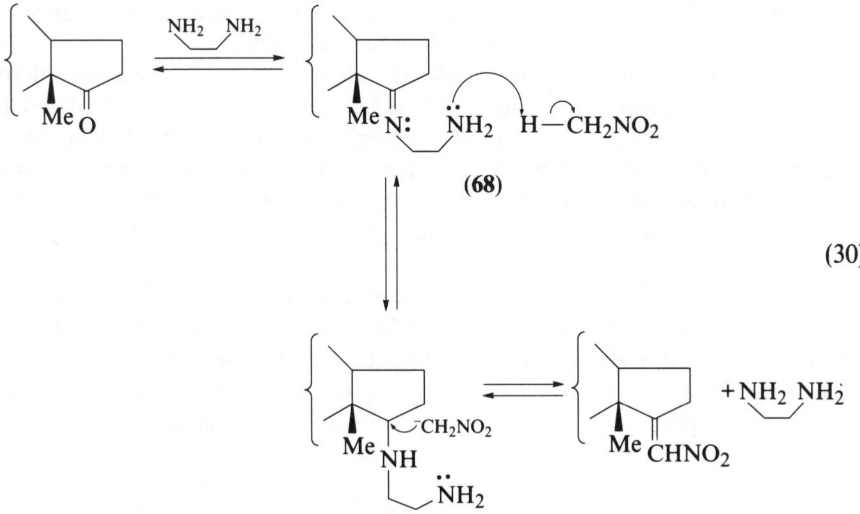

$$(30)$$

(68)

(69)   (70)

N,N-Dimethylethylenediamine has been found to be much more of a universal catalyst of ketone condensation with primary nitroalkanes [88]. It initiates the direct formation of allylic nitro compounds without α-nitroalkene intermediation. The authors propose a mechanism for the process. This condensation is the synthetic approach to various β,γ-unsaturated nitro compounds, such as (71) and (72) (equation 31).

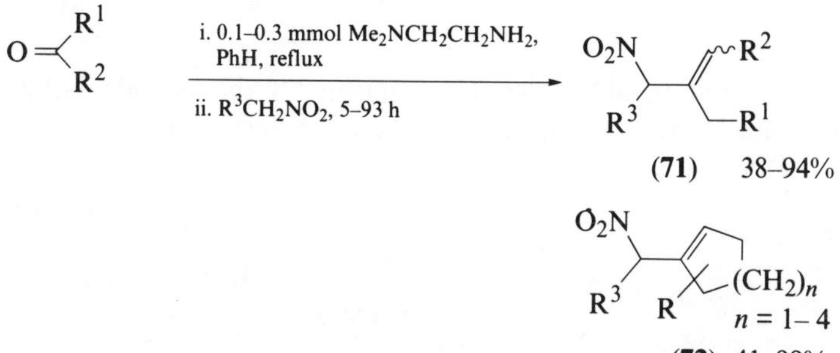

$$(31)$$

(71)   38–94%

(72) 41–99%

In contrast to aromatic aldehydes, their aliphatic analogues lead to formation of nitroalcohols under the action of an equivalent quantity of an alkali followed by acidification. The same result is achieved with basic catalysis. This process has been discussed in detail in several reviews [57–59]. There is a wide range of basic catalysts. For example, alkoxides provide good results in the reaction of aromatic aldehydes [89, 90]. Hydroxides, carbonates, tertiary amines [65] and ion-exchange resins [91, 92] have also been used successfully. Dehydration and acylation–deacylation processes are the most widely used for the transformation of alcohols into the corresponding nitroethenes.

Basic catalysis is inefficient for aliphatic aldehydes with chains longer than C_6, alicyclic aldehydes and citric aldehydes [93]. The reactions of these compounds can be carried out in the presence of KF and speeded up by a catalytic quantity of 18-crown-6. The same process was used twice in the multistep synthesis of stereoisomeric nitrocyclopentenes from D-glucose [94], the final target of which was the formation of the biologically active pseudo-nucleoside.

Reversibility of the reaction sometimes results in synthetic difficulties. For such cases, two methods that facilitate condensation were suggested by Seebach and Lehr [95] and Seebach *et al.* [96]. The reverse reaction may be suppressed by using dianions of primary nitro compounds. The reaction is stereospecific and the major products are the *threo*-nitroalcohols (**74a**; equation 32).

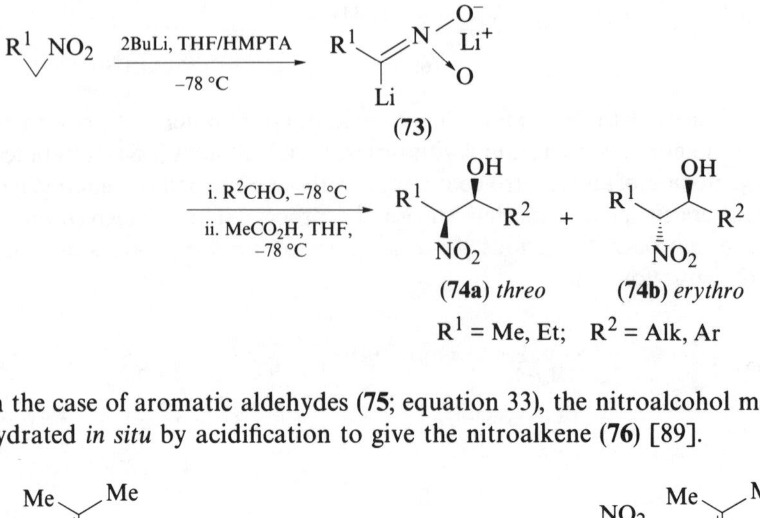

$$(73)$$

$$(32)$$

(**74a**) *threo*   (**74b**) *erythro*

R^1 = Me, Et; R^2 = Alk, Ar

In the case of aromatic aldehydes (**75**; equation 33), the nitroalcohol may be dehydrated *in situ* by acidification to give the nitroalkene (**76**) [89].

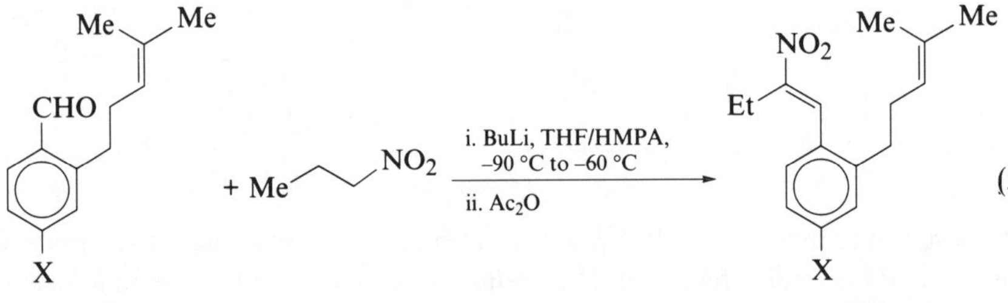

$$(33)$$

(**75**)   (**76**)

X = H, OMe

Another method suitable for a wide range of compounds is based on the condensation of aliphatic aldehydes (with chain lengths less than C_6) with silylnitroalkanes (silyl esters of the nitronic acids (**77**) in the presence of a catalytic amount of F^- [96–98]. In this case the dominant stereoisomeric products are the silyl esters of *erythro*-nitroaldols (**78**; equation 34) [96].

$$R^1C \underset{H}{\overset{O}{\diagdown}} \quad + \quad R^2CH = N \overset{OSiR_2^3R^4}{\underset{O}{\diagup}}$$

(**77**)

n-Bu$_4$NF/THF, –100 °C to –78 °C

$$R^4R_2^3SiO \underset{R^1}{\diagdown} NO_2 \quad \xrightarrow{H_2O} \quad HO \underset{R^1}{\diagdown} NO_2 \qquad (34)$$

(**78**) (**79**)

NaH/THF, 70 °C

$$R^1 \diagdown \overset{R^2}{\diagdown} NO_2 \qquad \xleftarrow{MsCl/Et_3N, CH_2Cl_2}$$

(**80**)

$R^1 = $ Alk, Ar, BuO; $R^2 = $ Me, Et, i-Pr, C$_5$H$_{11}$, C$_7$H$_{15}$;
$R^3 = R^4 = $ Me; $R^3 = $ Me; $R^4 = $ t-Bu

The resulting silyloxynitro compounds (**78**) undergo desilylation either under reflux with NaH in THF [98] or by hydrolysis followed by dehydration of the resulting nitroaldol in the presence of methanesulfonyl chloride [98]. The yields of the nitroalkenes (**80**) range from 60% to 82%. The (*E*) isomer is the major product in the reaction mixture. This method has been applied successfully for the chain extension of unsaturated nitrosugars under extremely mild conditions [99, 100]. Nitroaldol (**81**; equation 35) dehydration occurs as a result of acetylation followed by deacetylation.

The above modifications of the Henry reaction make it applicable to a wide range of nitroalkenes, but the modifications require rather expensive reagents and complex experimental facilities. Nitroalkene synthesis by dehydration of nitroalcohols and deacylation of nitroesters is usually rather simple. The elimination of water by various dehydrating agents depends upon the structure of the resulting nitroalkene. Acylated derivatives (acetic, benzoic) that accumulate readily in the reactions of nitroalcohols with chloroanhydrides or anhydrides of carboxylic acids are rather stable and eliminate the acidic elements without problems. Deacylation is therefore preferred to the dehydration of nitroalcohols.

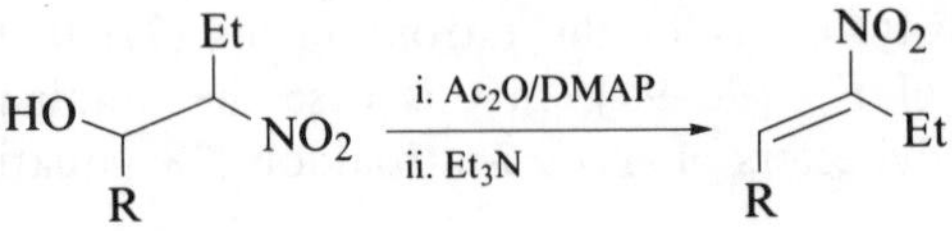

$$(35)$$

DMAP = N,N-dimethylaminopyridine

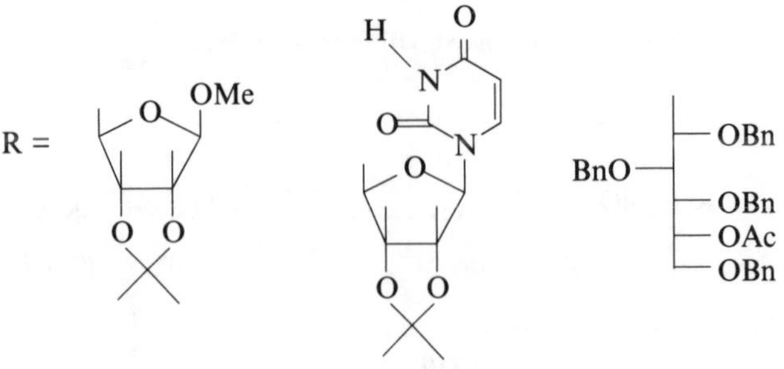

R =

48%   48%   60%

Some effective dehydrators are $ZnCl_2$, P_4O_{10}, HPO_3, $NaHSO_4$, $KHSO_4$, phthalic anhydride and acetic anhydride [101–103]. Such novel dehydrators as dicyclohexylcarbodiimide (DCC) [104], pivaloyl chloride, methanesulfonyl chloride [10, 105] and alumina allow dehydration to occur without heating.

Elimination of acetic acid is initiated by $NaHCO_3$, Na_2CO_3, CH_3CO_2Na, $Ca_3(PO_4)_2$, $Mg_3(PO_4)_2$ and others.

One of the most effective dehydrators is phthalic anhydride, which catalyses the process in high yield [102]. For example, phthalic anhydride facilitates the conversion of β-nitroethanol (83) into nitroethene (84) in 80% yield (equation 36).

$$HOCH_2CH_2NO_2 \xrightarrow[140-150\,°C,\,10^3\,Pa]{\text{phthalic anhydride}} CH_2{=}CHNO_2 \qquad (36)$$
$$(83) \hspace{5cm} (84)$$

Ion-exchange resin catalyses the condensation of cyclohexanone (85; equation 37) with nitromethane [91]. Reflux of the resultant nitroalcohol (86) with phthalic anhydride gives the product (87) [101].

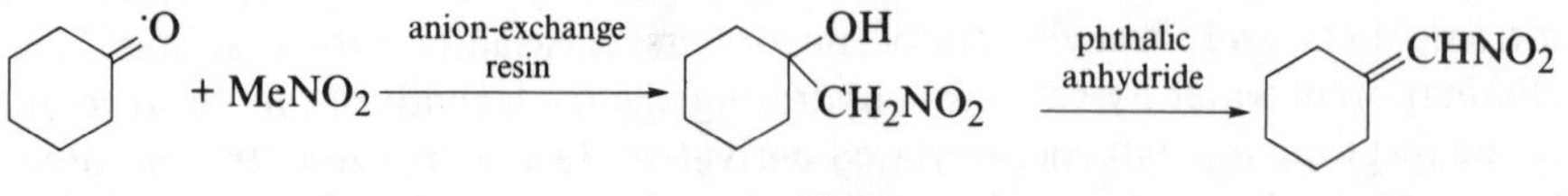

$$(37)$$

Dicyclohexylcarbodiimide (DCC) in combination with copper(I) chloride has opened the way to a series of nitroalkenes (equation 38) [104].

$$R^1-\underset{\underset{OH}{|}}{\overset{\overset{H}{|}}{C}}-\underset{\underset{H}{|}}{\overset{\overset{NO_2}{|}}{C}}-R^2 \xrightarrow[\text{ether/dioxan}]{\text{DCC, CuCl, room temperature}} \underset{H}{\overset{R^1}{\diagdown}}C=C\underset{R^2}{\overset{NO_2}{\diagup}} \qquad (38)$$

(88)   (89)

R^1 = H, Alk, alkenyl, 2-furyl; R^2 = H, Alk

A similar method has been used for the preparation of the cyclic nitroalcohol 2-nitro-2-cyclohexene-1-ol. This compound is an intermediate in the synthesis of nitroallylating agents.

Dehydration of 2-nitrocyclohexanols under the action of sodium hydride in THF leads to a number of nitrocyclohexenes [106]. Methanesulfonyl chloride in combination with TEA is a mild dehydrating agent. The process takes place in dichloromethane at 0 °C [105].

The ease of dehydration is determined by the transformation of the hydroxy group into a more nucleofuge leaving group. Such methods have been used in various syntheses [10, 54, 90, 98], such as the preparation of nitrovinyl-β-lactams from 4-formyl-β-lactams (equation 39) [90].

$$\xrightarrow[\substack{\text{MeNO}_2 \text{ (Solvent)},\\ 1-3\,h\\ \text{ii. MeSO}_2\text{Cl, CH}_2\text{Cl}_2, 3\,h}]{\text{i. R}^3\text{CH}_2\text{NO}_2, \text{Et}_3\text{N}} \qquad (39)$$

(90)   (91)

R^1 = MeO, PhO, phthalyl; R^2 = p-MeOC$_6$H$_4$, p-MeCOC$_6$H$_4$, CH$_2$CH(Ph)(OCOCH$_2$Cl), CH$_2$CH(Ph)OH; R^3 = H, Me, CH$_2$CO$_2$Me

Nitroalkenes are produced similarly from azetidine-2,3-diones and nitro-ethane. The condensation with nitromethane takes place in THF under the action of t-BuOK. Nitroalkenes synthesized in such a way have been used as starting materials for the preparation of β-lactam derivatives.

One of the steps in the production of the antibiotic (+)-thienamycin is the condensation of the azetidinone derivative (**92**; equation 40) with nitromethane under the action of 1,1,3,3-tetramethylguanidine (TMG), with subsequent transformation of the nitroalcohol to nitroalkene (**93**) [54].

Nitrodiol (**94**) is the starting material for the preparation of diastereomerically pure product (**95**) (90% yield) [92, 107]. The process takes place in accordance with equation (41).

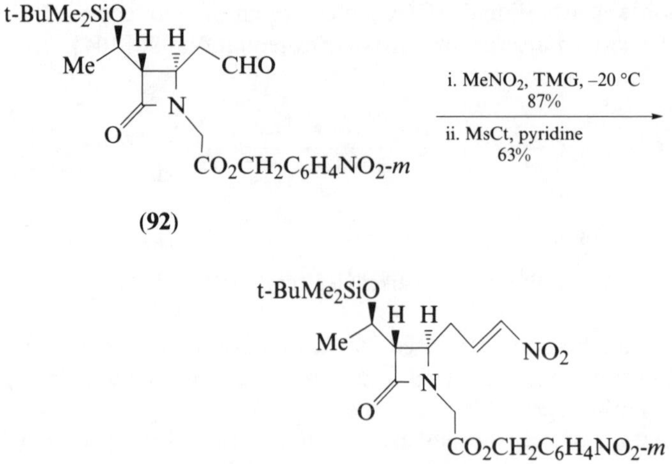

(92)

i. MeNO$_2$, TMG, −20 °C
87%

ii. MsCt, pyridine
63%

(40)

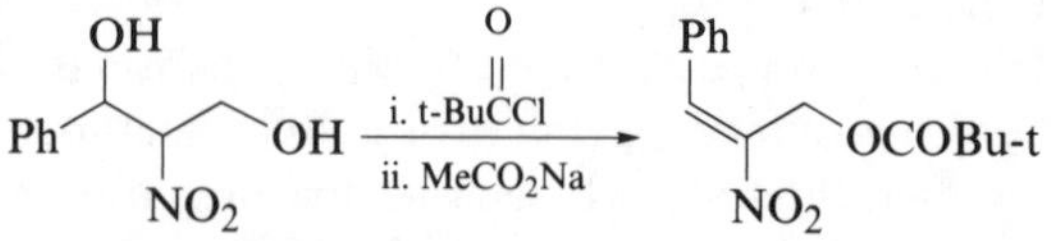

(93)

(41)

(94)   **(95)**

The same method is used for 2-nitropropene-3-il pivalate, which is known as a good multicoupling reagent [107].

Two methods of dehydration have been devised for disubstituted and trisubstituted nitroalkenes [100, 108, 109] and unsaturated nitrosugars [74], namely deacetylation *in situ* using (MeCO)$_2$O/DMAP (DMAP = 4-N,N-di-methylaminopyridine) or (CF$_3$CO)$_2$O/Et$_3$N. For example, the condensation of α,ω-alkenecarboxyaldehyde esters (**96**; equation 42) with nitroethane followed by acetylation and deacetylation leads to the corresponding nitroalkenes (**97**) which can be used for stereoselective syntheses of polycyclic structures [100].

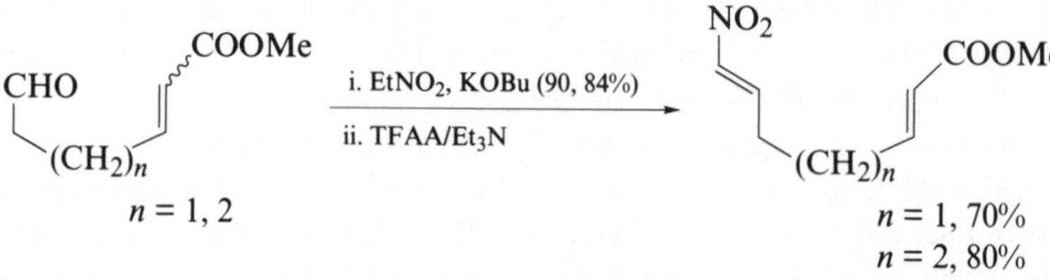

i. EtNO$_2$, KOBu (90, 84%)

ii. TFAA/Et$_3$N

(42)

$n = 1, 2$

$n = 1, 70\%$
$n = 2, 80\%$

(96)   **(97)**

A wider set of nitroalkenes of such kind is presented in the literature [108, 109].

Some functionalized nitroalkenes are not always prepared by common methods such as heating at high temperatures. A simple method for condensation [110] and subsequent chemoselective dehydration of β-nitroalcohols [111] on the surface of chromatographic alumina without a solvent has been described. This is the method of preparation for a large number of nitroalcohols, preferably from primary nitro compounds and aldehydes (68–80% yield) (equation 43) [110]. Phenylnitromethane undergoes condensation in a similar way [112].

$$R^1CHNO_2 + R^3CH{=}O \xrightarrow{Al_2O_3} R^1C(NO_2)CHOH \qquad (43)$$

$$\begin{array}{ccc} | & & | \quad | \\ R^2 & \mathbf{(99)} & R^2 \quad R^3 \end{array}$$

$$\mathbf{(98)} \hspace{4cm} \mathbf{(100)}\ 70\text{–}80\%$$

The yield of nitroalcohols in the condensation of γ-nitroaliphatic aldehydes with nitroalkanes is 80–89% [113]. The reaction takes place on an alumina surface in CH_2Cl_2. Some nitroalkenes are produced by condensation of 2-furylaldehydes with nitroalkanes [114]. It is recommended to use the latter method for the synthesis of polyfunctional nitro compounds that are sensitive to acids or bases [114]. The complex catalyst KF/Al_2O_3 (alumina-supported KF) speeds up the reaction which continues for 5–6 h and yields 55–79% of the nitroalcohol. In some cases, nitroalcohols derived from aromatic aldehydes undergo dehydration spontaneously during the preparation [115]. Various aliphatic nitroalcohols eliminate water under the action of basic Al_2O_3 in CH_2Cl_2 at 40 °C. The duration of the process is 7–50 h and the yields of nitroalkenes are 60–85% [111].

Nitroalkenes can also be prepared by deamination of β-nitroamines [32, 57, 58, 60]. Schiff bases depress the Michael reaction and thus support the condensation of aromatic aldehydes with nitro compounds [9].

It is impossible to isolate nitroamines formed in the reaction between azomethines (**101**; equation 44) and nitroalkanes owing to their immediate transformation into β-nitrostyrene derivatives [116].

$$\begin{array}{c} \hspace{6cm} R \\ \hspace{6cm} | \\ PhCH{=}NBu + H_2CNO_2 \xrightarrow[\text{ii. MeCO}_2\text{H}]{\text{i. PhH}} PhCH{=}CNO_2 \qquad (44) \\ \mathbf{(101)} \hspace{2cm} | \hspace{4cm} \mathbf{(102)} \\ \hspace{3.5cm} R \end{array}$$

α-Nitro-β-heterylstyrenes (**103**) can be synthesized in a similar way [117].

$$\mathbf{(103)}$$

$$X = O, S, NMe; \quad R = H, NO_2$$

Rather stable nitroamines (**105**; equation 45) are obtained if azomethines (**104**) are involved in the reaction. Deamination of such compounds takes place upon heating with mineral acids or the action of aliphatic amines to give the nitrostyrene (**106**) [118].

$$PhCH\!\!=\!\!NPh + MeNO_2 \longrightarrow \underset{\underset{(\mathbf{105})}{\overset{\mid}{NHPh}}}{PhCHCH_2NO_2} \xrightarrow[HCl]{-PhNH_3{}^+Cl^-} PhCH\!\!=\!\!CHNO_2$$

$$\underset{(\mathbf{104})}{} \qquad\qquad\qquad\qquad\qquad\qquad\qquad\qquad \underset{(\mathbf{106})}{}$$

$$(45)$$

Ferrocenylnitroalkenes are derived from formylferrocene and acetylferrocene under the action of morpholine. The predominant nitroamine product is deaminated using a mineral acid [68, 119].

The general method of synthesis for 3-substituted nitrocycloalkenes (**108**; equation 46) from 2-nitrocycloalkanones (via the hydrazone) includes reduction of the latter using $NaBH_4$ and subsequent deamination using acetic anhydride [120].

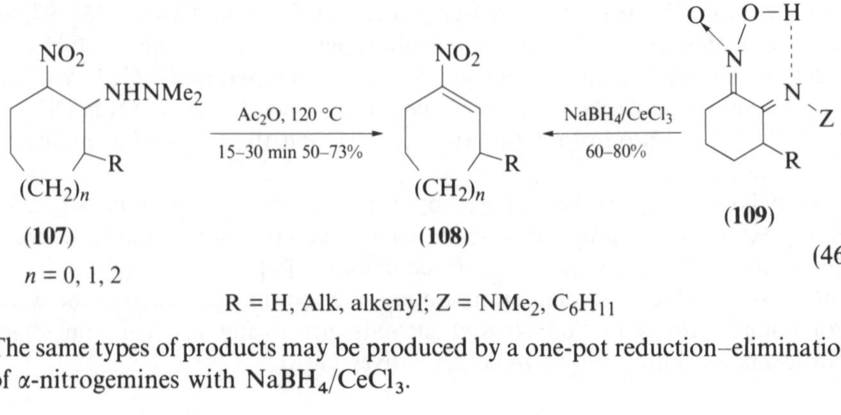

$$R = H, \text{ Alk, alkenyl}; \ Z = NMe_2, C_6H_{11}$$

$$(46)$$

The same types of products may be produced by a one-pot reduction–elimination of α-nitrogemines with $NaBH_4/CeCl_3$.

1.1.2.1 Geminally functionalized nitroalkenes

The prevailing approach to geminal alkoxynitroalkenes, cyanonitroalkenes, acetylnitroalkenes and benzoylnitroalkenes is the condensation of aldehydes (ketones) and their derivatives with methylene components containing the nitro group. α-Nitroalkenecarboxylic esters are the most interesting compounds among the geminally functionalized nitroalkenes as they are the unsaturated nitro precursors of α-amino acids.

α-Nitro-β-hydroxycarboxylic esters are prepared in the reaction of aliphatic aldehydes with nitroacetic esters under the action of sodium acetate. Acylation–deacylation of the ester leads to a number of aliphatic α-nitroacrylates [9, 121–123].

The same compounds can be synthesized by direct dehydration of α-nitro-β-hydroxycarboxylic esters diethylammonium salts with hydrochloric acid (equation 47) [124]. These salts are produced by condensation of aldehydes with nitroacetic esters induced by an equimolar amount of diethylamine [125].

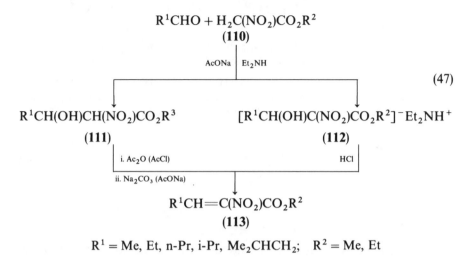

$$R^1CHO + H_2C(NO_2)CO_2R^2$$
(110)

AcONa | Et$_2$NH

(47)

$$R^1CH(OH)CH(NO_2)CO_2R^3 \qquad [R^1CH(OH)C(NO_2)CO_2R^2]^- Et_2NH^+$$
(111) **(112)**

i. Ac$_2$O (AcCl)   HCl
ii. Na$_2$CO$_3$ (AcONa)

$$R^1CH{=}C(NO_2)CO_2R^2$$
(113)

R^1 = Me, Et, n-Pr, i-Pr, Me$_2$CHCH$_2$; R^2 = Me, Et

All attempts to isolate the simplest α-nitroacrylate from α-nitro-β-hydroxy-propionate or its acylated derivative failed. The formation of the compound was determined by isolation of the corresponding adducts of the diene. The reactions between β-nitroalcohols and cyclopentadiene or anthracene were performed in ampoules with heating [126].

There are only a few examples of nitroacetic ester alkenylation by aromatic aldehydes under the action of bases [127]. In many cases, the substituted derivatives of 2,4-dinitroglutaric esters, isoxazoline N-oxides or β-amino-α-nitropropionates [128, 129] have been isolated.

The reaction of Schiff base heteroanalogues of aldehydes with nitroacetic esters also results in the formation of 2,4-dinitroglutaric esters, i.e. the products of nitroacetic ester addition to nitroacrylic intermediates [130]. Nevertheless, acetic anhydride induces acylation of the first-formed nitroamine (equation 48), which then easily eliminates acetamide to form the nitro compound. The latter does not undergo condensation with CH acids.

$$R^1CH(R^2NH)CH(NO_2)CO_2R^3 \xrightarrow{Ac_2O} [R^1CH(R^2NCOMe)CH(NO_2)CO_2R^3]$$
(114)

$$\xrightarrow[-R^2NHCOMe]{} R^1CH{=}C(NO_2)CO_2R^3 \qquad (48)$$
(115)

R^1 = Ph, p-ClC$_6$H$_4$, p-MeOC$_6$H$_4$, 3-indolyl;
R^2 = Me, Bu, Ph; R^3 = Me, Et

Substituted α-nitroacrylic esters ((E) and (Z) isomers) are prepared by condensation of benzalamines, furfuralamines or indalamines and their derivatives with nitroacetic esters (equation 49) [131–134].

$$R^1CH{=}NR^2 + H_2C(NO_2)CO_2R^3 \xrightarrow{Ac_2O} R^1CH{=}C(NO_2)CO_2R^3 \quad (49)$$

$$\textbf{(116)} \qquad\qquad\qquad\qquad\qquad\qquad\qquad\qquad \textbf{(117)}$$

$$R^1 = Ph,\ p\text{-}ClC_6H_4,\ p\text{-}MeOC_6H_4,\ 2\text{-furyl, 3-indolyl};$$
$$R^2 = Me,\ n\text{-}Bu,\ Ph;\quad R^3 = Me,\ Et$$

Besides Schiff bases, ymmonium salts have also been used in the reactions with nitroacetic esters (and nitroalkanes). Indole-type conjugated nitro compounds have been successfully synthesized from the chlorides **118** (equation 50) [135].

$$
\begin{array}{c}
\text{(image of reaction)}
\end{array}
$$

$$(50)$$

$$R = H,\ Me;\quad X = H,\ Me,\ Et,\ CO_2Me$$

The reaction between acetals and nitroacetic ester is a convenient method for other complicated synthetic approaches to β-substituted α-nitroacrylates. For example, this method was used to produce α-nitro-β,β-dimethylacrylic esters that are not obtainable by other methods (**121**; equation 51) in acetic anhydride [136]. Acetone does not undergo chemical transformation under the same conditions.

$$Me_2C(OEt)_2 + H_2C(NO_2)CO_2Me \xrightarrow[\substack{AcONa, \\ heat}]{Ac_2O} Me_2C{=}C(NO_2)CO_2Me \quad (51)$$

$$\textbf{(120)} \qquad\qquad\qquad\qquad\qquad\qquad\qquad \textbf{(121)}$$

Subsequently this method was extended to aromatic acetals and some substituted α-nitrocinnamic esters (**122**) have been prepared [137].

$$R^1\!\!-\!\!\underset{\underset{R^2\quad R^3}{}}{\bigcirc}\!\!-CH{=}C(NO_2)CO_2Me$$

$$\textbf{(122)}$$

$$R^1 = H,\ OMe,\ NMe_2;\quad R^2 = H;\quad R^3 = H,\ OMe;\quad R^1, R^2 = OCH_2O$$

Acetals of dimethylaminobenzaldehyde and 3-indalaldehyde react with nitroacetic esters in toluene without acetic anhydride participation.

Titanium tetrachloride in pyridine induces alkenylation of nitroacetic esters (**110**) in the presence of some aliphatic, aromatic and heterocyclic aldehydes to give α,β-unsaturated α-nitrocarboxylic esters (equation 52) [138].

$$RCHO + H_2C(NO_2)CO_2Et \xrightarrow[\substack{THF, 0–22\,°C}]{TiCl_4/\text{pyridine}} RCH{=}C(NO_2)CO_2Et \quad (52)$$

$$\textbf{(110)} \qquad\qquad\qquad\qquad\qquad\qquad\qquad \textbf{(123)}$$

$$R = Ph,\ p\text{-}Me_2NC_6H_4,\ p\text{-}MeOC_6H_4,\ p\text{-}ClC_6H_4,\ 2\text{-furyl, 2-thienyl, i-Pr, t-Bu}$$

Alkenylation of nitroacetonitrile (124) with aldehydes under basic catalysis (equation 53) gives rise to a number of aromatic and heterocyclic cyanonitroethenes [9, 139].

$$RCHO + H_2C(NO_2)CN \xrightarrow[Na_2CO_3/MeNH_2 \cdot HCl]{Et_2NH \text{ or}} RCH{=}C(NO_2)CN \qquad (53)$$

$$\textbf{(124)} \qquad\qquad\qquad\qquad \textbf{(125)}$$

$$R = Ph, \ o\text{-}HOC_6H_4, \ p\text{-}ClC_6H_4, \ 2\text{-furyl}$$

Nitroacetone (127; equation 54) reacts with azomethines (126) in a similar manner to nitroacetic esters [9, 140] under the action of Ac_2O. Fatty aromatic unsaturated nitroketones are formed in this case [128].

$$R^1CH{=}NR^2 + H_2C(NO_2)COMe \xrightarrow[R^2NHAc]{Ac_2O} R^1CH{=}C(NO_2)COMe \qquad (54)$$

$$\textbf{(126)} \qquad\quad \textbf{(127)} \qquad\qquad\qquad\qquad \textbf{(128)}$$

$$R^1 = Ph, \ p\text{-}ClC_6H_4, \ p\text{-}MeOC_6H_4, \ p\text{-}O_2NC_6H_4, \ 2\text{-furyl}; \quad R^2 = Bu$$

The more active methylene fragment of nitroacetophenone [141] condenses with benzalmethylamine in acetic anhydride without heating to give α-nitrochalcone (48% yield). Various authors [142–145] have worked out a common synthetic method for α-nitroethenes (129, 130; equation 55) that uses various electron-withdrawing substituents such as CN, CO_2Alk, COMe, COPh and others located in the geminal position to the nitro group. So, CH acids (nitroacetic esters, nitroacetonitrile, nitroacetone, nitroacetophenone) that contain the active NO_2 group are alkenylated by aromatic and heterocyclic aldehydes under the action of acid catalysis ($POCl_3$, $SOCl_2$, HCl, etc.). Nitroacetonitrile undergoes this kind of reaction without a catalyst.

$$R^1CH{=}C(NO_2)CN \xleftarrow{H_2C(NO_2)CN} R^1CHO$$

$$\textbf{(129)}$$

$$\xrightarrow[(110, 127)]{H_2C(NO_2)CO_2R^2} R^1CH{=}C(NO_2)CO_2R^2 \qquad (55)$$

$$\textbf{(130)}$$

$$R^1 = Ar, \ 2\text{-furyl}, \ 2\text{-thienyl}, \ 2\text{-pyrrolyl}, \ 3\text{-indolyl}; \quad R^2 = Me, Ph, OMe, OEt$$

1.1.3 OXIDATION OF NITROGEN-CONTAINING COMPOUNDS

Nitroalkenes (133; equation 56) can be produced through α,β-unsaturated ketoxime oxidation by trifluoroperacetic acid [146]. This is a synthetic approach to aliphatic and carbocyclic hydroxynitroalkenes (75–88% yield) [146].

The synthesis of active dienophilic 3-nitrocycloalkenones devised by Corey and Estreicher [147] includes the oxidation of oxime (134) to nitroalcohol (135) and further oxidation by pyridinium chlorochromate or by a mixture of chromic and sulfuric acids (equation 57).

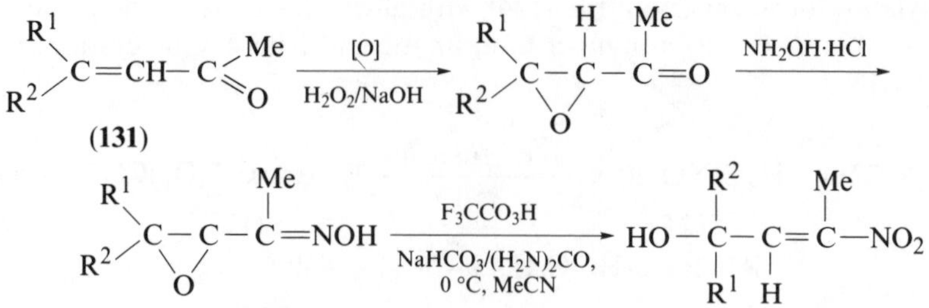

$$
\underset{\textbf{(131)}}{\overset{R^1}{\underset{R^2}{>}}C=CH-C\overset{Me}{\underset{O}{<}}}\xrightarrow[H_2O_2/NaOH]{[O]}\underset{}{\overset{R^1}{\underset{R^2}{>}}C-\overset{H}{\underset{O}{C}}-\overset{Me}{C}=O}\xrightarrow{NH_2OH\cdot HCl}
$$

$$
\underset{\textbf{(132)}}{\overset{R^1}{\underset{R^2}{>}}C-\overset{Me}{\underset{O}{C}}-C=NOH}\xrightarrow[\substack{NaHCO_3/(H_2N)_2CO,\\0\,°C,\ MeCN}]{F_3CCO_3H}\underset{\textbf{(133)}}{HO-\overset{R^2}{\underset{R^1}{C}}-\overset{Me}{\underset{H}{C}}=C-NO_2}\qquad(56)
$$

$$
R^1 = H,\ R^2 = Me \quad 84\%
$$
$$
R^1 = R^2 = Me \quad\quad 88\%
$$

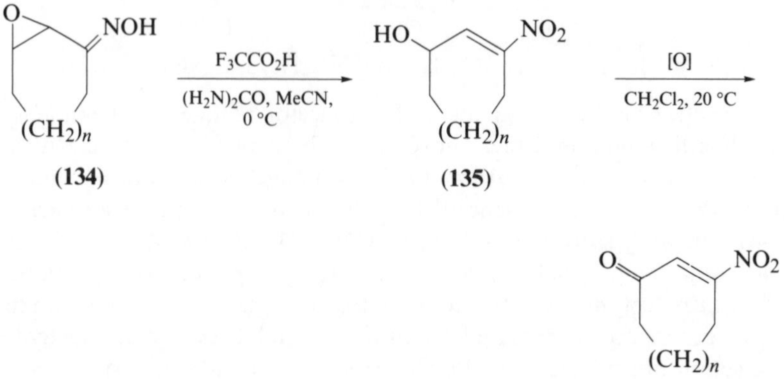

$$
\xrightarrow[\substack{(H_2N)_2CO,\ MeCN,\\0\,°C}]{F_3CCO_2H}
$$

$$
\xrightarrow[CH_2Cl_2,\ 20\,°C]{[O]}
$$

(134)   **(135)**

$$(57)$$

(136)

Oxidation and further dehydrohalogenation of α-haloketoximes lead to cyclic and acyclic nitroalkenes (**138**; equation 58) in 31–66% yield [148].

$$
\underset{\substack{|\\Hal\\ \textbf{(137)}}}{R^1CH\overset{R^2}{\underset{}{C}}=NOH}\xrightarrow[MeCN,\ TFAA\ +\ 30\%\ H_2O_2]{F_3CCO_3H}\underset{\textbf{(138)}}{R^1CH=C(R^2)NO_2}\qquad(58)
$$

(TFAA = trifluoroacetic anhydride)

Oxidation of alicyclic β,γ-nitroalkenes (equation 59) by m-chloroperbenzoic acid (MCPBA) leads to 2-nitromethylene-1-cycloalkanoles (60–83% yield) [149].

Nucleophilic nitromethylation of N-pyridinium salts (equation 60) gives (**141**) and (**143**), and subsequent oxidation produces (**142**) and (**144**). The location of the nitromethylene pattern is defined by the structure of the initial salt. If position 4 is vacant, the products (**142**) are obtained. If position 4 is occupied compounds (**144**) are produced [150].

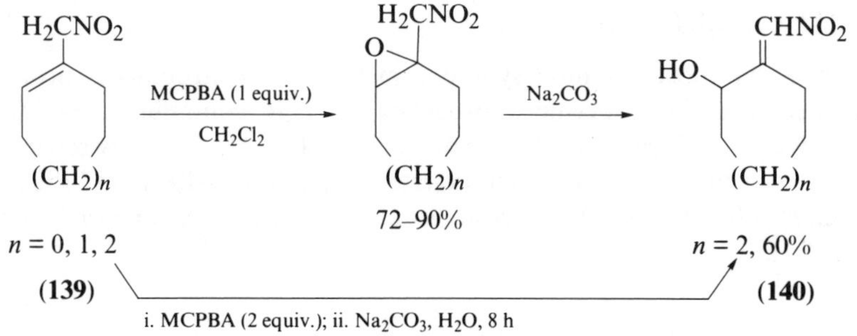

$$(59)$$

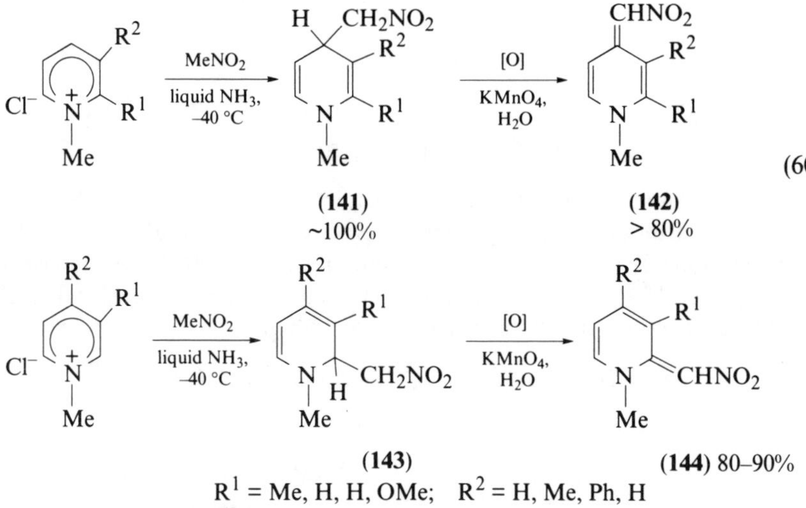

n = 0, 1, 2

(139)

72–90%

n = 2, 60%

(140)

i. MCPBA (2 equiv.); ii. Na$_2$CO$_3$, H$_2$O, 8 h

$$(60)$$

(141)
~100%

(142)
> 80%

(143)

(144) 80–90%

R^1 = Me, H, H, OMe; R^2 = H, Me, Ph, H

It is probable that nitromethylation proceeded via π-complex formation. The oxidation products as indicated by PMR are in the (E) form. The (Z) form is found for compound (142) with R^2 = H. This kind of reaction has been developed for more complicated structures with the pyridine fragment. The aim has been to synthesize heterocyclic, chromophoric systems of the type (145) [151].

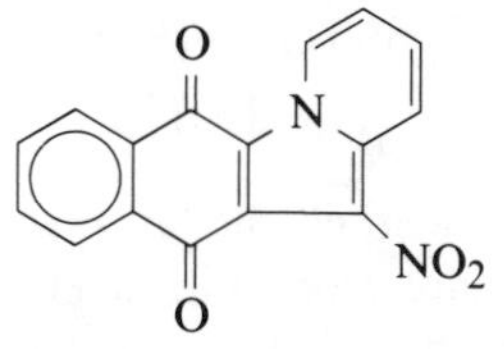

(145)

1.1.4 OTHER METHODS

Apart from the most common synthetic methods for nitroethenes presented above there are some specific approaches to these compounds. α-Nitro-β-indolylacrylates (**147**; equation 61) are prepared by the reactions between indoles and β-alkoxy-substituted α-nitroacrylic esters (**146**) in 66–84% yield [152]. The compounds (**146**) are synthesized easily from nitroacetic esters and orthoformic ethers [153].

$$\text{(146)} \quad + R^2OCH=C(NO_2)CO_2R^4 \longrightarrow$$

$$\text{(147)} \quad CH=C(NO_2)CO_2R^4 \tag{61}$$

$$R^1, R^2 = H, Me; \quad R^3, R^4 = Me, Et$$

The synthesis of polyfunctional aliphatic and alicyclic nitroalkenes by reacting acetoxy, alkylsulfonic and alkylthionyl derivatives of conjugated nitroethenes (**148**; equation 62) with copper–zinc compounds (RCu(CN)ZnI) is described in the literature [154]. This is an addition–elimination process which takes place at $-55\,^{\circ}C$ and completes within a few minutes in high yield (for example, **150**).

$$PhSO_2CH=CHNO_2 + NC(CH_2)_3Cu(CN)ZnI \xrightarrow[5\,min]{THF, -55\,^{\circ}C}$$
$$\text{(148)} \qquad\qquad\qquad \text{(149)}$$

$$NC(CH_2)_3CH=CHNO_2 \tag{62}$$
$$\text{(150) } 84\%$$

A very simple synthesis of nitroethene (**155**; equation 63) involves industrially produced, cheap compounds as starting materials [155].

$$HOCH_2CH_2OH \xrightarrow[20\%\ NaOH]{PhSH, reflux, 1\,h} PhSHCH_2CH_2OH \xrightarrow[room\ temperature]{PBr_3} PhSCH_2CH_2Br$$
$$\text{(151)} \qquad\qquad\qquad \text{(152)} \qquad\qquad\qquad \text{(153)}$$

$$\xrightarrow[\substack{PhH, 12\,h, room\ temperature}]{\substack{i.\ NaNO_2, DMSO, room\ temperature, 8\,h,\\ ii.\ 4\text{-}t\text{-}BuPhIO_2/TFA,}} Ph\overset{+}{\underset{O^-}{S}}CH_2CH_2NO_2 \xrightarrow{PhH, reflux} CH_2=CHNO_2 \tag{63}$$
$$\text{(154)} \qquad\qquad\qquad \text{(155)}$$

Zavlin and Efremov [156] worked out an elegant synthesis of nitroketene (**156**) by elimination of ethanol from nitroacetic ester under the action of phosphoric anhydride (equation 64). The structure of the nitroketene has been proven by its chemical transformations and using physicochemical methods.

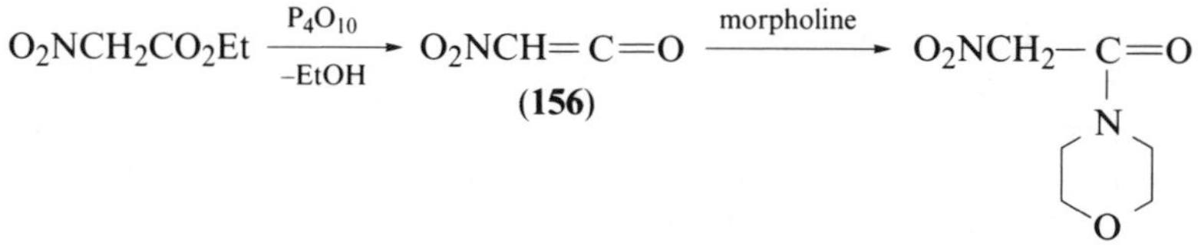

$$O_2NCH_2CO_2Et \xrightarrow[-EtOH]{P_4O_{10}} O_2NCH=C=O \xrightarrow{\text{morpholine}} O_2NCH_2-C=O \quad (64)$$

(156)

(157)

1.2 MONONITRODIENES

Buckley [157] was the first to describe nitrodienes with conjugated double bonds, e.g. 3-nitro-1,3-pentadiene and 2-methyl-3-nitro-1,3-butadiene.

A common synthetic approach to 2-nitroalkadienes (**159**; equation 65) includes the condensation of nitroallyls (**158**) with formaldehyde and subsequent dehydration of the corresponding nitroalcohol salts [158a].

$$R^1CH=C(R^2)CH_2NO_2 \xrightarrow{CH_2O(MeONa)} R^1CH=C(R^2)C(CH_2OH)=NOO^-Na^+$$

(158)

$$\xrightarrow{HCl} R^1CH=C(R^2)C(NO_2)=CH_2 \quad (65)$$

(159)

$$R^1 = R^2 = (CH_2); \quad R^1 = H; \quad R^2 = Me$$

Acetylation of (**161**; equation 66) followed by elimination of $MeCO_2H$ leads to 3-nitrodienes (**163**) [158b].

$$R^1CH=CR^2CH_2NO_2 \xrightarrow{MeCHO, base} R^1CH=CR^2CH(NO_2)CHMe$$

$$\text{(160)} \qquad\qquad \text{(161)} \qquad\qquad \overset{|}{OH}$$

$$\xrightarrow{MeCOCl} R^1CH=CR^2CH(NO_2)CH(Me)OCOMe$$

(162)

$$\xrightarrow[-MeCO_2H]{MeCO_2Na} R^1CH=CR^2C(NO_2)=CHMe \quad (66)$$

(163)

$$R^1 = R^2 = H; \quad R^1 = Me, R^2 = H; \quad R^1 = H, R^2 = Me; \quad R^1 = R^2 = Me$$

Another way of building up the nitrodiene system is through candensation of unsaturated aldehydes with nitroalkanes. For example, 1-nitro-1,3-pentadiene is prepared by acylation–deacylation of 1-nitro-3-pentene-2-ol, which is produced by the reaction between nitromethane and crotonic aldehyde [159] or acetaldol [160].

Condensation of cinnamic aldehyde, its derivatives [161] and heteroarylacrylic aldehydes [162–164] with nitromethane followed by neutralization of the salt produces 4-aryl-1-nitro-1,3-butadienes, 4-heteroaryl-1-nitro-1,3-butadienes and 1-nitro-1,3,5-hexatrienes. The same aldehydes as well as methyl styryl ketone condense in absolute ethanol with nitroacetonitrile without a catalyst or with nitroacetone in the presence of acid to give geminally functionalized nitrodienes (**165**; equation 67) in good yield [144a, 145a, 165, 166].

$$R^1CH{=}CHC(O)R^2 + H_2C(X)NO_2 \xrightarrow{EtOH} R^1CH{=}CHC(R^2){=}C(X)NO_2 \quad (67)$$

(164)   **(165)** 64–80%

$R^1 = $ Ph, 2-furyl, 2-thienyl; $R^2 = $ H, Alk; X = COMe, CN

It is possible to produce conjugated nitrodienes of the type (**167**; equation 68) by dehydrohalogenation [167] of (**166**).

$$NO_2CH_2CHX(CH{=}CH)_nCO_2Et \xrightarrow[-HX]{} O_2N(CH{=}CH)_{n+1}CO_2Et \quad (68)$$

(166)   **(167)**

$$X = Cl, Br; \quad n = 0\text{--}2$$

A synthesis of 3-nitropentadienes (**169**; equation 69) has been devised by Kvitko *et al.* [168–170] consisting of the condensation of sodium nitromalonic dialdehyde (**168**) with CH acids under the action of basic catalysis.

$$[O_2NC(CHO)_2]^-Na^+ + H_2CR^1R^2 \xrightarrow{C_5H_5N} [R^1R^2C{=}CHCCH{=}CR^1R^2]^-Na^+$$

(168) $\underset{}{\overset{|}{NO_2}}$

$$\xrightarrow{H^+} R^1R^2C{=}CHC(NO_2){=}CHCHR^1R^2 \quad (69)$$

(169)

$$R^1 = NO_2, R^2 = CN; \quad R^1 = CO_2Et, R^2 = CN; \quad R^1 = R^2 = CN$$

In contrast with the methylene-containing aliphatic compounds, heterocyclic molecules with the same active fragment (e.g. iodomethylates of α-picoline, N-methyl-2-methylbenzimidazole, etc.) (**170**; equation 70) react only with one aldehyde group of dialdehyde (**168**).

(168)   **(170)**   **(171)**

Nitromalonic dialdehyde (**172**) and its salt (**173**; equation 71) are unique bifunctional systems owing to the combination of electrophilic and nucleophilic centres [168]. The reaction of (**173**) with primary aliphatic amines leads to (**174**).

Condensation of these products with CH acids results in the formation of novel, substituted dienes (175).

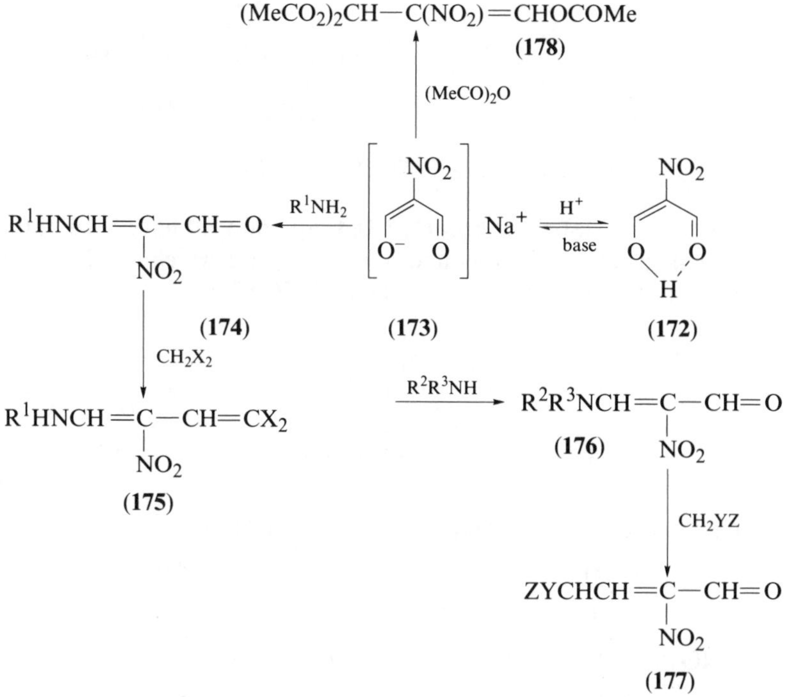

$$(MeCO_2)_2CH-C(NO_2)=CHOCOMe$$
(178)

$(MeCO)_2O$

$$R^1HNCH=C-CH=O \xleftarrow{R^1NH_2} \left[\begin{array}{c} NO_2 \\ \diagup \diagdown \\ O^- \quad O \end{array}\right] Na^+ \underset{base}{\overset{H^+}{\rightleftharpoons}} \begin{array}{c} NO_2 \\ \diagup \diagdown \\ O \quad O \\ H \end{array}$$

(71)

$R^1HNCH=C-CH=O$ with NO_2 below

(174)   (173)   (172)

CH_2X_2

$$R^1HNCH=C-CH=CX_2 \xrightarrow{R^2R^3NH} R^2R^3NCH=C-CH=O$$

with NO_2 below left; (176) with NO_2 below

(175)

CH_2YZ

$$ZYCHCH=C-CH=O$$

with NO_2 below

(177)

R^1 = Me, Ph; R^2, R^3 = Alk, Ar;   X = Y = CN;   Z = CN, NO_2, CO_2Et

2-Nitropropanedial (172) (a highly polarized *cis*-monoenol) is stabilized by intramolecular hydrogen bonding. A specific property of this compound in comparison to the anion (173) is its ability to form aminonitroacroleins (176) in the reaction with secondary amines [169]. Further reaction of compounds (176) with CH acids leads to the formation of substituted 2-nitrobutenals (177). In accordance with infrared (IR) and nuclear magnetic resonance (NMR) data, the structures of N-monoalkylaminonitroacroleins (174) and N,N-dialkylamino-nitroacroleins (176) and the corresponding aryl derivatives are nearly dipolar [169].

One possible synthetic approach to nitrodienes is the direct introduction of the nitro group into a diene system. Coe and Doumani [171] declared the direct nitration of butadiene by nitric acid to give nitrobutadiene, but subsequently this was disproved.

The addition of NO_2X to dienes (179; equation 72) followed by elimination of HX is a common method for preparing nitrodienes (180) which gives access to a large variety of 1-nitrodienes, but usually in rather low yield.

$$R^1CH=CR^2CR^3=CHR^4 + NO_2X \longrightarrow R^1\underset{\underset{NO_2}{|}}{C}HCR^2=CR^3\underset{\underset{X}{|}}{C}HR^4$$

$$(179)$$

$$\xrightarrow[-Hx]{base} R^1\underset{\underset{NO_2}{|}}{C}=CR^2CR^3=CHR^4 \qquad (72)$$

$$(180)$$

$$X = NO_2, [172, 173], ONO_2 [174, 175]$$

A number of nitro-1,3-alkadienes (**183, 184**; equation 73) have been obtained by dehydrohalogenation of compounds (**181**) and (**182**), which are the products of the addition of iodine and N_2O_4 to dienes [176, 177].

$$R^1CH=CR^2CH=CH_2$$

$$\downarrow {\scriptstyle N_2O_4 + I_2}$$

$$(73)$$

$$R^1CH(NO_2)CR^2=CHCH_2I \qquad\qquad R^1CH(I)CR^2=CHCH_2NO_2$$

$$(181) \qquad\qquad\qquad\qquad\qquad (182)$$

$$\downarrow {\scriptstyle Pb(OAc)_2} \qquad\qquad\qquad\qquad\qquad \downarrow {\scriptstyle Pb(OAc)_2}$$

$$R^1C(NO_2)=CR^2CH=CH_2 \qquad\qquad R^1CH=CR^2CH=CHNO_2$$

$$(183) \qquad\qquad\qquad\qquad\qquad (184)$$

Very high yields of some aliphatic and alicyclic nitrodienes have been achieved in the nitrofluoroacetoxylation of dienes. For example, trifluoroacetyl nitrate combines with 1,3-butadiene to produce (**186a**) and (**186b**). Subsequent elimination gives nitrobutadiene (**187**; equation 74) [178, 179].

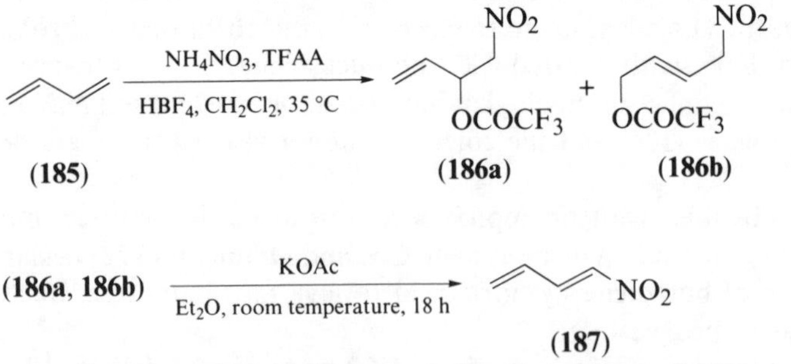

$$(185) \xrightarrow[\text{HBF}_4, \text{CH}_2\text{Cl}_2, 35\,°C]{\text{NH}_4\text{NO}_3, \text{TFAA}} (186a) + (186b)$$

$$(74)$$

$$(186a, 186b) \xrightarrow[\text{Et}_2\text{O, room temperature, 18 h}]{\text{KOAc}} (187)$$

This is the synthetic approach to a series of nitrobutadiene derivatives (**188**), as well as to 1-nitrocyclohexa-1,3-diene (70% yield) and 1-nitrocycloocta-1,3-

diene (70% yield) [178, 179]. The process takes place under the action of $MeCO_2K$, NaH, Et_3N and $EtCO_2K$.

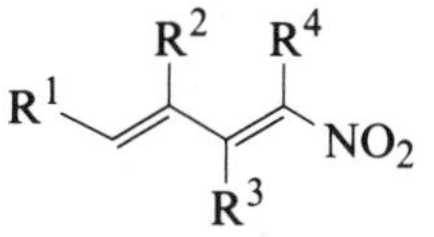

	R^1	R^2	R^3	R^4	Yield (%)
	H	H	Me	H	55
	H	Me	Me	H	75
(188)	Me	H	H	Me	84
	CO_2Et	H	H	H	35
	CO_2Me	H	H	Me	94

A highly selective method for the preparation of functionalized nitroalkenes involving addition of Cu–Zn–organic compounds to β-(alkylthio)nitroalkenes and β-(phenylsulfonyl) nitroalkenes followed by elimination of the sulfur-containing group [154] has been used successfully for the synthesis of substituted 1-nitro-1,3-dienes (**191, 193**; equation 75) [180].

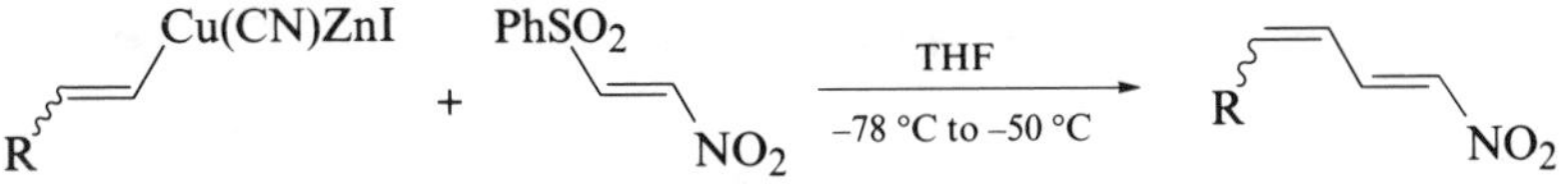

(**189**) (**190**) (**191**)

R = Bu, 82%, (Z, E)
R = C_6H_{13}, 90%, (E, E)   (75)

(**189**) + (**192**) → (**193**) 90%

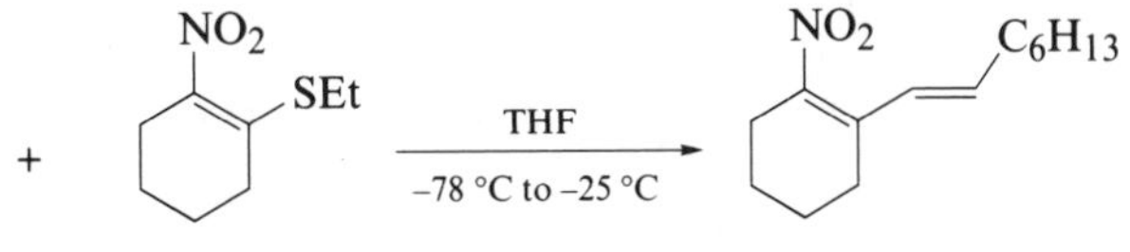

Attempts to synthesize 2-nitro-1,3-dienes by conjugated diene nitroselenation followed by oxidative deselenation have resulted in only the products of the second transformation, namely 3,4-epoxy-3-nitroalkenes (**195, 196**; equation 76), probably owing to the high activity of 2-nitro-1,3-dienes [181]. Subsequently pure nitrodienes (**197**) and (**198**) have been isolated in the deselenation of compounds of type (**194**) using dimethylaminopyridine (DMAP) [182a].

Deselenation of other compounds under similar conditions gives unstable 2-nitrodienes, which have been used in heterodiene synthesis *in situ*. The same problem has been observed for 3-substituted 2-nitro-1,3-dienes. The synthesis of these compounds (equation 77) is described in the literature [182b].

An original synthetic method for substituted 2-nitro-1,3-butadienes based on desulfonylation of 3-nitro-3-thiolene-1,1-dioxides has been suggested by Berestovitskaya *et al.* [183] (see Section 3.5).

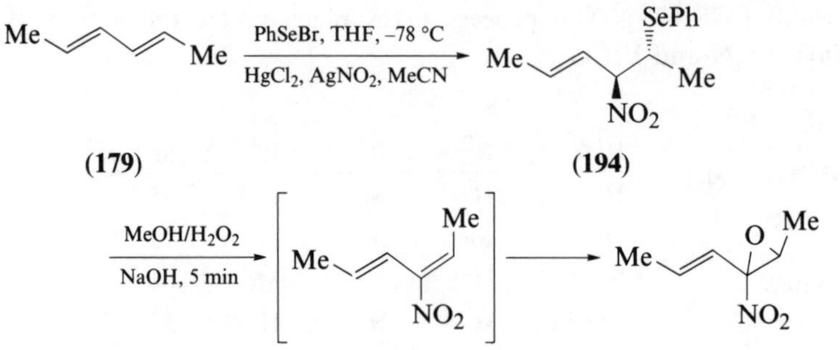

(179)   (194)

$$(76)$$

(195)
40–51%

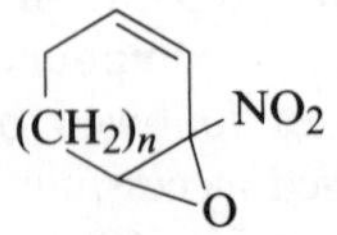

(196)

n = 1 (68%), 2 (41%)

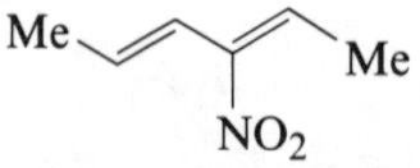

(197) 45% 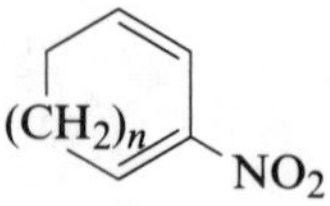   (198) n = 1 (38%), 2 (75%)

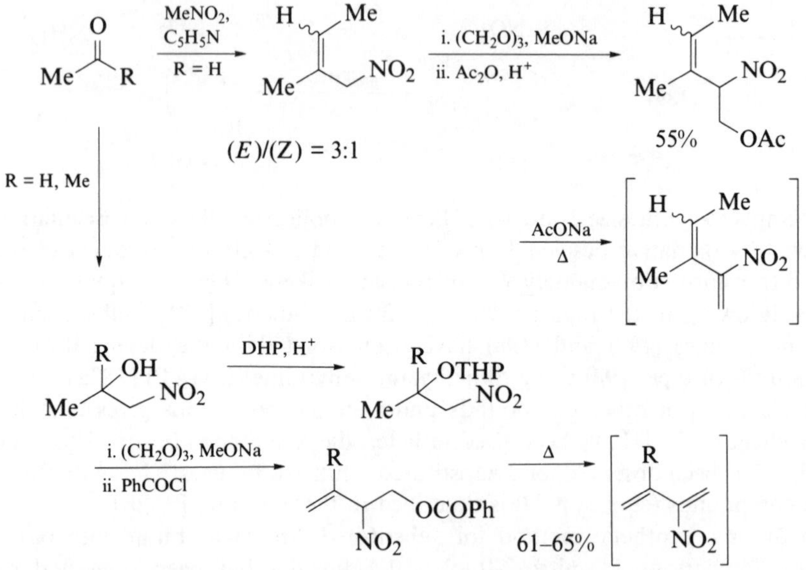

(E)/(Z) = 3:1   55%

$$(77)$$

61–65%

(DHP = 2,3-dihydro-4H-pyrane, THP = α-tetrahydropyranyl)

1.3 DINITROALKENES

Addition of N_2O_4 to the triple bond of acetylene derivatives is a preparative method for 1,2-dinitroalkenes [184–186], as is the oxidation of 1,2-dinitroalkanes produced by alkene nitration [187]. The N_2O_4 addition method is restricted by the instability of some products because of partial oxidation of the acetylene by N_2O_4, particularly the monosubstituted acetylenes. The products of the reactions of acetylene with N_2O_4 have not been isolated.

Nitration of 3-hexyne (**199**; equation 78) leads to the formation of stereoisomeric dinitroalkenes (**200a**) and (**200b**) as well as some other products (**201–203**) [184, 185].

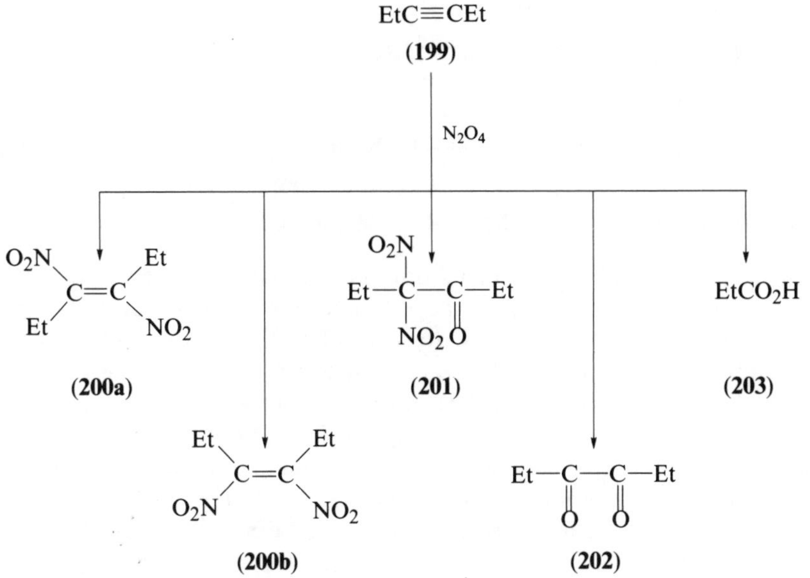

$$EtC{\equiv}CEt$$
(199)

N_2O_4

(78)

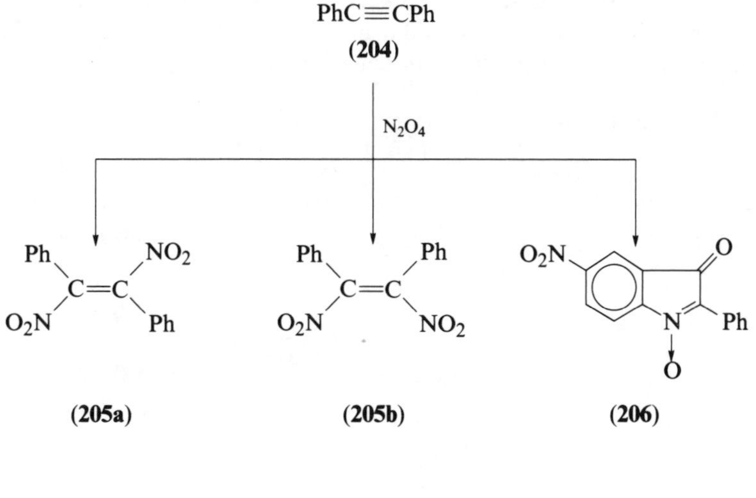

(200a)   **(201)**   **(203)**

(200b)   **(202)**

$$PhC{\equiv}CPh$$
(204)

N_2O_4

(79)

(205a)   **(205b)**   **(206)**

The nitration of only one monosubstituted alkyne has been reported, i.e. phenylacetylene, which gave the corresponding 1,2-dinitroalkene [188].

The action of N_2O_4 upon tolan (204; equation 79) in ether also results in the formation of the stereoisomeric products of addition (205a, 205b), along with (206) [189].

(Z)-Dinitrostilbene (205b) is obtainable by nitration of (E)-isomeric nitrocarboxy-α-stilbenes and methoxycarbonyl-α-stilbenes [190]. (Z)-1,2-Dinitro-1-aryl-propenes are prepared in a similar way [190].

Oxidation methods of the C=C bond formation in dianions (208; equation 80) has been used by Lipina et al. [187] for the preparation of dinitroalkenes (209) and 1,4-dinitro-1,3-alkadienes.

$$O_2NCHR^1CR^2HNO_2 \xrightarrow{\text{2MeONa}} {}^-OON{=}CR^1CR^2{=}NOO^- \cdot 2Na$$

$$\text{(207)} \qquad\qquad\qquad\qquad \text{(208)}$$

$$\xrightarrow[-2e^-]{\text{Br}_2} O_2NCR^1{=}CR^2NO_2 \qquad\qquad (80)$$

$$\text{(209)}$$

$$R^1 = R^2 = H; \quad R^1 = R^2 = Me; \quad R^1 = Me, R^2 = Et; \quad R^1 = R^2 = Ph$$

The simplest vicinal dinitroalkene 1,2-dinitroethene (209) has been synthesized in low yield (22%) by oxidation of the 1,2-dinitroethene disodium salt using bromine at low temperature ($-50\,°C$ to $-60\,°C$) [191]. The unstable dinitroethene (209) was frozen out of the reaction mixture at $-70\,°C$. Besides this product, 1,4-dinitro-1,3-butadiene (9%) and 1,2-dibromo-1,2-dinitroethane (5%) were isolated by means of chromatography on silica [191].

The reaction of 2,3-dinitrobutane with MeONa and Br_2 leads to 2,3-dinitro-2-butene (24%) and 2-bromo-2-nitro-3-methoxybutane (18%). 2,3-Dinitro-2-pentene (20%) and (Z)-1,2-dinitrostilbene and (E)-1,2-dinitrostilbene (33%) have been prepared in a similar way [187]. However, this method is inapplicable to the production of 1,2-dinitrocycloalkenes because of the facile denitration of the corresponding salts. Recently, the new 1,2-dinitrocyclohexene (213) was synthesized [192] in accordance with equation (81).

(210)   (211)   (212) 42%

(213) 58%   (81)

Two nitroalkenes (**215**; equation 82) have been prepared by oxidative coupling of 1-nitro-1-haloalkanes (**214**) in the presence of NaOH [193].

$$\underset{\underset{(214)}{\overset{|}{\text{Hal}}}}{\text{AlkCHNO}_2} \xrightarrow{\text{NaOH}} \underset{\underset{(215)}{\overset{|}{\text{O}_2\text{N}}\ \overset{|}{\text{NO}_2}}}{\text{AlkC}=\text{CAlk}} \qquad (82)$$

Alk = Me, Et

1.3.1 GEMINAL DINITROALKENES

The simplest, highly reactive, geminal dinitroalkene (**217**; equation 83) is formed by spontaneous dehydration of dinitroethanol (**216**). Despite numerous attempts (**217**) has not been isolated in a free state, and its formation has only been implied by the products of diene synthesis (**219**) [194] or 1,3-dipolar cyclo-addition [195].

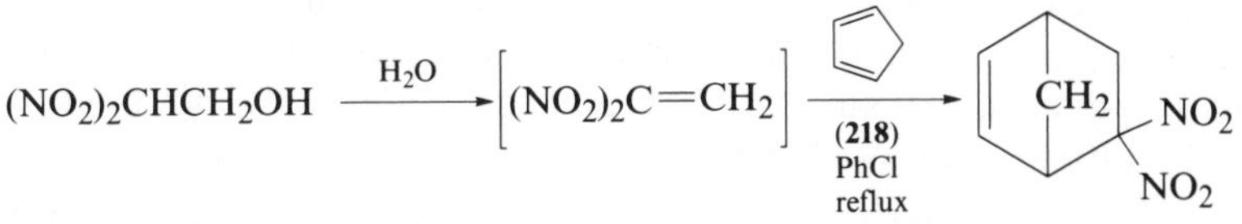

$$\underset{(216)}{(\text{NO}_2)_2\text{CHCH}_2\text{OH}} \xrightarrow{\text{H}_2\text{O}} \underset{(217)}{\left[(\text{NO}_2)_2\text{C}=\text{CH}_2\right]} \xrightarrow[\substack{(218) \\ \text{PhCl} \\ \text{reflux}}]{} \underset{(219)}{\text{CH}_2\ \text{NO}_2 \ \text{NO}_2} \qquad (83)$$

1,1-Dinitroethene forms as an intermediate in some other reactions of geminal nitroethane derivatives [195, 196].

Stable 1,1-dinitrostyrenes (**221**; equations 84) and 1,1-dinitro-2,2-diarylethenes (**223**) have been obtained as a result of nitrostyrene and diarylethene nitration [9, 197].

$$\underset{(220)}{\text{ArCH}=\text{CHNO}_2} \xrightarrow{\text{N}_2\text{O}_4} \underset{(221)}{\text{ArCH}=\text{C(NO}_2)_2}$$

$$(84)$$

$$\underset{(222)}{\text{Ar}_2\text{C}=\text{CHR}} \xrightarrow{\text{N}_2\text{O}_4} \underset{(223)}{\text{Ar}_2\text{C}=\text{C(NO}_2)_2}$$

R = H, CO$_2$H

1,1-Dinitro-2,2-diphenylethene (**225**; equation 85) has been prepared by nitration of 1,1-diphenylethane (**224**) using dilute HNO$_3$ [198].

$$\underset{(224)}{\text{Ph}_2\text{CHCH}_3} \xrightarrow[70-80\,^{\circ}\text{C}]{\text{HNO}_3} \underset{(225)\ 53-80\%}{\text{Ph}_2\text{C}=\text{C(NO}_2)_2} \qquad (85)$$

The reaction of iodotrinitromethane (**226**; equation 86) (as well as bis-trinitro-methylmercury) with diphenyldiazomethane or 9-diazofluorene leads to the 1,1-dinitro-2,2-diarylethene (**228**) [196, 199].

$$IC(NO_2)_3 + R_2CN_2 \xrightarrow[-I_2, -NO_2, -N_2]{Et_2O, 25\,°C} R_2C{=}C(NO_2)_2 \qquad (86)$$

$$\textbf{(226)} \qquad \textbf{(227)} \qquad\qquad\qquad\qquad \textbf{(228)}$$

$$R = Ph\ (87\%),\ Ph{-}Ph\ (85\%)$$

Geminal dinitrostyrenes are prepared in good yield (60–80%) by the reaction of tetranitromethane, bromotrinitromethane, dibromodinitromethane, dinitro-cyanocarboethoxymethane or triazolyltrinitromethane with phenyldiazo-methane [200].

1.4 DINITRODIENES

There are two main synthetic approaches to conjugated 1,4-dinitro-1,3-buta-dienes, namely condensation of 1,2-dinitroethane with aldehydes and conden-sation of nitroalkanes with α,β-dicarbonyl compounds. Catalytic condensation in the presence of base according to the first method is unsuccessful because of the instability of 1,2-dinitroethane under these conditions. However, glyoxal combines with nitromethane [201, 202] and some other simple nitroalkanes to form 1,4-dinitro-1,3-dienes (**233**; equation 87).

$$2RCH_2NO_2 + O{=}CHCH{=}O \longrightarrow RCH(NO_2)CH(OH)CH(OH)CHNO_2R$$

$$\textbf{(229)} \qquad\qquad \textbf{(230)} \qquad\qquad\qquad\qquad \textbf{(231)}$$

$$\longrightarrow RCH(NO_2)CH(OCOMe)CH(OCOMe)CH(NO_2)R$$

$$\textbf{(232)}$$

$$\longrightarrow RC(NO_2){=}CHCH{=}C(NO_2)R \quad (87)$$

$$\textbf{(233)}$$

$$R = H,\ Alk$$

Another synthetic method involves elimination of the elements of difunctiona-lized 1,4-dinitroalkanes.

The simplest representative of the 1,4-dinitro-1,3-dienes is also obtainable according to the reactions in equation (88) [203].

$$O_2NCH_2CH{=}CHCH_2NO_2 \xrightarrow{Cl_2} O_2NCH_2\underset{\underset{Cl}{|}}{C}H\underset{\underset{Cl}{|}}{C}HCH_2NO_2 \xrightarrow[-HCl]{Pb(OAc)_2}$$

$$\textbf{(234)} \qquad\qquad\qquad\qquad \textbf{(235)}$$

$$O_2NCH{=}CHCH{=}CHNO_2 \quad (88)$$

$$\textbf{(236)}$$

A similar method is used to obtain 2-methyl-1,4-dinitrobutadienes and 2,3-dimethyl-1,4-dinitrobutadienes under the action of sodium acetate. It is possible to produce two stereoisomers of the former product [204, 205].

Halogenation of 1,4-dinitro-1,3-butadienes followed by HHal elimination gives 1,4-dinitro-1,4-dihalo-1,3-butadienes [206, 207].

Nitration of 6-nitrocholesta-3,5-diene by NOCl produces 3,6-dinitrocholesta-3,5-diene (237). The oxidation of the latter by lead(IV) acetate leads to the conjugated steroid dinitrotriene (238) [208, 209].

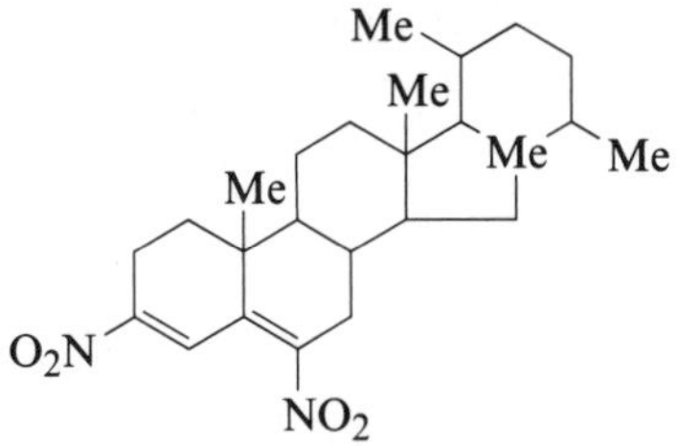

(237) 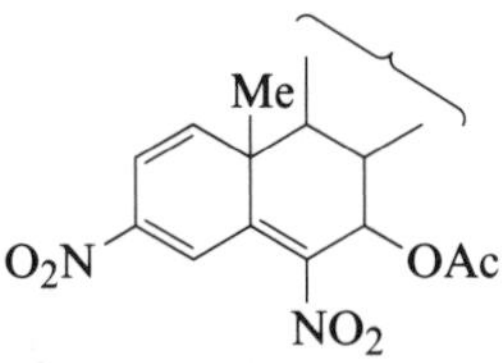   (238)

Oxidation of 3-hexene-2,5-dione dioximes by Pb(OAc)$_4$ gives 2,4-dinitro-2,4-hexadienes (11%), along with a number of heterocyclic products [210].

Desulfonylation of 2,5-dinitro-3-thiolene 1,1-dioxide (239; equation 89) at 140–180 °C represents an easy route to the corresponding substituted 1,4-dinitro-1,3-diene (see Section 3.5) [211].

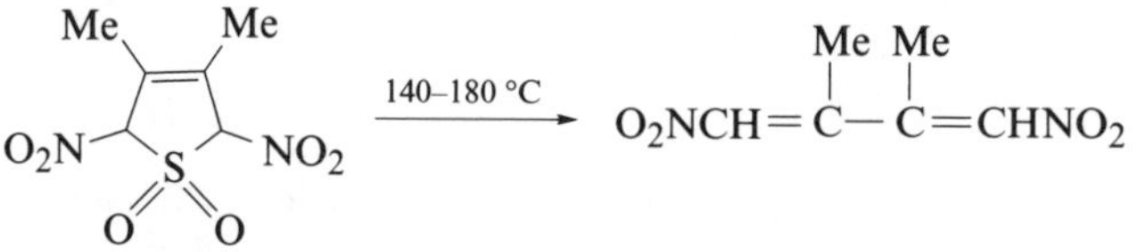

(239)   (240)   (89)

The synthesis of 2.3-dinitro-1,4-diamino-1,3-butadienes (241; equation 90) from 3,4-dinitrothiophene and amines is based on the ring-opening reaction. Reaction of (241) with Grignard reagents leads to alkyl-substituted and aryl-substituted 2,3-dinitro-1,3-butadienes (242a, 242b) [212].

The principally new synthetic approach to 1,4-dinitro-1,3-butadienes (245; equation 91) [205, 213] consists in the oxidation of 1,4-dinitro-2-butene dianions (244). A similar method has been used to prepare 1,2-dinitroalkenes [187]. Bromine has been found to be a universal oxidizer for these anions though FeCl$_3$, KMnO$_4$, AgNO$_3$, K$_2$S$_2$O$_8$ and other oxidizing agents may be successfully used for the oxidation of dinitro compounds that are stable in aqueous solution. It has been shown by redox titration that the reaction is based on the two-electron oxidation of the conjugated dianion [214].

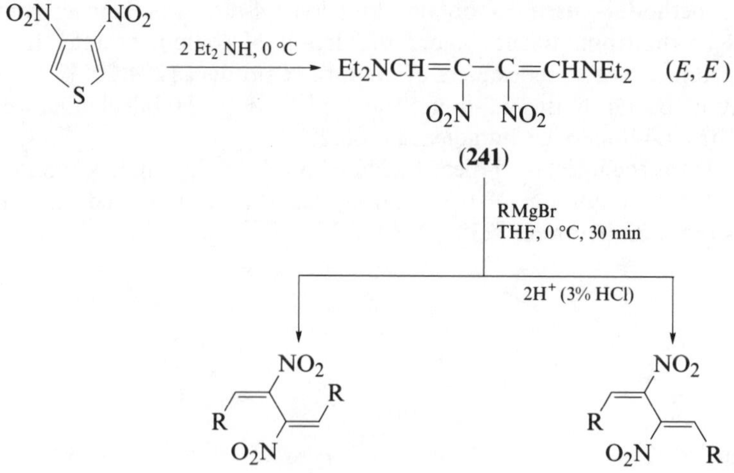

$$\underset{(241)}{Et_2NCH=\underset{O_2N}{\overset{|}{C}}-\underset{NO_2}{\overset{|}{C}}=CHNEt_2} \quad (E, E)$$

via

$$\begin{array}{c} O_2N \quad NO_2 \\ \diagup \diagdown \\ S \end{array} \xrightarrow{\text{2 Et}_2\text{NH, 0 °C}}$$

RMgBr
THF, 0 °C, 30 min

2H⁺ (3% HCl)   (90)

(E, E) (242a)   (E, Z) (242b)

R	Yield (%)	(E, E)/(E, Z)
Et	94	92:8
Cyclohexyl	90	85:15
Ph	88	100:0

$$\underset{(243)}{O_2NCHR^1CR^2=CR^3CR^1HNO_2} \xrightarrow{\text{NaOMe}} \underset{(244)}{\left[{}^-O_2N=CR^1CR^2=CR^3CR^1=NO_2{}^- \right]}$$

$$2Na^+ \xrightarrow[-2e^-]{\text{'Br}_2\text{'}} \underset{(245)}{O_2NCR^1=CR^2-CR^3=CR^1NO_2} \quad (91)$$

$R^1 = R^2 = R^3 = H;$ $R^1 = H, R^2 = R^3 = Ph;$

$R^1 = R^3 = H, R^2 = Me;$ $R^1 = Ph, R^2 = R^3 = H;$

$R^1 = H, R^2 = R^3 = Me$ $R^1 = CH_2OH, R^2 = R^3 = H$

Compound (243) was prepared by 1,3-butadiene nitration [205]. This method also gives rise to 1,4-dinitrocyclohexadiene (246) and 1,4-dinitrocyclooctatetraene (247).

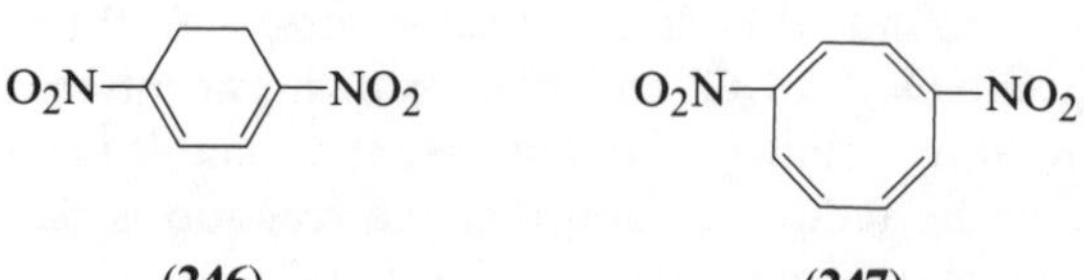

$$O_2N-\bigcirc-NO_2 \qquad\qquad O_2N-\bigcirc-NO_2$$

(246)   (247)

1.5 MONONITROTRIENES AND DINITROTRIENES

The first preparation of fatty aromatic nitrotrienes (**250**; equation 92) was carried out by Severin and Ipach [162] through the dehydration of unsaturated

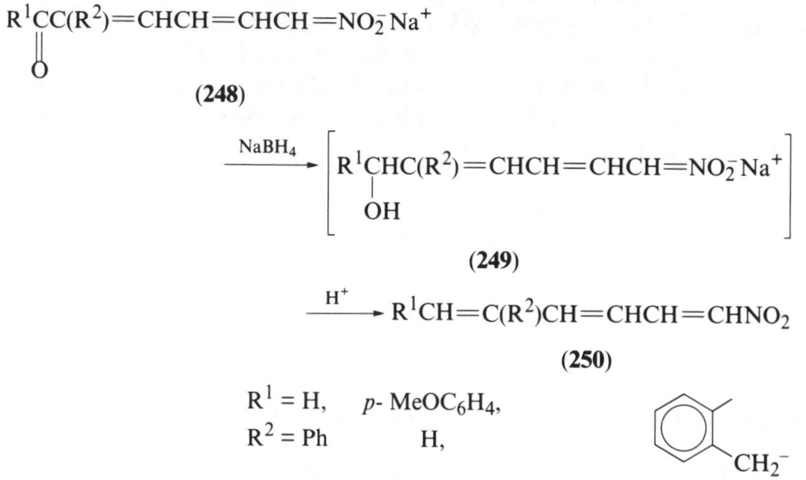

$$R^1CC(R^2)=CHCH=CHCH=NO_2^-Na^+$$
$$\underset{O}{\|}$$

(248)

$$\xrightarrow{NaBH_4} \left[\underset{OH}{R^1CHC(R^2)}=CHCH=CHCH=NO_2^-Na^+ \right]$$

(249)　　　　　　　　　　　　(92)

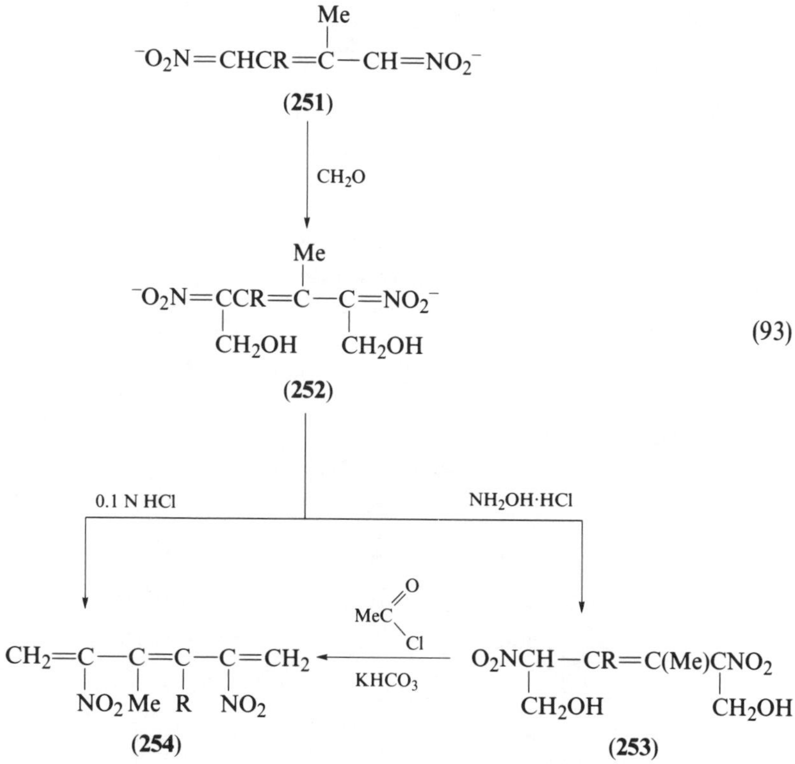

$$\xrightarrow{H^+} R^1CH=C(R^2)CH=CHCH=CHNO_2$$

(250)

R^1 = H,　　p- MeOC$_6$H$_4$,

R^2 = Ph　　　H,

$$\underset{\text{(251)}}{\overset{\overset{\displaystyle Me}{|}}{{}^-O_2N=CHCR=C-CH=NO_2^-}}$$

$\downarrow$ CH$_2$O

$$\underset{\underset{\text{(252)}}{\overset{|}{CH_2OH}\quad\overset{|}{CH_2OH}}}{\overset{\overset{\displaystyle Me}{|}}{{}^-O_2N=CCR=C-C=NO_2^-}}$$　　　　　(93)

0.1 N HCl　　　　　　　　　　　　　　NH$_2$OH·HCl

$$\underset{\underset{\text{(254)}}{\overset{|}{NO_2}\ \overset{|}{Me}\ \ R\ \overset{|}{NO_2}}}{CH_2=C-C=C-C=CH_2} \xleftarrow[\text{KHCO}_3]{\underset{Cl}{\overset{O}{\diagup\!\!\diagup}}{MeC}} \underset{\underset{\text{(253)}}{\overset{|}{CH_2OH}\qquad\overset{|}{CH_2OH}}}{O_2NCH-CR=C(Me)CNO_2}$$

nitroalcohol salts (249). The alcohols (249) were prepared by reduction of the corresponding ketones (248). The latter are the products of reaction between fatty aromatic carbonyl compounds and 1-nitro-4-dimethylamino-1,3-alkadienes.

Formation of compounds (253) and (254) results from the acidification of salt (252; equation 93). Dinitroglycol (253) may be transformed to the 2,5-dinitro-1,3,5-hexatriene (254) under the appropriate conditions [213].

Fatty aromatic dinitrotrienes (255) with E,E,E-structure are prepared by condensation of 1,4-dinitro-2-butenes with aromatic aldehydes in the presence of ammonium acetate or via Schiff bases [215].

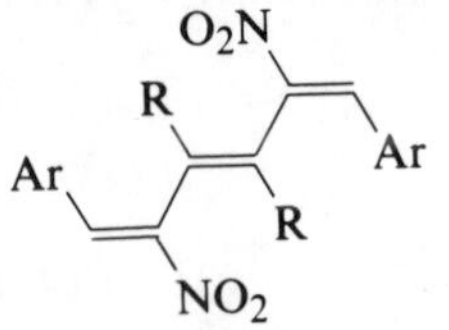

(255)

1.6 NITROALKYNES AND NITROENYNES

Attempts to produce p-nitrophenylnitroacetylene by nitration of β-nitrostyrene have so far been unsuccessful [216, 217].

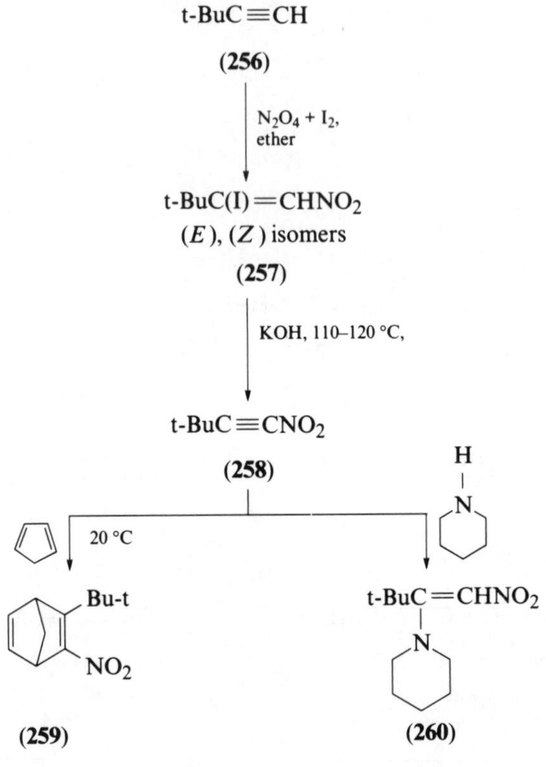

(94)

In 1969, Jager and Viehe [218] synthesized alkylnitroacetylenes by the action of KOH on vicinal halonitroethenes. For example, steaming of (257) over solid KOH with heating in vacuum results in the formation of the nitroacetylene (258; equation 94) (96–97%). The purity of the isolated product (258) is controlled by GLC and is around 98–100%. Compound (258) participates readily in the reactions of hydration, bromination, hydrohalogenation, alkoxylation, amination and diene synthesis [219, 220].

Jager et al. [221] have performed several substitutional nitrations of (261; equation 95).

$$R^1C\equiv CSn(R^2)_3 \xrightarrow[\substack{\text{i. } N_2O_5, \text{ ether, } -60\ ^\circ C \\ \text{ii. } NO_2BF_4, CH_2Cl_2, -20\ ^\circ C \\ \text{iii. } N_2O_4, -60\ ^\circ C;\ \text{iv. } [O]}]{} R^1C\equiv CNO_2 \qquad (95)$$

$$(261) \qquad\qquad\qquad\qquad\qquad\qquad\qquad (262)$$

R^1	R^2	Yield (%)		
		i	ii	iii, iv
t-Bu	Me	40	70	40
i-Pr	n-Bu	35	40	30
n-Bu	Me	–	10	–
i-Pr	Me	30	50	20

The method of C–Sn bond cleavage in compounds of type (261) using N_2O_4 and N_2O_5 has made it possible to widen the range of nitroacetylenes $XC\equiv CNO_2$ obtainable with $X = Ph, Me_3Si, Pr, i\text{-}Pr$, and others [222, 223]. Nitration–desilylation of (263) under the action of nitronium hexafluorophosphate (NHFP) gives rise to an extended number of trialkylsilylnitroacetylenes (264; equation 96) [223].

$$Me_3SiC\equiv CSiR^1R^2R^3 \xrightarrow[\text{1 h, 20 } ^\circ C]{\text{NHFP/MeCN}} R^1R^2R^3SiC\equiv CNO_2 + Me_3SiC\equiv CNO_2$$

$$(263) \qquad\qquad\qquad\qquad\qquad (264) \qquad\qquad\qquad\qquad (96)$$

$$R^1R^2R^3Si = Me_3Si, i\text{-}PrMe_2Si, t\text{-}BuMe_2Si, i\text{-}Pr_3Si$$

A convenient synthetic approach to trimethylsilylnitroacetylene (266; equation 97) is through the reaction between compound (265) and NO_2BF_4 in the presence of CsF and an ammonium salt [224].

$$Me_3SiC\equiv CSiMe_3 + NO_2BF_4 \xrightarrow[CH_2Cl_2,\ 2\ h,\ 20\ ^\circ C]{CsF/n\text{-}Bu_4N^+BF_4^-} Me_3SiC\equiv CNO_2 \qquad (97)$$

$$(265) \qquad\qquad\qquad\qquad\qquad\qquad\qquad (266)\ (70\%)$$

The chemical activity of these compounds in the cycloaddition reaction [225] is illustrated in equation (98).

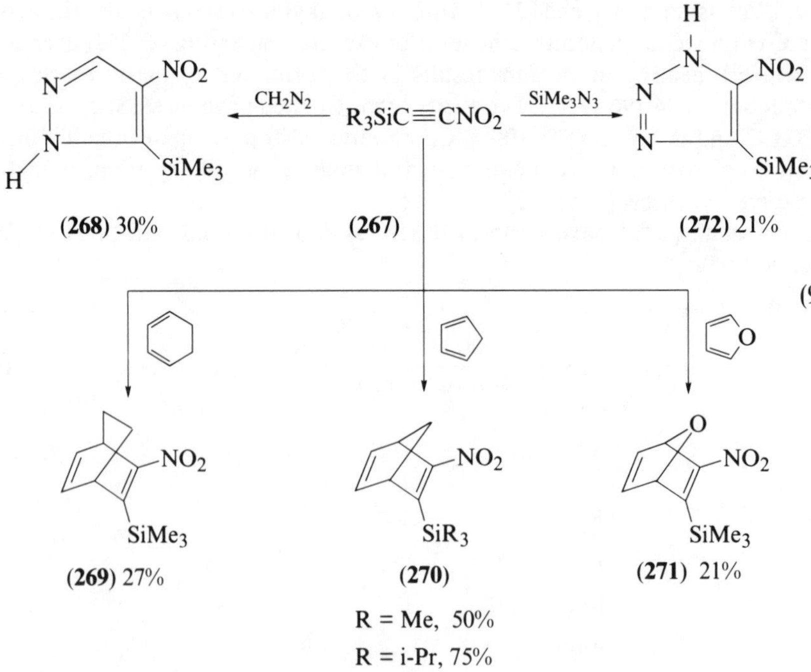

(98)

Nitroacetylenes containing the NO_2 group in the β or γ position to the acetylene fragment have been extensively documented [220].

A number of alkylnitrovinylacetylenes (275; equation 99) have been prepared by dehydrohalogenation of (274) under the action of $Pb(OAc)_2$ in glacial acetic acid [176, 226].

$$RC{\equiv}CCH{=}CH_2 + N_2O_4 + I_2 \longrightarrow [RC{\equiv}CCHICH_2NO_2]$$

(273)   (274)

$$\xrightarrow[\text{AcOH}]{\text{Pb(OAc)}_2} RC{\equiv}CCH{=}CHNO_2 \quad (99)$$

(275)

R = H, Me, Et, Pr, t-Bu

The same method has been used for the preparation of elementoorganic derivatives of (275) (R = Me_3Si, Et_3Si, Et_3Ge) [227].

1.7 TETRANITROETHENE AND TETRANITROBUTADIENE

Synthesis of such strong π acids as pernitroalkenes was the natural development of polynitroalkene chemistry. The sterically overloaded and unstable tetranitroethene (277; equation 100) was synthesized by Griffin and Baum [228] through mild pyrolysis of hexanitroethane (276); it was isolated in the form of the products of diene synthesis (278, 279).

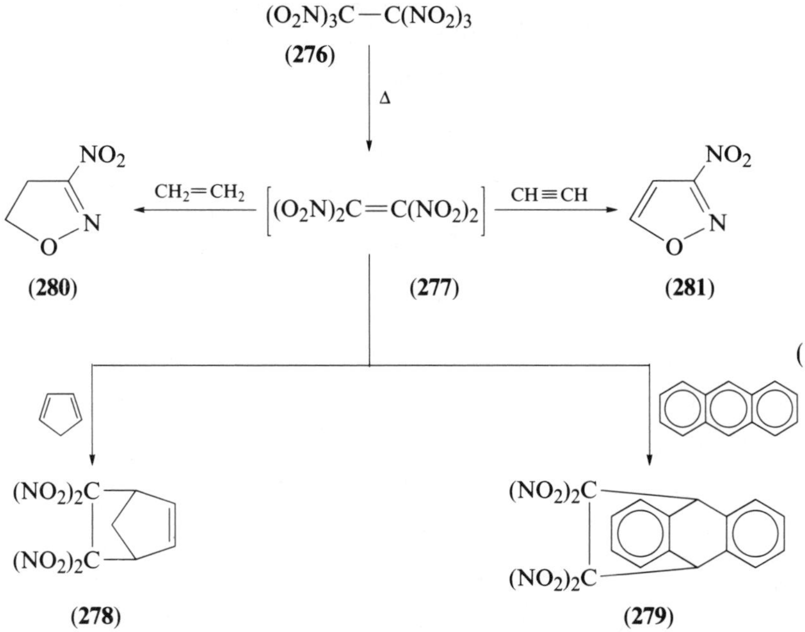

$$(O_2N)_3C-C(NO_2)_3$$

(276)

$$\Delta$$

(280) $CH_2=CH_2$ $[(O_2N)_2C=C(NO_2)_2]$ $CH\equiv CH$ (281)

(277)

(100)

$(NO_2)_2C$
$(NO_2)_2C$

(278)

$(NO_2)_2C$
$(NO_2)_2C$

(279)

Subsequently, tetranitroethene was isolated in 50% yield by flash pyrolysis as a solid, yellow-green substance which could be purified by sublimation. This compound is stable in vacuum at room temperature but its physicochemical data are not yet known. Compound (277) was allowed to react with ethene, acetylene and their derivatives to form 3-nitroisoxazoles and 3-nitroisoxazolines, respectively. It has been pointed out [229] that tetranitroethene is more active than tetracyanoethene in cycloaddition reactions. Tetranitroethene formed by heating hexanitroethane in alcohol gives the products of alkoxylation directly [230].

In working out the possibilities for 1,1,4,4-tetranitro-1,3-butadiene synthesis, Lipina et al. [231] studied the properties of 1,1,4,4-tetranitro-2,3-butanediol diacetate in various media. It was found that deacylation takes place easily under normal conditions in the presence of a base such as OH^-, MeO^- or aniline. Tetranitroethene forms the products of bis addition with these bases in situ (284–286; equation 101). Formation of the intermediate (283) was proven through the formation of charge transfer complexes (CTCs) with various donors (pyrene, anthracene). The CTC slowly transforms into the product (284) in the presence of base. Formation of unstable tetranitrobutadiene (283) was also detected by spectroscopic methods after storing the diacetate (282) in chloroform solution or after keeping the corresponding diol in concentrated sulfuric acid for long periods [231].

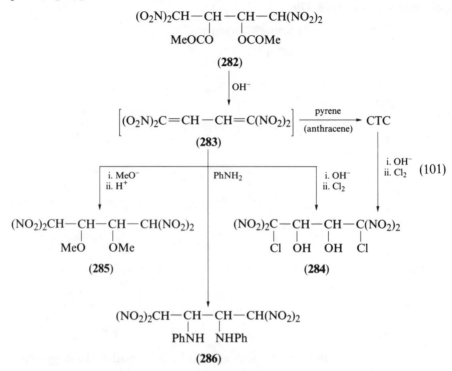

1.8 REFERENCES

[1] Haitinger, L. *Liebigs Ann. Chem.*, 1878, **193**, 366; *Monatsh. Chem.*, 1881, **2**, 286.
[2] Konovalov, M. *Zh. Obshch. Khim.*, 1893, **25**, 472; 1899, **31**, 57.
[3] Hass, H. B., Hodge, E. B., and Vanderbilt, B. M. *Ind. Eng. Chem.*, 1936, **28**, 339.
[4] Topchiev, A. V. *The Nitration of Hydrocarbons and other Compounds*, Akad. Nauk SSSR, Moscow, 1956, p. 296; *Chem. Abstr.*, 1957, **51**, 12 126d.
[5] Bagal, L. I., Stotskii, A. A., and Novitskaya, N. I. *Zh. Org. Khim.*, 1968, **4**, 1380; 1967, 3, 1201.
[6] Stotskii, A. A., Kirichenko, V. V., and Bagal, L. I. *Zh. Org. Khim.*, 1973, **9**, 2464.
[7] Shin, C., Masaki, M., and Ohta, M. *Bull. Chem. Soc. Jpn*, 1970, **43**, 3219.
[8] Shin, C., Yonezawa, Y. Narukawa, H., Nanjo, K., and Yoshimura, J. *Bull Chem. Soc. Jpn*, 1972, **45**, 3595.
[9] Benhaoua, H., Piet, J.-C., Danion-Bougot, R., Toupet, L., and Carrie, R. *Bull. Soc. Chim. Fr.*, 1987, 325.
[10] Azaro, F., Pitacco, G., and Valentin, E. *Tetrahedron*, 1987, **43**, 3279.
[11] Gupta, A., Ahmad, F., and Siddiqui, M. S. *Indian J. Chem.*, Sect. B, 1987, **26**, 870.
[12] Van der Lee, J. *Recl Trav. Chim. Pays-Bas*, 1928, **47**, 920.
[13] Van der Lee, J. *Recl Trav. Chim. Pays-Bas*, 1926, **45**, 674.
[14] Erdmann, H. *Ber.*, 1891, **24**, 2771.

[15] Posner, T. *Liebigs Ann. Chem.*, 1912, **389**, 1.
[16] Tinsley, S. W. *J. Org. Chem.*, 1961, **26**, 4723.
[17] Yandovskii, V. N., Ryabinkin, I. I., and Tselinskii, I. V. *Zh. Org. Khim.*, 1980, **16**, 2084.
[18] Bordwell, F. G. and Garbisch, I. E. W. *J. Am. Chem. Soc.*, 1960, **82**, 3588.
[19] Bordwell, F. G. and Garbisch, I. E. W. *J. Org. Chem.*, 1962, **27**, 2322; Fukunaga, K. *Yuki Gosei Kagaku Kyokai Shi*, 1976, **34**, 980; *Chem. Abstr.*, 1977, **87**, 167 640.
[20] Peterson, J. R., Do, H. D., and Dunham, A. J. *Can. J. Chem.*, 1988, **66**, 1670.
[21] Kovac, J. and Brezny, R. *Collect. Czech. Chem. Commun.*, 1977, **42**, 1880.
[22] Sitkin, A. I., Klimenko, V. I., and Fridman, A. L. *Zh. Org. Khim.*, 1975, **11**, 2452; 1977, **13**, 648, 2623; *Chem. Abstr.*, 1975, **82**, 111 874a; *Chem. Abstr.*, 1977, **87**, 22 903v; *Chem. Abstr.*, 1978, **88**, 894 576.
[23] Sitkin, A. I., Klimenko, V. I., and Fridman, A. L. *Khim. Geterotsikl. Soedin.*, 1977, 1604; Sitkin, A. I. and Klimenko, V. I. *Khim. Geterotsikl. Soedin.*, 1979, 41; *Chem. Abstr.*, 1978, **88**, 120 900g; *Chem. Abstr.*, 1979, **90**, 151 903f.
[24] Schmidt, E. and Fischer, H. *Chem. Ber.*, 1920, **53**, 1529.
[25] Masnovi, J. M. and Kochi, J. K. *Recl Trav. Chim. Pays-Bas*, 1986, **105**, 286.
[26] Altukhov, K. V. and Perekalin, V. V. *Usp. Khim.*, 1976, **45**, 2050.
[27] Penozek, S., Jagur-Jradzinski, J., and Szwara, M. *J. Am. Chem. Soc.*, 1968, **90**, 2174.
[28] Corey, E. J. and Estreicher, H. *Tetrahedron Lett.*, 1980, **21**, 1113.
[29] Levi, N., and Rose, I. D. *Q. Rev. I*, 1947, **1**, 363; Levi, N., Scaife, C. W., and Wilder-Smith, A. E. *J. Chem. Soc.*, 1948, 52.
[30] Seifert, W. K. *J. Org. Chem.*, 1963, **28**, 125.
[31] Boberg, F. *Liebigs Ann. Chem.*, 1959, **626**, 71.
[32] Perekalin, V. V. *Unsaturated Nitro Compounds*, Israel Program for Scientific Translations, Jerusalem, 1964; Perekalin, V. V., Sopova, A. S., and Lipina, E. S. *Unsaturated Nitro Compounds*, Khimiya, Moscow, 1982, pp. 6–51.
[33] Schaarschmidt, A. and Hoffmeier, H. *Chem. Ber.*, 1925, **58B**, 1047.
[34] Bryant, D. K., Challis, B. C., and Iley, J. *J. Chem. Soc., Chem. Commun.*, 1989, 1027.
[35] Rockenbauer, A., Gyow, M., and Tudos, F. *Tetrahedron Lett.*, 1986, **27**, 3425.
[36] Powell, J. L., Ridd, J. H., and Sandall, J. P. B. *J. Chem. Soc., Chem. Commun.*, 1990, 402.
[37] Nikolaev, A. D., Sitkin, A. I., and Kamai, G. Kh. *Zh. Org. Khim.*, 1968, **4**, 1411; *Chem. Abstr.*, 1968, **69**, 86 568a.
[38] Gold, M. H., *J. Am. Chem. Soc.*, 1946, **68**, 2544.
[39] Birchenbach, L., Goubeau, J., and Berninger, E. *Chem. Ber.*, 1932, **65**, 1339.
[40] Kabalka, G. W. and Varma, R. S. *Org. Prep. Proced. Int.*, 1987, **19**, 283.
[41] Szarek, W. A., and Lance, D. G. *Carbohydr. Res.*, 1970, **13**, 75.
[42] Szarek, W. A., Lance, D. G., and Beach, R. L. *J Chem. Soc., Chem. Commun.*, 1968, 356.
[43] (a) Hassner, A., Kropp, J. E., and Kent, G. J. *J. Org. Chem.*, 1969, **34**, 2628.
(b) Sy, W.-W. and By, A. W. *Tetrahedron Lett.*, 1985, **26**, 1193.
(c) Jew, S.-S., Kim, H.-d., Cho, J.-s. and Cook, C.-h. *Chem. Lett.*, 1986, 1747.
[44] Stevens, T. E. and Emmons, W. D. *J. Am. Chem. Soc.*, 1958, **80**, 338.
[45] Kunai, A., Yanagi, Y., and Sasaki, K. *Tetrahedron Lett.*, 1983, **24**, 4443.
[46] Corey, E. J. and Estreicher, H. *J. Am. Chem. Soc.*, 1978, **100**, 6294.
[47] Vankar, Y. D. and Bawa, A. *Synth. Commun.*, 1985, **15**, 1253.
[48] Hayama, T., Tomoda, S., Takeuchi, Y., and Nomura, Y. *Chem. Lett.*, 1982, 1109.
[49] Hayama, T., Tomoda, S., Takeuchi, Y., and Nomura, Y. *Tetrahedron Lett.*, 1982, **28**, 4733.
[50] Hayama, T., Tomoda, S., Takeuchi, Y., and Nomura, Y. *J. Org. Chem.*, 1984, **49**, 3235.

[51] Sakakibara, T., Takai, I., Ohara, E., and Sudoh, R. *J. Chem. Soc., Chem. Commun.,* 1981, 261.
[52] Sakakibara, T., Ikuta, S., and Sudoh, R. *Synthesis,* 1982, 261.
[53] Vankar, Y. D., Bawa, A., and Kumaravel, G. *Tetrahedron,* 1991, **47**, 2027.
[54] Hanessian, S., Desilets, D., and Bennani, Y. L. *J. Org. Chem.,* 1990, **55**, 3098.
[55] Seebach, D., Calderari, G., and Knochel, P. *Tetrahedron,* 1985, **41**, 4861.
[56] Muller, E. (Ed.) *Houben-Weyl Methoden der Organische Chimie,* Georg Thieme Verlag, Stuttgart, 1971, Vol. X(I), pp. 9–462.
[57] Baer, H. H. and Urbas, L. in *The Chemistry of the Nitro and Nitroso Groups,* Part 2 (Ed. M. Feuer), Interscience, New York, 1970, pp. 63–147.
[58] Novikov, S. S., Shvekhgeimer, G. A., Sevostyanova, V. V., and Shliapochnicov, V. A. *Chemistry of Aliphatic and Alicyclic Nitro Compounds,* Khimiya, Moscow, 1974, pp. 56–101.
[59] Kozlov, L. M., and Burmistrov, V. I. in *Nitroalcohols and Their Derivatives* (Ed. S. S. Novikov), Kazan, Moscow, 1960, p. 181.
[60] Seebach, D. *Chimia,* 1979, **13**, 1.
[61] Koremura, M., Oku, H., Shoho, T., and Nakanishi, T. *Takamine Kenkyusho Nempo,* 1861, **13**, 198; *Chem. Abstr.,* 1962, **57**, 16451d.
[62] Kada, R., Knoppova, V., Kovac, J., and Malnakova, I. *Collect. Czech. Chem. Commun.,* 1984, **49**, 2496.
[63] Egorova, V. S., Ivanova, V. N., and Putokhin, N. I. *Zh. Obshch. Khim.,* 1964, **34**, 4084; King, W. J. and Nord, F. F. *J. Org. Chem.,* 1949, **49**, 405.
[64] Hamdan, A. and Wasley, J. W. F. *Synth. Commun.,* 1985, **15**, 71; *Chem. Abstr.,* 1985, **103**, 71142d.
[65] Sakakibara, T., Koezuka, M., and Sudon, R. *Bull. Chem. Soc. Jpn,* 1978, **51**, 3095.
[66] Simonov, A. M. and Dalgatov, D. D. *Zh. Obshch. Khim.,* 1964, **34**, 3052.
[67] Baer, H. H. and Achmatowicz, B. *J. Org. Chem.,* 1964, **29**, 3180; Campbell, R. D. and Pitzer, C. L. *J. Org. Chem.,* 1959, **24**, 1531.
[68] Shiga, M., Kono, H., Motoyama, I., and Hata, K. *Bull. Chem. Soc. Jpn,* 1968, **41**, 1897.
[69] Oreshko, G. V. and Eremenko, L. T. *Izv. Akad. Nauk SSSR, Ser. Khim.,* 1989, 1107.
[70] Klager, K. *Monatsh. Chem.,* 1965, **96**, 1.
[71] Gairaud, G. B. and Lappin, G. R. *J. Org. Chem.,* 1953, **18**, 1.
[72] Wessling, M. and Schafer, H. *Chem. Ber.,* 1991, **124**, 2303.
[73] Agafonov, N. E., Sedishev, I. P., and Zhulin, V. M. *Izv. Akad. Nauk SSSR, Ser. Khim.,* 1989, 2153; Agafonov, N. E., Sedishev, I. P., Dudin, A. V., Kutin, A. A., Stashina, G. A., and Zhulin, V. M. *Izv. Akad. Nauk SSSR, Ser. Khim.,* 1991, 426
[74] McDonald, E. and Martin, R. T. *Tetrahedron Lett.,* 1977, **15**, 1317.
[75] Daunzonne, O. and Royer, R. *Synthesis,* 1984, 1054.
[76] Miller, R. B. and Tsang, T. *Tetrahedron Lett.,* 1988, **29**, 6715.
[77] Goldfarb, Ya. L., Kalik, M. A., and Zavialova, V. K. *Zh. Org. Khim.,* 1979, **15**, 1540.
[78] Bulatova, N. N. and Suvorov, N. N. *Khim. Geterotsikl. Soedin.,* 1969. 813.
[79] Rodina, O. A. and Mamaev, V. P. *Zh. Obshch. Khim.,* 1964, **34**, 2146.
[80] Zhilina, O. D., Shagalov, L. B., Vasiliev, A. M., and Suvorov, N. N. *Khim. Geterotsikl. Soedin.,* 1980, 949.
[81] (a) Mayer, A. A. and Murphy, B. P. *Synth. Commun.,* 1985, **15**, 423; *Chem. Abstr.,* 1985, **103**, 141542k.
(b) Kubo, A., Saito, N., Kawakami, N., Matsuyama, Y., and Miwa, T. *Synthesis,* 1987, 824.
[82] Karmarkar, S. N., Kelkar, M. S., and Waden, M. S. *Synthesis,* 1985, 510; Young, T. E. and Beidler, W. T. *J. Org. Chem.,* 1985, **50**, 1182.

[83] Rizzacasa, M. A., Sargent, M. V., Skelton, B. W., and White, A. H. *Aust. J. Chem.*, 1990, **43**, 79.

[84] Young, E. H. P. *J. Chem. Soc.*, 1958, 3493.

[85] Heinzelman, R. V., Anthony, W. C., Lyttle, D. A., and Szmuszkovicz, J. *J. Org. Chem.*, 1960, **25**, 1548.

[86] Yakhontov, L. N., Uritskaya, M. Ya., and Rubtsov, M. V. *Zh. Org. Khim.*, 1965, **1**, 2040.

[87] Barton, D. H. R., Motherwell, W. B., and Zard, S. Z. *J. Chem. Soc., Chem. Commun.*, 1982, 551; *Bull. Soc. Chim. Fr.*, 1983, **2**, 61.

[88] Tamura, R., Sato, M., and Oda, D. *J. Org. Chem.*, 1986, **51**, 4368.

[89] Denmark, S. E., Kesler, B. S., and Moon, Y.-C. *J. Org. Chem.*, 1992, **57**, 4912.

[90] Palomo, C., Aizpurua, J. M., Cossio, F. P., Garcia, J. M., Lopez, M. C., and Oiarbide, M. *J. Org. Chem.*, 1990, **55**, 2070.

[91] Astle, M. J. and Abbott, F. P. *J. Org. Chem.*, 1956, **21**, 1228.

[92] Knochel, P. and Seebach, D. *Tetrahedron Lett.*, 1982, **23**, 3897.

[93] Wollenberg, R. H. and Miller, S. J. *Tetrahedron Lett.*, 1978, 3219.

[94] Okaichi, J., Cha, B. C., and Kitagawa, I. *Tetrahedron*, 1990, **46**, 7459.

[95] Seebach, D. and Lehr, F. *Helv. Chim. Acta*, 1979, **62**, 2239.

[96] Seebach, D., Beck, A. K., Mukhopadhyay, T., and Thomas, E. *Helv. Chim. Acta*, 1982, **65**, 1101.

[97] Colvin, E. W., Beck, A. K., and Seebach, D. *Helv. Chim. Acta*, 1981, **64**, 2264.

[98] Lee, K. and Oh, D. Y. *Synth. Commun.*, 1989, **19**, 3055.

[99] Martin, O. R., Khamis, F. E., El-Shenawy, H. A., and Rao, S. P. *Tetrahedron Lett.*, 1989, **30**, 6139, 6143.

[100] Denmark, S. E., Senanayake, C. B. W., and Ho, G.-D. *Tetrahedron*, 1990, **46**, 4857.

[101] Topchiev, A. V., Alaniya, V. P., and Vagin, M. F. *Dokl. Akad. Nauk SSSR*, 1963, **151**, 350.

[102] Ranganathan, D., Rao, C. B., Ranganathan, S., Mehrotra, A. K., and Iyengar, R. *J. Org. Chem.*, 1980, **45**, 1185; Yamashita, T. and Namba, K. *Kogyo Kagaku Kyokaishi*, 1963, **23**, 293; *Chem. Abstr.*, 1963, **59**, 8577b.

[103] Busklev, G. D. and Scalfe, C. W. *J. Chem. Soc.*, 1947, 1471.

[104] Knochel, P. and Seebach, D. *Synthesis*, 1982, 1017.

[105] Melton, J. and McMurry, J. E. *J. Org. Chem.*, 1975, **40**, 2138.

[106] Dampawan, P. and Zajac, W. W. *Tetrahedron Lett.*, 1982, **23**, 135.

[107] Seebach, D. and Knochel, P. *Helv. Chim. Acta*, 1984, **67**, 261.

[108] Denmark, S. E., Moon, Y.-C., and Senanayake, C. B. W. *J. Am. Chem. Soc.*, 1990, **112**, 311.

[109] Denmark, S. E., Moon, Y.-C., Cramer, C. J., Dappen, M. S., and Senanayake, C. B. W. *Tetrahedron*, 1990, **46**, 7373.

[110] Rosini, G., Ballini, R., and Sorrenti, P. *Synthesis*, 1983, 1014.

[111] Ballini, R., Castagnani, R., and Petrini, M. *J. Org. Chem.*, 1992, **57**, 2160.

[112] Asaro, F., Pitacco, G., and Valentin, E. *Tetrahedron*, 1987, **43**, 3279.

[113] Bowman, W. R. and Jackson, S. W. *Tetrahedron*, 1990, **46**, 7313.

[114] Rosini, G., Ballini, R., Petrini, M., and Sorrenti, P. *Synthesis*, 1985, 515.

[115] Melot, J.-N., Texier-Boullet, F., and Fousaund, A. *Tetrahedron Lett.*, 1986, **27**, 493.

[116] Worral, D. E. *J. Am. Chem. Soc.*, 1934, **56**, 1556; Crowell, T. and Ramirez, F. *J. Am. Chem. Soc.*, 1951, **73**, 2268.

[117] Arcoria, A., Bottino, F. A., and Sciotto, D. *J. Heterocycl. Chem.*, 1977, **14**, 1353.

[118] Mayer, M. C. *Bull. Soc. Chim. Fr.*, 1905, **33**, 395.

[119] Kono, H., Shiga, M., Motoyama, I., and Hata, K. *Bull. Chem. Soc. Jpn*, 1969, **42**, 3267.

[120] Denmark, S. E., Sternberg, J. A., and Lueoend, R. *J. Org. Chem.*, 1988, **53**, 1251.
[121] Ivanova, I. S., Konnova, Yu. V., and Novikov, S. S. *Izv. Akad, Nauk SSSR, Ser. Khim.*, 1962, 1677.
[122] Babievskii, K. K., Belikov, V. M., and Tikhonova, N. A. *Izv. Akad. Nauk SSSR, Ser. Khim.*, 1965, 89.
[123] Umezawa, S. and Zen, S. *Bull. Chem. Soc. Jpn*, 1963, **36**, 1143.
[124] Bocharova, L. A., Polyanskaya, A. S., and Perekalin, V. V. *USSR*, 1964, **164**, 271; *Chem. Abstr.*, 1965, **62**, 451g; Bocharova, L. A. and Polyanskaya, A. S. *Tr. Pedvuzov (Pedagog. Vissh. Ucheb. Zaved.) Dalnego Vostoka, Khabarovsk (USSR)*, 1966, 74; *Chem. Abstr.*, 1969, **70**, 86 975b.
[125] Dornow, A. and Frese, A. *Liebigs Ann. Chem.*, 1953, **581**, 211.
[126] Babievskiii, K. K., Belikov. V. M., and Tikhonova, N. A. *Dokl. Akad. Nauk SSSR*, 1965, **160**, 103.
[127] Stefanovic, G., Vandyel, V., and Boyanovic, J. *Glasnik Khem. Drushtva, Belgrad*, 1955, **20**, 545; *Chem. Abstr.*, 1958, **52**, 16 2781i.
[128] Shipchandler, M. T. *Synthesis*, 1979, 666.
[129] Dornow, A. and Wiehler, G. *Liebigs Ann. Chem.*, 1952, **578**, 113.
[130] Dornow, A. and Frese, A. *Liebigs Ann. Chem.*, 1952, **578**, 122.
[131] Dornow, A. and Menzel, H. *Liebigs Ann. Chem.*, 1954, **588**, 40.
[132] Babievskii, K. K., Belikov, V. M. Vinogradova, A. I., and Latov, V. K. *Zh. Org. Khim.*, 1973, **9**, 1700.
[133] Aboskalova, N. I., Polyanskaya, A. S., and Perekalin, V. V. *Metody Sint., Str. Khim. Prevraschen Nitrosoedin., Gertsenovskie Chteniya, Leningrad Gos. Pedagog. Inst., Leningrad*, 1978, 58; *Chem. Abstr.*, 1979, **91**, 39 246p.
[134] Brown, R. F. C. and Meehan, C. V. *Aust. J. Chem.*, 1968, **21**, 1581; Yamamura, K., Watarai, S. and Kinugasa, T. *Bull. Chem. Soc. Jpn*, 1971, **44**, 2440; Aboskalova, N. I., Polyanskaya, A. S., and Perekalin, V. V. *Dokl. Akad. Nauk SSSR*, 1967, **176**, 829.
[135] Babievskii, K. K., Kochetkov, K. A., and Belikov, V. M. *Izv. Akad. Nauk SSSR, Ser. Khim.*, 1977, 2310.
[136] Kochetkov, K. A., Garbalinskaya, N. S., Babievskii, K. K., and Belikov, V. M. *USSR Pat. 632 686* (Cl. C07C79/43), 1978; *Chem. Abstr.*, 1979, **90**, 54 687f.
[137] Kochetkov, K. A., Babievskii, K. K., Belikov, V. M., Garbalinskaya, N. S., and Bakhmutov, V. I. *Izv. Akad. Nauk SSSR, Ser. Khim.*, 1980, 639.
[138] Lehnert, W. *Tetrahedron*, 1972, **28**, 663.
[139] Ried, W. and Kohler, E. *Liebigs Ann. Chem.*, 1956, **598**, 145.
[140] Dornow, A. and Sassenberg, W. *Liebigs Ann. Chem.*, 1957, **602**, 14.
[141] Dornow, A., Muller, A., and Lipfert, S. *Liebigs Ann. Chem.*, 1955, **594**, 191.
[142] Aboskalova, N. I., Polyanskaya, A. S., Perekalin, V. V., Demireva, Z. I., and Sokolova, L. N. *Zh. Org. Khim.*, 1972, **8**, 1332.
[143] Polyanskaya, A. S., Perekalin, V. V., Aboskalova, N. I., Demireva, Z. I., Sokolova, L. N., and Abdulkina, Z. A. *Zh. Org. Khim.*, 1979, **15**, 2057.
[144] (a) Shadrin, V. Yu. *Nitroacetonitrile in Reactions with Carbonyl Compounds*, dissertation abstract, Leningrad State Pedagogical Institute, Leningrad, 1987.
(b) Sokolova, L. N., Polyanskaya, A. S., Aboskalova, N. I., Demireva, Z. I., Abdulkina, Z. A., and Pyatnitskaya, I. A. *XXX Gertsenovskie Chteniya*, Khimiya, Moscow, 1977, p. 59; *Chem. Abstr.*, 1978, **89**, 43 024r.
[145] (a) Aboskalova, N. I. and Shadrin, V. Yu. *Sint. Issled. Nitrosoedin., Aminokislot., Leningrad Gos. Pedagog. Inst., Leningrad*, 1983, 8; *Chem. Abstr.*, 1984, **100**, 85 534h.
(b) Polyanskaya, A. S., Perekalin, V. V., Aboskalova, N. I., and Sokolova, L. N. *USSR Pat. 335 246* (Cl. CO7d), 1972; *Chem. Abstr.*, 1972, **77**, 34 322m.

(c) Aboskalova, N. I., Babievskii, K. K., Belikov, V. M., Perekalin, V. V., and Polyanskaya, A. S. *Zh. Org. Khim.*, 1973, **9**, 1058.

[146] Takamoto, T., Ikeda, Y., Tachimori, Y., Seta, A., and Sudoh, R. *J. Chem. Soc., Chem. Commun.*, 1978, 350.

[147] Corey, E. J. and Estreicher, H. *Tetrahedron Lett.*, 1981, **22**, 603.

[148] Sakakibara, T., Ikeda, Y., and Sudoh, R. *Bull. Chem. Soc. Jpn*, 1982, **55**, 635.

[149] Sakakibara, T., Manandhar, M., Ohkita, N., and Ishido, Y. *Bull. Chem. Soc. Jpn*, 1987, **60**, 3425.

[150] Buurman, D. J., Veldhuizen, A., and Van der Plas, H. C. *J. Org. Chem.*, 1990, **55**, 778.

[151] Ayyangar, N. R., Lugade, A. G., and Rajadhyaksha, M. N. *Indian J. Chem., Sect. B*, 1986, **25**, 1126.

[152] Babievskii, K. K., Belikov, V. M., and Tikhonova, N. A. *Khim. Geterotsikl. Soedin. Sb. 1, Azotsoderzh. Heterocykli*, 1967, 46; *Chem. Abstr.*, 1969, **70**, 78 343d.

[153] Babievskii, K. K., Belikov, V. M., and Tikhonova, N. A. *Izv. Akad. Nauk SSSR, Ser. Khim.*, 1967, 877.

[154] Jubert, C. and Knochel, P. *J. Org. Chem.*, 1992, **57**, 5431.

[155] Ranganathan, S., Ranganathan, D., and Singh, S. K. *Tetrahedron Lett.*, 1987, **28**, 2893.

[156] Zavlin, P. M. and Efremov, D. A. *Zh. Nauchn. Prikl. Fotogr. Kinematogr.*, 1992, 128.

[157] Buckley, G. D. and Charlish, J. L. *J. Chem. Soc.*, 1947, 1472.

[158] (a) Nikolinski, R. and Spacov, T. *God. Khim. Tekhu. Inst., Sofia*, 1936, **3**(2), 94, 1956.

(b) Perekalin, V. V., Mladenov, I., Lipina, E. S., and Aleksiev, D. I. *God. Vissh. Khimkotekhnol. Inst., Burgas, Bulgaria*, 1973, **10**, 595; *Chem. Abstr.*, 1975, **82**, 169 910a.

[159] Fort, G. and Melean, A. *J. Chem. Soc.*, 1948, 1907.

[160] Novikov, S. S., Burmistrova, M. S., Gorelik, V. P., and Chkhikvadze, V. P. *Izv. Akad. Nauk SSSR, Ser. Khim.*, 1961, 695.

[161] Kochetkov, N. K. and Dudykina, N. V. *Zh. Obshch. Khim.*, 1958, **28**, 2399.

[162] Severin, T. and Ipach, I. *Chem. Ber.*, 1978, **111**, 692.

[163] Moldenhauer, O., Irion, W., Mastaglio. D., Pfluger, R., and Marwitz, H. *Liebigs Ann. Chem.*, 1953, **583**, 46.

[164] Alaniya, V. P. and Sokolov, N. A. *Zh. Org. Khim.*, 1965, **1**, 2245.

[165] Shadrin, V. Yu, Polyanskaya, A. S., Berkova, G. A., Paperno, T. Ya., and Aboskalova, N. I. *Sint. Issled. Nitrosoedin., Aminokislot., Leningrad Gos. Pedagog. Inst., Leningrad*, 1986, 20.

[166] Pechenko, L. M., Polyanskaya, A. S., Perekalin, V. V., Aboskalova, N. I., and Demireva, Z. I. *Metody Sint., Str. Khim. Prevraschen Nitrosoedin., Gertsenovskie Chteniya, Leningrad Gos. Pedagog. Inst., Leningrad*, 1978, 63; *Chem. Abstr.*, 1979, **91**, 55 611j.

[167] Yanovskaya, L. A., Stepanova, R. N., and Kucherov, V. F. *Izv. Akad. Nauk SSSR, Ser. Khim.*, 1964, 2093.

[168] Kvitko, S. M., Perekalin, V. V., and Maksimov, Yu. V. *Zh. Org. Khim.*, 1973, **9**, 2228.

[169] Kvitko, S. M., Maksimov, Yu. V., Paperno, T. Ya., and Perekalin, V. V. *Zh. Org. Khim.*, 1973, **9**, 471.

[170] Kvitko, S. M., Slovokhotova, N. A., Perekalin, V. V., Vasilieva, V. N., and Bobvich, Ya. S. *Dokl. Akad. Nauk SSSR*, 1962, **143**, 345.

[171] Coe, C. S. and Doumani, T. F. *US Pat. 2 478 243*, 1949; *Chem. Abstr.*, 1950, **44**, 1128g.

[172] Padwa, A., Crumrine, D., Hartman, R., and Layton, R. *J. Am. Chem. Soc.*, 1967, **89**, 4435.
[173] Lipina, E. S., Perekalin, V. V., and Bobovich, Ya. S. *Zh. Obshch. Khim.*, 1964, **34**, 3640.
[174] Sulimov, I. G., Samoilovich, T. I., Perekalin, V. V., Polyanskaya, A. S., and Usik, N. V. *Zh. Org. Khim.*, 1972, **8**, 1343.
[175] Polyanskaya, A. S. and Perekalin, V. V. *Dokl. Chem. (Engl. Transl.)*, 1974, **217**, 586.
[176] Petrov, A. A., Rall, K. B., and Vildavskaya, A. I. *Zh. Org. Khim.*, 1965, **1**, 240.
[177] Samoilovich, T. I., Polyanskaya, A. S., and Perekalin, V. V. *Dokl. Akad. Nauk SSSR*, 1974, **217**, 1335; Samoilovich, T. I., Polyanskaya, A. S., and Perekalin, V. V. *Zh. Org. Khim.*, 1967, **3**, 1532.
[178] Bloom, A. J. and Mellor, J. M. *Tetrahedron Lett.*, 1986, **27**, 873.
[179] Bloom, A. J. and Mellor, J. M. *J. Chem. Soc., Perkin Trans. 1*, 1987, 2737.
[180] Retherford, C. and Knochel, P. *Tetrahedron Lett.*, 1991, **32**, 441.
[181] Najera, C., Yus, M., Karlson, V., Gogoll, A., and Backvall, J.-E. *Tetrahedron Lett.*, 1990, **31**, 4199.
[182] (a) Backvall, J.-E. Karlsson. U., and Chinchilla, P. *Tetrahedron Lett.*, 1991, **32**, 5607.
(b) Barco, A., Benetti, S., Pollini, G. P., Spalluto, G., and Zanirato, V. *Tetrahedron Lett.*, 1991, **32**, 2517.
[183] Berestovitskaya, V. M., Speranskii, E. M., and Perekalin, V. V. *Zh. Org. Khim.*, 1976, **12**, 2256; Berestovitskaya, V. M., Speranskii, E. M., Sulimov, I. G., and Trukhin, E. V. *Zh. Org. Khim.*, 1972, **8**, 1763.
[184] Freeman, J. P. and Emmons, W. D. *J. Am. Chem. Soc.*, 1957, **79**, 1712.
[185] Mead, T. E. and Clapp, L. B. *J. Org. Chem.*, 1958, **23**, 921.
[186] Nikolaeva, A. D. and Kashevarova, L. V. *Izv. Vuzov Khimikotekhnol.*, 1975, **18**, 1550; Nikolaeva, A. D., Mikhailovskii, D. I., and Kirsanov, A. P. *Zh. Org. Khim.*, 1975, **11**, 691.
[187] Lipina, E. S., Pavlova, Z. F., Prikhodko, L. V., Paperno, T. Ya., and Perekalin, V. V. *Dokl. Akad. Nauk SSSR*, 1970, **192**, 810.
[188] Wieland, H. and Blumlich, E. *Liebigs Ann. Chem.*, 1921, **424**, 100.
[189] Campbell, R. N., Shavel, J., and Campbell, B. K. *J. Am. Chem. Soc.*, 1953, **75**, 2400.
[190] Dore, J.-C. and Viel, C. *C. R. Hebd. Seances Acad. Sci., Ser. C*, 1973, **276**, 1675.
[191] Lipina, E. S., Pavlova, Z. F., and Perekalin, V. V. *Zh. Org. Khim.*, 1969, **5**, 1312.
[192] Boyer, J. H. and Pagoria, P. F. *Org. Prep. Proced. Int.*, 1986, **18**, 363.
[193] Bisgrove, D. E., Brown, J. F., and Clapp, L. B. in *Organic Syntheses* (Ed. N. Rabjoher), Interscience, New York, 1963, Coll. Vol. 4, p. 372.
[194] Gold, M. H. and Klager, K. *Tetrahedron*, 1963, **19**(1), 77; Klager, K., Kispersky, J. P., and Hamel, E. *J. Org. Chem.*, 1961, **26**, 4368.
[195] Fridman, A. L., Gabitov, F. A., and Surkov, V. D. *Zh. Org. Khim.*, 1972, **8**, 2457.
[196] Kamlet, J. and Dacons, C. *J. Org. Chem.*, 1961, **26**, 3005; Thyagarajan, B. S. and Gopalakrishnan, P. V. *Tetrahedron*, 1964, **20**, 1051.
[197] Sitkin, A. I., Safiulina, O. Z., and Fridman, A. L. *Sb. Nauchn. Tr. Kuzbas. Politekh. Inst.*, 1976, **81**, 142; *Chem. Abstr.*, 1977, **86**, 16 195d.
[198] Tsukervanik, I. P. and Sokolnikova, M. D. *Zh. Obshch. Khim.*, 1954, **24**, 1435.
[199] Gabitov, F. A., Fridman, A. L., and Nikolaeva, A. D. *Zh. Org. Khim.*, 1969, **5**, 2245.
[200] Ladyzhnikova, T. D., Tyrkov, A. G., Soloviev, N. A., and Altukhov, K. V. *Zh. Org. Khim.*, 1989, **25**, 444; Manuel, D. V., Soloviev, N. A., Ladyzhnikova, T. D., and Altukhov, K. V. *Gertsenovskie Chteniya, Leningrad Pedagog. Inst., Leningrad (USSR)*, 1989, 133; *Chem. Abstr.*, 1991, **114**, 5745d.

[201] Novikov, S. S., Korsakova, I. S., and Babievskii, K. K. *Izv. Akad. Nauk SSSR, Ser Khim.*, 1960, 944.

[202] Durden, J. A., Heywood, D. L., Sousa, A. A., and Spurr, H. W. *J. Agric. Food Chem.*, 1970, **18**, 50.

[203] Perekalin, V. V. and Lerner, O. M. *Dokl. Akad. Nauk SSSR*, 1959, **129**, 1303.

[204] Zonis, E. S., Lerner, O. M., and Perekalin, V. V. *Zh. Prikl. Khim.*, 1961, **34**, 711; Lipina, E. S. and Perekalin, V. V. *Acta Phys. Chem.*, 1973, **19**, 125.

[205] Prikhodko, L. V., Lipina, E. S., and Perekalin, V. V. *Zh. Org. Khim.*, 1970, **6**, 1748.

[206] Carroll, F. I., Kerbow, S. C., and Wall, M. E. *Can. J. Chem.*, 1966, **44**, 2115.

[207] Lipina, E. S. and Perekalin, V. V. *Zh. Obshch. Khim.*, 1964, **34**, 3544.

[208] Shafiullah, I. and Husain, S. *J. Chem. Res. (S)*, 1983, 255.

[209] Husain, S., Agarwal, V., and Gupta, K. C. *Indian J. Chem., Sect. B*, 1988, **27**, 852.

[210] Ohsawa, A., Arai, H. J., Gets, H., Akimoto, T., Tsuji, A., and Iitaka, Y. *J. Org. Chem.*, 1979, **44**, 3524.

[211] Berestovitskaya, V. M., Titova, M. V., and Perekalin, V. V. *Zh. Org. Khim.*, 1980, **16**, 891.

[212] Dell Erba, C., Mele, A., Novi, M., Petrillo, G., and Stagnaro, P. *Tetrahedron Lett.*, 1990, **31**, 4933.

[213] Zagibalova, L. Ya., Lipina, E. S., Berkova, G. A., Paperno, T. Ya., Perekalin, V. V., and Pozdnyakov, V. P. *Zh. Org. Khim.*, 1981, **17**, 2302.

[214] Lipina, E. S., Stepanov, N. D., Bagal, I. L., and Bodina, R. I. *Zh Org. Khim.*, 1980, **16**, 2404.

[215] Lipina, E. S., Perekalin, V. V., and Bobvich, Ya. S. *Zh. Obshch. Khim.*, 1964, **34**, 3635.

[216] Novikov, S. S., Belikov, V. M., Demyanenko, V. F., and Lapshina, L. V. *Izv. Akad. Nauk SSSR, Ser. Khim.*, 1960, 1295.

[217] Viehe, H. G., Jager, V. V., and Compernolle, F. *Angew. Chem.*, 1969, **81**, 999.

[218] Jager, V. V. and Viehe, H. G. *Angew. Chem.*, 1969, **81**, 259.

[219] Derycke, C., Jager, V., Putzeys, J. P., Meersscke, M., and Viehe, H. G. *Angew. Chem.*, 1973, **85**, 447.

[220] Rall, K. B., Vildavskaya, A. I., and Petrov, A. A. *Usp. Khim.*, 1975, **44**, 744.

[221] Jager, V. V., Motte, J.-C., and Viehe, H. G. *Chimia*, 1975, **29**, 516.

[222] Petrov, A. A., Zavgorodnii, V. S., Rall, K. B., Vildavskaya, A. I., and Bogoradovskii, E. T. *Zh. Obshch. Khim.*, 1978, **48**, 943.

[223] Schmitt, R. J., Bottaro, J. C., Malhotra, R., and Bedford, C. D. *J. Org. Chem.*, 1987, **52**, 2294.

[224] Schmitt, R. J. and Bedford, C. D. *Synthesis*, 1986, 132.

[225] Bottaro, J. C., Schmitt, R. J., Bedford, C. D., Gilardi, R., and George, C. *J. Org. Chem.*, 1990, **55**, 1916.

[226] Petrov, A. A., Rall, K. B., and Vildavskaya, A. I. *Zh. Obshch. Khim.*, 1964, 34, 3513; Vildavskaya, A. I. and Rall, K. B. *Zh. Org. Khim.*, 1968, **4**, 959.

[227] Vildavskaya, A. I., Rall, K. B., and Petrov, A. A. *Zh. Obshch. Khim.*, 1971, **41**, 1279.

[228] Griffin, T. S. and Baum, K. *J. Org. Chem.*, 1980, **45**, 2880; 1981, **46**, 4811.

[229] Baum, K. and Tzeng, D. *J. Org. Chem.*, 1985, **50**, 2736.

[230] Tzeng, D. and Baum, K. *J. Org. Chem.*, 1983, **48**, 5384.

[231] Nekrasova, G. V., Lipina, E. S., Pozdnyakov, V. P., and Perekalin, V. V. *Zh. Org. Khim.*, 1984, **20**, 2502.

2 CHEMICAL TRANSFORMATIONS OF UNSATURATED NITRO COMPOUNDS

2.1 REDUCTION

The reaction of reduction is a widely investigated process in the chemistry of unsaturated nitro compounds. Researchers have worked out a good number of approaches to nitroalkene reduction that lead to a wide range of organic compounds, such as nitroalkanes, N-substituted hydroxylamines, aldoximes and ketoximes, amines and ketones [1, 2a, 2b]. The reduction reaction is a simple synthetic approach to different biologically active amines, for example, α-amino acids, tryptamine, benzedrine, mescaline, and others. In many cases, reduction results in heterocyclization. It is reasonable to discuss this reaction with respect to the nature of the reducing agent.

Sodium or aluminium amalgam induced reduction of α-nitroalkenecarbonic esters (**1**; equation 1) leads to the corresponding oximes (**2**) and unsaturated α-amino acid alkylates (**3**) [3].

$$R^1CR^2C{=}C(NO_2)CO_2R^3 \xrightarrow[\text{Et}_2\text{O, room temperature}]{\text{Al/Hg}}$$
(1)

$$\rightarrow R^1CHR^2C(CO_2R^3){=}NOH$$
(2)

$$R^1 = H; \quad R^2 = Et, i\text{-}Pr, n\text{-}Pr;$$
$$R^3 = Me, Et$$

$$\hookrightarrow R^1CR^2{=}C(NH_2)CO_2R^3$$
(3)

$$R^1 = R^2 = R^3 = Me \qquad (1)$$

Zinc dust in ether/acetic acid or alcohol/acetic acid mixtures initiates oxime formation. Nitroalkenes with carbon chains from C_6 to C_{10} are reduced to the corresponding oximes (**5**; equation 2) in 50–60% yield [4].

$$RCH{=}CHNO_2 \xrightarrow[\text{reflux}]{\text{Zn + MeCO}_2\text{H (25\%) + Et}_2\text{O}} RCH_2CH{=}NOH \qquad (2)$$

$$\textbf{(4)} \qquad\qquad\qquad\qquad\qquad\qquad\qquad\qquad \textbf{(5)}$$

$$R = Alk$$

Under similar conditions, as well as under the action of zinc dust in a methanolic solution of ammonia, 6-nitro-5-cholestenes transform into acylated enamines (**6**) and 6-keto and 6-oximino derivatives (**7**) [5, 6].

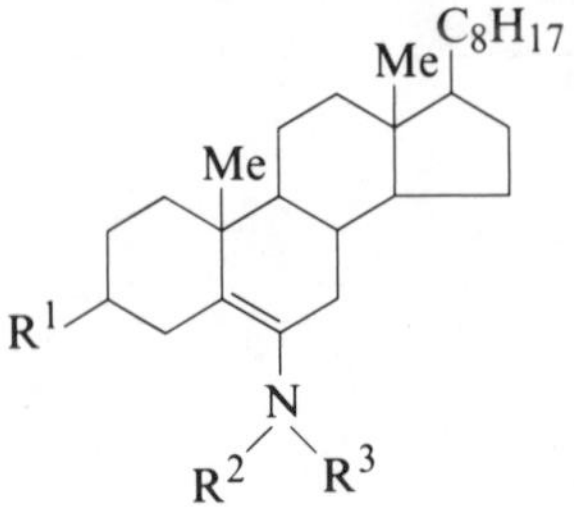

$$R^1 = H, Cl, OH, OAc; \quad R^2, R^3 = H, COMe, OCOMe; \quad R^4 = O, NOH$$

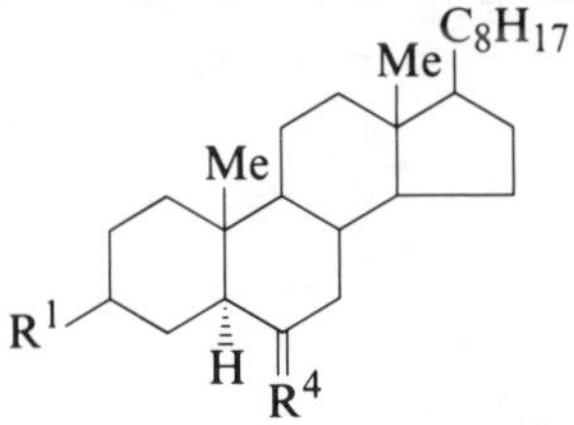

Under the action of iron in hydrochloric acid, unsaturated nitro compounds transform into ketoximes (**9**; equation 3), and further into ketones (**10**) [7].

$$R^1CH{=}CR^2NO_2 \xrightarrow{\text{Fe + HCl}} R^1CH_2C(R^2){=}NOH \xrightarrow{\text{HCl}}$$

$$\textbf{(8)} \qquad\qquad\qquad\qquad \textbf{(9)}$$

$$R^1CH_2C(O)R^2 \qquad (3)$$

$$\textbf{(10)}$$

$$R^1 = Ph, \text{ } p\text{-MeOC}_6\text{H}_4, \text{ } m\text{-MeOC}_6\text{H}_4, \text{ 2-furyl}; \quad R^2 = Me, Et, Pr$$

Nitro and β-nitrovinyl groups in the *ortho* position of the benzene ring promote reductive heterocyclization and the formation of indole derivatives [8, 9]. The substituted 2,β-dinitrostyrenes (**11**; equation 4) can be reduced by iron powder in 80% acetic acid to produce the corresponding indoles (**12**).

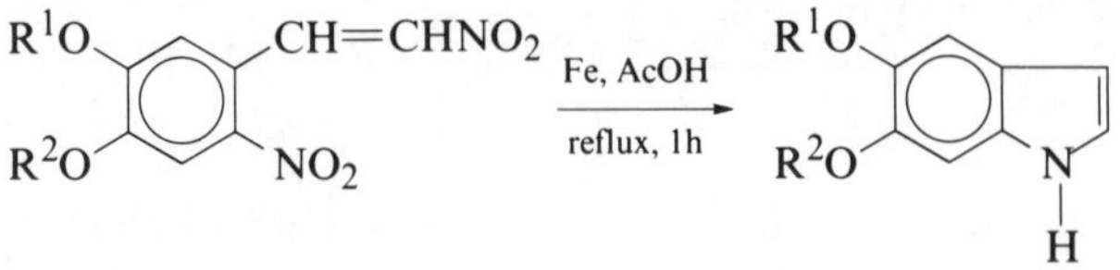

$$(4)$$

$$\textbf{(11)} \qquad\qquad\qquad\qquad \textbf{(12)}$$

$$R^1 = R^2 = PhCH_2;$$
$$R^1 = Me, R^2 = PhCH_2;$$
$$R^1 = PhCH_2, R^2 = Me$$

The combination lead/acetic acid in dimethylformamide (DMF) reduces 1-nitroalkenes to aldoximes or ketoximes in very good yields [10].

In acidic media, tin(II) chloride transforms nitroalkenes into α-substituted oximes and the corresponding saturated nitro compounds or their mixtures [11, 12]. The same reducing agent in acetone and ethyl acetate initiates aldoxime and ketoxime formation [2, 13, 14], and in alcohols or thiols (equation 5) compounds of type (14) are formed [15].

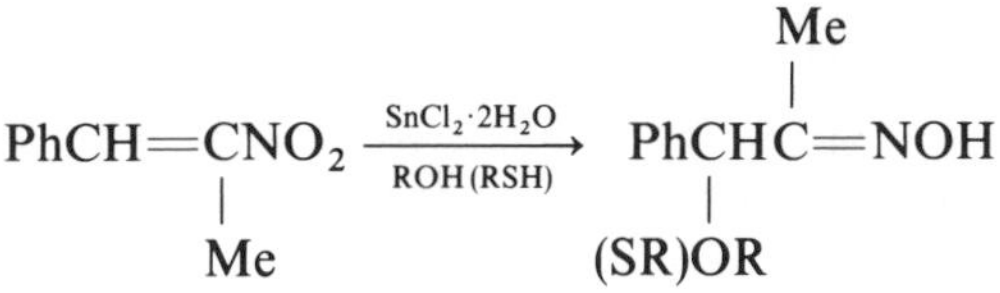

$$PhCH\!\!=\!\!CNO_2 \xrightarrow[\text{ROH(RSH)}]{\text{SnCl}_2 \cdot 2\text{H}_2\text{O}} \underset{(SR)OR}{\overset{Me}{PhCHC\!\!=\!\!NOH}} \qquad (5)$$

$$\underset{Me}{|}$$

(13)   (14)

R = Alk

6-Nitro-Δ^5-steroids are reduced to compounds of type (16) under the action of $SnCl_2 \cdot 2H_2O$ in THF (equation 6) [16].

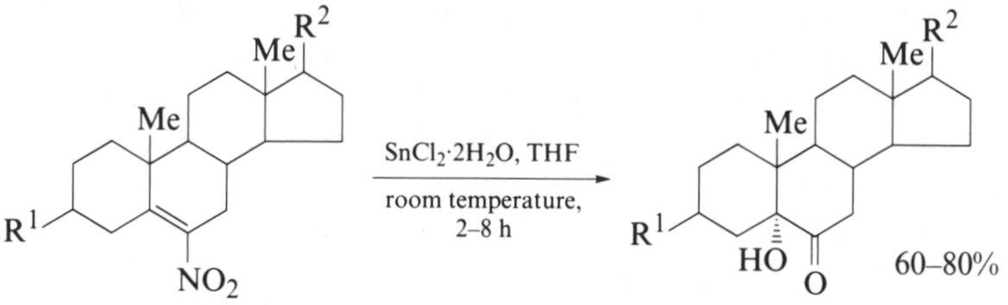

$$(6)$$

(15)   (16)

R^1 = H, OH, OAc, OMe, Cl; $R^2 = C_8H_{17}$, OAc

The hydrogenation of nitrochromenes in the presence of $SnCl_2$ proceeds via the oxime to give 3-chromanones [17].

Under reflux, solutions of 2-nitromethylene-3-methylbenzothiazoline and 1-methyl-1,2-dihydroquinoline in concentrated hydrochloric acid with spongy tin produce hydrates of chloromethylates and 2-aminomethylbenzothiazole and quinoline, respectively [18].

Nitroalkene reduction by tin(II) chloride in basic media (sodium stannite) gives ketoximes [19] in high yields. No aldoximes are formed under these conditions.

Chromium(II) chloride readily reduces β-aryl-α,β-unsaturated nitroalkenes (17; equation 7) to oximes (18) [20] or ketones [21]. Sometimes, under mild conditions, the reaction stops at the α-hydroxyoxime.

The reduction of 3-nitroflavenes under the action of $CrCl_2$ has become a new and effective synthetic method for flavones [22].

Steroid nitroalkenes (19; equation 8) are hydrogenated to the corresponding α-hydroxyoximes (20) under the action of $CrCl_2$ [23, 24].

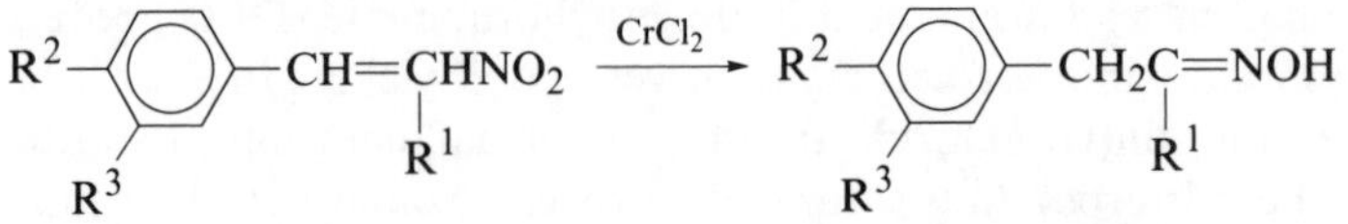

$$R^2-\text{C}_6\text{H}_3(R^3)-CH=CHNO_2 \xrightarrow{\text{CrCl}_2} R^2-\text{C}_6\text{H}_3(R^3)-CH_2C=NOH \quad (7)$$

with R^1 substituents on the $CHNO_2$ / $C=NOH$ carbons.

(17)   **(18)**

$R^1 = R^2 = R^3 = H;$ $R^1 = Me, R^2 = Br, R^3 = H;$
$R^1 = Me, R^2 = R^3 = H;$ $R^1 = Me, R^2 = R^3 = OEt$

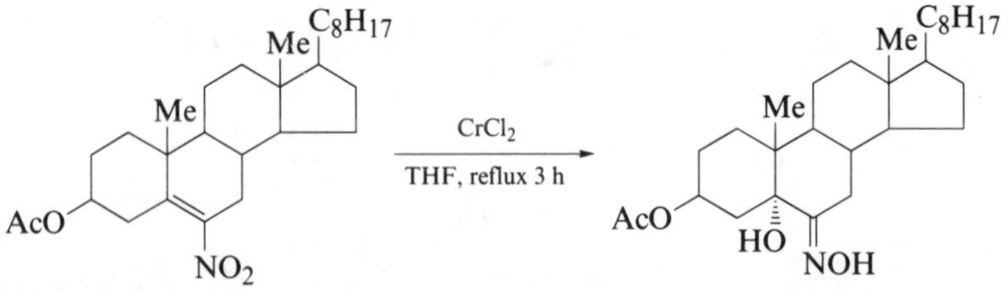

$$ (8) $$

(19)   **(20)** 80%

A wide range of such compounds as oximes, ketones, dimers, pyrroles and others can be prepared by nitrostyrene reduction under the action of titanium(III) chloride (six-fold to 18-fold excess) at pH 6.5 and 0–35 °C. The simplicity of the method and the mechanism of pyrrole formation are discussed in the literature [25].

Tributyltin hydride reduces α,β-nitroalkenes to give nitroalkanes in CH_2Cl_2 via the tin complex (**21**; equation 9) [26, 27]; the complex ($R^2 = H$) can be treated with hydrofluoric acid to produce the nitroalkane (**22**). In the presence of an oxidizing agent (MCPBA or ozone) the complex (**21**) transforms into the ketone (**10**).

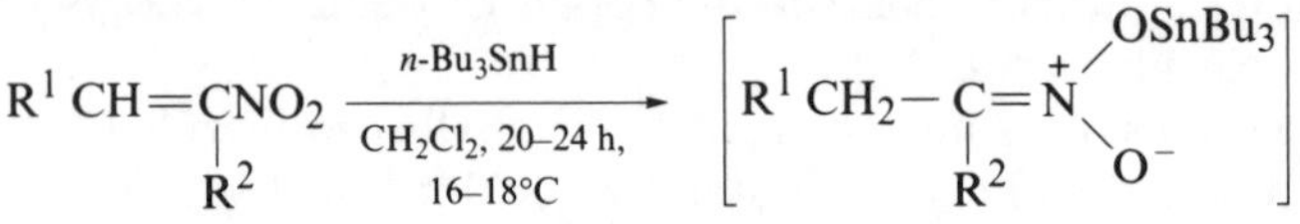

$$R^1CH=CNO_2(R^2) \xrightarrow[\substack{CH_2Cl_2, 20-24\ h, \\ 16-18°C}]{n\text{-Bu}_3SnH} \left[R^1CH_2-\overset{+}{C}(R^2)=N \overset{OSnBu_3}{\underset{O^-}{\diagup}} \right] \quad (9)$$

(8)   **(21)**

$$\text{MeOH/H}_2\text{F}_2 \swarrow \qquad \searrow \text{MCPBA or O}_3$$

$R^1CH_2CHNO_2(R^2)$ $R^1CH_2C=O\,(R^2)$

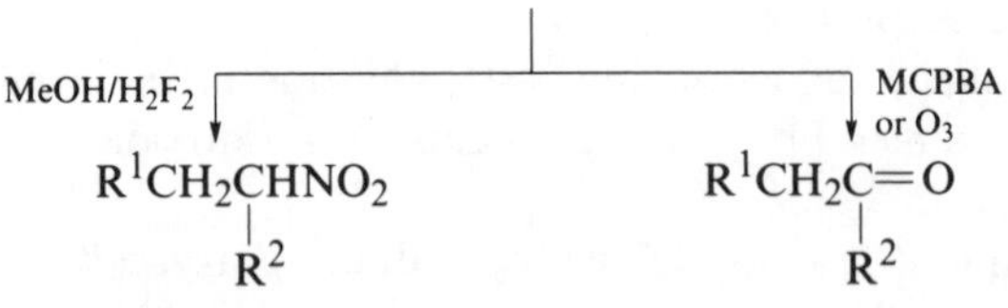

(22)   **(10)**

$R^1 = Ar, R^2 = H$ $R^1 = Ar, R^2 = H, Me$

Nitroalkene reduction by lithium aluminium hydride (LAH), lithium borohydride, sodium trimethoxyborohydride has been widely investigated.

A double bond that is normally inert towards lithium aluminium hydride is easily reduced when it is conjugated with electrophilic groups. That is why such a method is used for the preparation of nitroalkanes, oximes and amines. The product structure depends on the conditions of hydrogenation, particularly the temperature [28]. In many cases, low-temperature ($-40\,°C$ to $-70\,°C$) hydrogenation does not involve the nitro group, and only the double bond is reduced to give the mono nitroalkane or dinitroalkane. The reduction at $0\,°C$ leads to oximes; amines are isolated in boiling ether or in THF or dioxan at room temperature (equation 10) [29–35].

$$R^1CH_2CHR^2NH_2 \xleftarrow[\text{Et}_2\text{O, THF}]{25\text{–}35\,°C,\ \text{LiAlH}_4} R^1CH{=}CR^2NO_2 \xrightarrow{0\,°C,\ \text{LiAlH}_4} R^1CH_2CR^2{=}NOH$$

$$\textbf{(23)} \qquad\qquad\qquad\qquad \textbf{(8)} \qquad\qquad\qquad\qquad \textbf{(24)}$$

$$\Big| \quad \text{LiAlH}_4,\ -50\,°C\ \text{to}\ -70\ °C$$

$$R^1CH_2CHR^2NO_2 \qquad\qquad R^1CH_2CR^2(NO_2)CHR^1CHR^2NO_2$$

$$\textbf{(22)} \qquad\qquad\qquad\qquad \textbf{(25)}$$

$$R^1, R^2 = \text{H, Alk, Ar, Het} \qquad\qquad\qquad (10)$$

Many amines have been isolated in the form of chlorohydrates, oxalates or picrates [36, 37].

Primary amines are obtained in the reduction of 1-nitro-1-alkene-3-ynes by LAH [38].

Substituted 3-aminoalkylazaindoles (**27**; equation 11) have been synthesized in high yields by LAH hydrogenation of the corresponding starting materials (**26**) [39].

$$\textbf{(26)} \qquad\qquad\qquad\qquad\qquad \textbf{(27)} \qquad\qquad\qquad (11)$$

$$R^1 = R^3 = \text{H}, R^2 = \text{Me}; \quad R^1 = \text{Cl}, R^2 = \text{Me}, R^3 = \text{Me, Et}$$

The hydration of heteroarylnitroethenes containing dibenzofuran or dibenzothiophene rings leads to amines that can be used for the synthesis of analogues of alkaloids of the ellipticine group [40].

The action of LAH on 1-phenyl-4-nitro-1,3-butadiene results entirely in the reduction of the nitrovinyl fragment [41].

1,2-Dinitroalkenes are inert towards LAH.

Sodium borohydride and sodium trimethoxyborohydride have been used successfully for the reduction of alkylnitroalkenes, arylnitroalkenes and heteroarylnitroalkenes to produce the corresponding nitroalkanes in THF, dioxan, alcohols or acetonitrile (equation 12). The reaction is promoted by ion-exchange resins or silica gel [2, 27, 42–49].

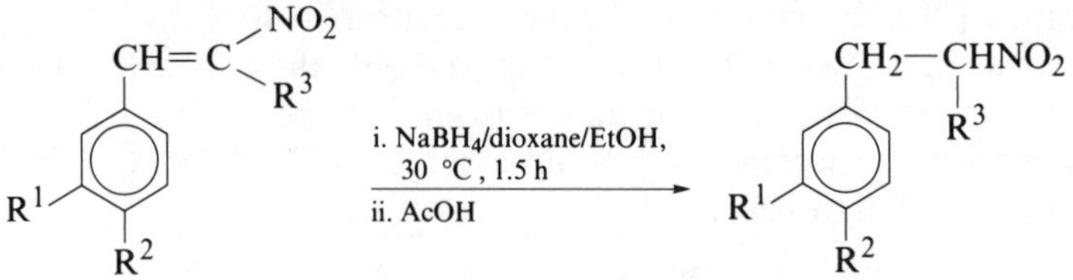

$$
\begin{array}{l}
\text{i. NaBH}_4/\text{dioxane/EtOH,} \\
\quad 30\ ^\circ\text{C}, 1.5\ \text{h} \\
\text{ii. AcOH}
\end{array}
\tag{12}
$$

$$
R^1 = \text{H, H, H, H, OMe, OCH}_2\text{Ph, OMe}
$$
$$
R^2 = \text{H, H, OMe, NMe, OCH}_2\text{Ph, OMe, OMe}
$$
$$
R^3 = \text{H, Me, H, H, H, H, H}
$$

Variation in the experimental conditions leads to the formation of dimers and polymers [47] or carbonyl compounds [27].

The reduction of unsaturated nitrosugars by $NaBH_4$ is selective: only the double bond is reduced. This is the synthetic approach to saturated nitrosugars that are the intermediates in aminosugar preparation [50, 51]. Such a method is closely linked to the quantitative transformation of steroid nitroalkenes into the corresponding nitroalkanes [52].

The selective reduction of the double bond in arylnitroalkenes by $NaBH_4$ has been used for the synthesis of 3-(β-nitroethyl) indoles (**29**; equation 13). Such indoles are the starting materials for the production of ergot alkaloid ($\mp$)-aurantioclavin [53].

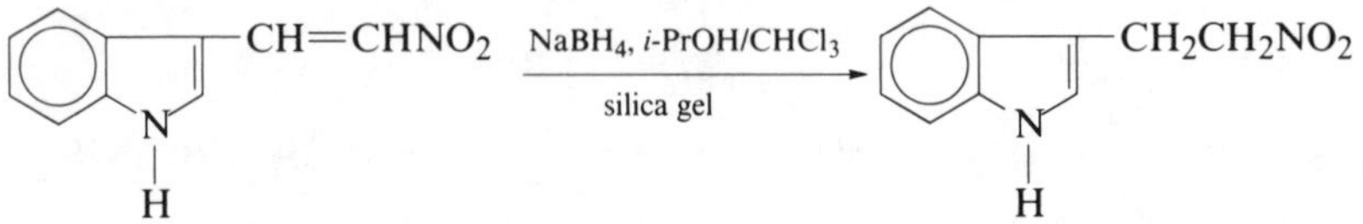

$$
\text{NaBH}_4,\ i\text{-PrOH/CHCl}_3 \quad \xrightarrow{\text{silica gel}}
\tag{13}
$$

(**28**)   (**29**) 84%

Sodium borohydride has been widely used for the selective reduction of substituted 3-nitrochromenes (**30**; equation 14) to 3-nitrochromanes (**31**) [54]. These compounds are of particular interest owing to the structural similarity with flavones.

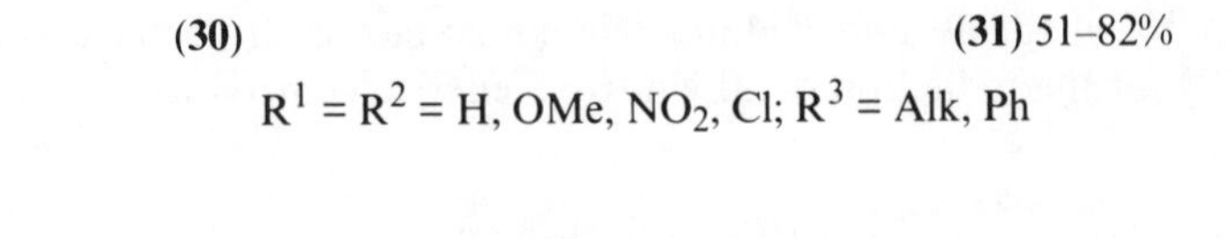

$$
\text{NaBH}_4,\ \text{MeOH/THF} \quad \xrightarrow{\text{room temperature}}
\tag{14}
$$

(**30**)   (**31**) 51–82%

$$
R^1 = R^2 = \text{H, OMe, NO}_2,\ \text{Cl};\ R^3 = \text{Alk, Ph}
$$

The reduction of (**32**) with $NaBH_4$ (equation 15) followed by heterocyclization of the intermediate halonitroalkane provides a general and convenient synthesis of substituted nitrodihydrobenzofurans (**33**) [55].

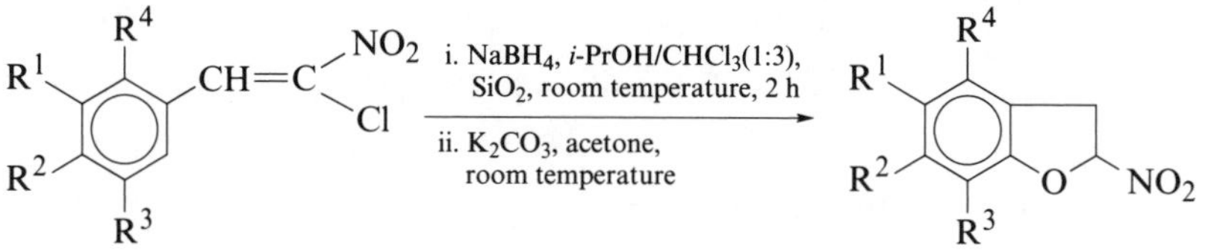

(15)

(**32**) (**33**)

R^1 = H, Br, Cl, OMe, NO_2; R^2, R^3, R^4 = H, OMe, Br

A mixture of $NaBH_4$ and $NiCl_2$ (catalytic amount) in methanol is an effective reducing agent for nitrocyclohexenes and aliphatic nitro compounds to produce the amines. The Ni_2B formed *in situ* is an active hydrogenation catalyst [56].

It is known [57] that the salts of nitro compounds (nitronates) are easily reduced by borane complexes to give the hydroxylamines. It is probable that the reaction proceeds via the intermediate complex $[CHCH{=}N(O)OBH_3]^-$ M^+, which can be easily hydrolysed to the nitroalkane or reduced by a borane complex to the hydroxylamine.

Sodium borohydride catalyses the reduction of conjugated nitroalkenes with borane complexes to give the hydroxylamines (**34**; equation 16) in high yield [2, 58, 59].

$$R^1C(R^2){=}C(R^3)NO_2 \xrightarrow[\text{ii. } H_2O,\, H^+,\, 60\text{–}65\,^{\circ}C,\, 2\,h]{\text{i. } BH_3,\, THF,\, NaBH_4\,(\text{catalyst}),\ \text{room temperature, } 1\,h} R^1CH(R^2)CH(R^3)NHOH \qquad (16)$$

(**34**)

Prolongation of the reaction period from one hour up to six days results in the formation of the corresponding amines in about 90% yield [2].

A subsequent modification of this method [59, 60] involves the formation of borane *in situ* from sodium borohydride and boron trifluoride etherate in THF. Such a process makes it possible to exclude the initial borane.

Zinc borohydride has been used recently for the selective reduction of unsaturated nitro compounds (equation 17). According to the method, monosubstituted nitroalkenes give nitroalkanes and disubstituted nitroalkenes give oximes. The mechanism of the process is discussed in the literature [61, 62].

A similar method has been used for the preparation of a number of cycloalkanone (C_5–C_7) oximes. The conditions of the process are mild and it is useful for the reduction of nitro compounds that are sensitive to basic or acidic media due to the neutral nature of $Zn(BH_4)_2$.

$$R^1R^2C{=}CR^3(NO_2) \xrightarrow[\substack{0-25\,°C,\,0.75-12\,h}]{Zn(BH_4)_2,\,DME} \begin{cases} \xrightarrow{R^1=R^3=H} R^1CH_2CH_2NO_2 \\ \qquad\qquad \textbf{(36)}\ \ 88-93\% \\ \\ \longrightarrow R^1R^2CHC(R^3){=}NOH \\ \qquad\qquad \textbf{(37)}\ \ 73-83\% \end{cases}$$

$$\textbf{(35)}$$

$$(17)$$

$$R^1 = Ar;\quad R^2, R^3 = H,\ Alk;\quad DME = 1,2\text{-dimethoxyethane}$$

Nitroalkene reduction catalysed by platinum, palladium or nickel in various media is selectively directed towards the formation of nitroalkanes, oximes, amines, or nitrogen-containing heterocycles [63–65]. For example, hydrogenation of nitroalkenes over palladium black produces nitroalkanes or, in acidic media, oximes [63, 65].

Effective reduction of arylnitroalkenes to amines takes place in a mixture of sulfuric and acetic acids under the action of palladium black [66–69].

The catalytic reduction of β-nitrostyrenes over a Pd/C catalyst in an EtOH/HCl mixture gives amines (**39**; equation 8) that are the precursors of 1,2,3,4-tetrahydroquinolines (**40**) [170].

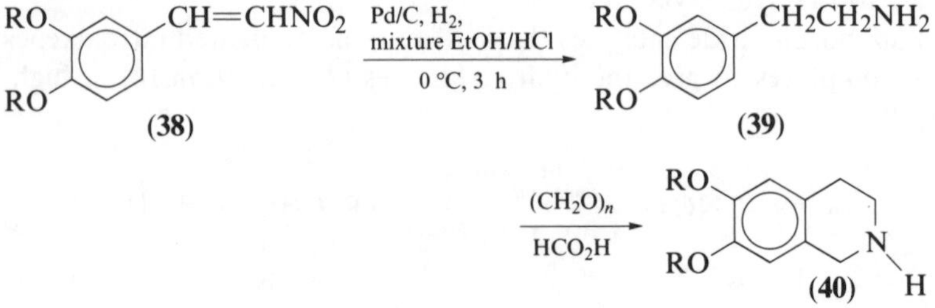

$$R = H,\ Me,\ CH_2Ph$$

This synthetic approach to compounds of type (**39**) is more convenient for their industrial production than reduction of β-nitrostyrenes by LiAlH$_4$ [71] or borane [60].

The catalytic hydrogenation of nitroalkenes over a palladium/charcoal catalyst is selective and depends upon the medium, giving oximes, ketones or amines [72a, 72b].

A convenient method for nitroalkene hydration using ammonium formate in the presence of 5% Pd/C in methanol/THF 1:1 by volume has been proposed [73]. The process occurs at room temperature and produces the corresponding oximes in high yields. Such a reaction can be accelerated and made to yield amines if NaBH$_4$ and 10% Pd/C in THF are used [74].

1,2-Dinitroalkenes are catalytically reduced under high pressure over platinum oxide or nickel to form diamines [75].

The catalytic ($Pd/BaSO_4$) hydration of 1-nitro-4-phenylbutadiene leads to the unsaturated amine [41].

The reduction of substituted α-nitroacrylic esters in the presence of various catalysts is of practical importance in connection with the synthesis of α-amino acids, such as D,L-phenylalanine, D,L-leucine D,L-tryptophan and others [76–79]. For example, the palladium catalysed hydration of α-nitro-β-isopropylacrylic ester (41; equation 19) in a water/alcohol solution under normal pressure [76] gives D,L-leucine (42). The same result is achieved with a skeletal nickel catalyst in methanol under high pressure [78].

$$Me_2CHCH{=}C{-}CO_2Alk \xrightarrow[\text{ii. H}_2O(\text{H}^+)]{\text{i. H}_2(\text{Pd, H}^+ \text{ or Raney nickel}}} Me_2CHCH_2CHCO_2H \quad (19)$$

$$\underset{\text{NO}_2}{|} \qquad\qquad\qquad\qquad\qquad\qquad \underset{\text{NH}_2}{|}$$

(41)   **(42)**

D,L-α-Aminobutyric acid (44; equation 20) is prepared from the methyl ester of α-nitrocrotonic acid (43) by hydration over PtO_2 in ethylacetate at 45 °C followed by hydrolysis in a 15% solution of hydrochloric acid [80].

The high-pressure hydrogenation of (43) in methanol over a skeletal nickel catalyst proceeds violently and does not give α-aminobutyric acid as it was supposed, but gives a mixture of the *allo* and *threo* forms of D,L-*o*-methylthreonine (45) [78]. The formation of the methoxy derivative takes place via methanol addition to the nitrocrotonic ester under the action of traces of alkali which are present in the skeletal nickel.

$$
\begin{array}{c}
\qquad\qquad\qquad \xrightarrow[\text{ii. HOH(H}^+)]{\text{i. PtO}_2, \text{EtOAc, 45 °C}} MeCH_2CHCO_2H \\
\qquad\qquad\qquad\qquad\qquad\qquad\qquad\qquad \underset{\text{NH}_2}{|} \\
MeCH{=}C(NO_2)CO_2Me \xrightarrow{\text{H}_2} \quad\qquad\qquad \textbf{(44)} \qquad\qquad\qquad (20) \\
\textbf{(43)} \\
\qquad\qquad\qquad \xrightarrow[\text{ii. HOH(H}^+)]{\text{i. Raney nickel, MeOH}} MeCHCHCO_2H \\
\qquad\qquad\qquad\qquad\qquad\qquad\qquad\quad \underset{\text{MeO NH}_2}{|\ \ |} \\
\qquad\qquad\qquad\qquad\qquad\qquad\qquad\qquad\quad \textbf{(45)}
\end{array}
$$

The reduction of α-nitrocinnamic ester (46; equation 21) by Pd/Al_2O_3 at room temperature and normal pressure is a new synthetic method for D,L-phenylalanine (47) [77].

$$PhCH{=}C(NO_2)CO_2Alk \xrightarrow[\text{ii. H}_2O(\text{H}^+)]{\text{i. H}_2(\text{Pd/Al}_2\text{O}_3)} PhCH_2CH(NH_2)CO_2H \quad (21)$$

(46)   **(47)**

The kinetic study of this process [81a] exposed the difference between the hydration of compound (46) and that of α-nitroalkenes. In the former case, along with compound (47) there were also formed α-nitro-β-phenylpropionates, α-hydroxylamino-β-phenylpropionates and the oxime of β-phenylpyruvate. The maximum yield of methylphenylalanine is 69%.

The stepwise reduction of substituted α-nitroacrylates with sodium borohydride on 10% Pd/C followed by the action of hydrochloric acid gives rise to a series of aromatic α-amino acid chloroanhydrides [81b].

Reduction of ester (48) over palladium black in methanol (equation 22) leads to the methyl esters of the corresponding saturated β-nitro acid (49) and unsaturated β-amino acid (50) [82].

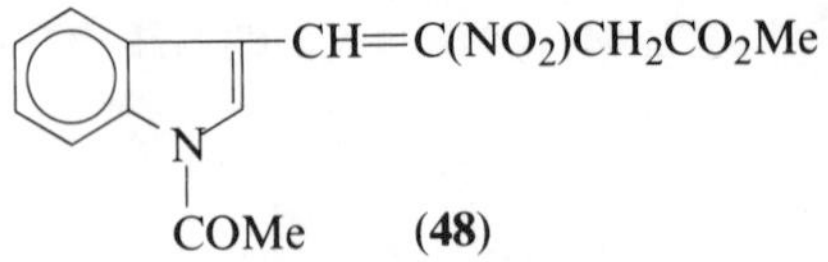

$$\text{(indole)}-CH=C(NO_2)CH_2CO_2Me$$

COMe   (48)

$$\xrightarrow{\text{H}_2,\ \text{Pd}}$$

$$\text{(indole)}-CH_2CH(NO_2)CH_2CO_2Me$$
COMe   (49) 40%

$$\text{(indole)}-CH_2C=CHCO_2Me$$
                    $|$
                    NH$_2$
COMe   (50) 27%

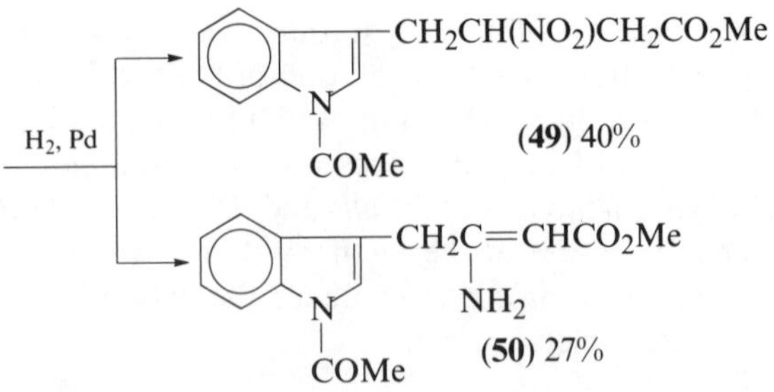

(22)

The importance of the D,L-tryptophan synthesis has created a strong interest in effective methods of nitroindolylacrylic ester reduction. D,L-Tryptophan is obtained in the two-step reduction (equation 23) of ester (51). The first step involves the action of palladium black and the second step requires an autoclave under 6–7 MPa pressure and skeletal nickel. Methyl D,L-tryptophan ester is hydrolysed by 10% NaOH to the amino acid [79, 83].

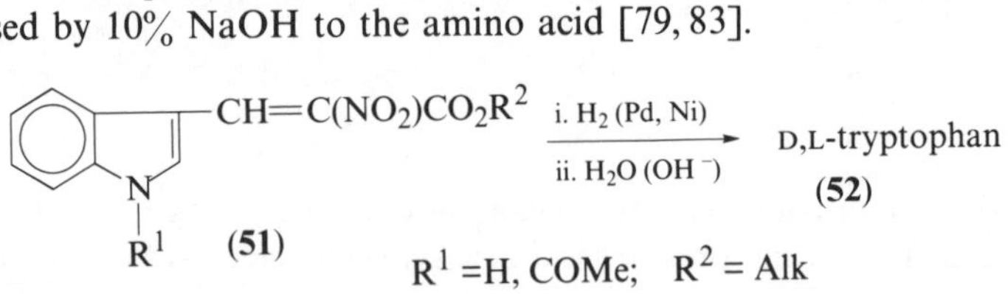

$$\text{(indole)}-CH=C(NO_2)CO_2R^2 \xrightarrow[\text{ii. H}_2O\ (OH^-)]{\text{i. H}_2\ (Pd,\ Ni)} \text{D,L-tryptophan}$$

R^1 (51)   (52)

R^1 = H, COMe;   R^2 = Alk

(23)

Raney nickel can be used to prepare D,L-tryptophan from ethyl α-nitro-β-(1-acetylindolyl-3)acrylate [84].

The reduction of α-nitro-β-(6-methylindolyl-3)acrylates under the action of 10% Pd/C in ethyl acetate or SnCl$_2$ in acidic medium gives rise to unsaturated α-amino acid esters [85].

The stepwise hydrogenation of substituted 2-β-dinitrostyrenes (**11**; equation 24) gives rise to indole derivatives (**12**) [86].

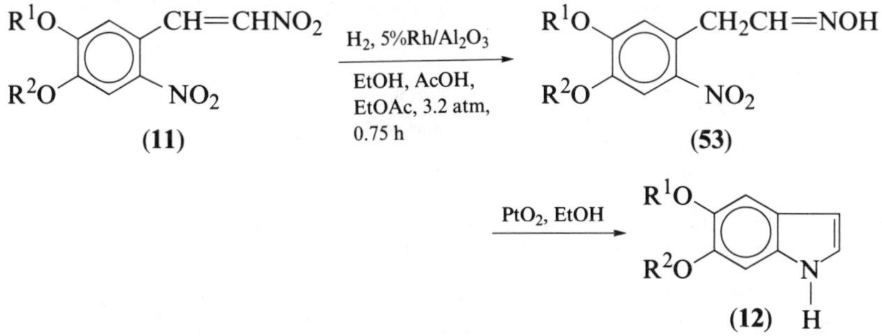

$$R^1 = Me, R^2 = Me, CH_2Ph; \quad R^1 = R^2 = Me; \quad R^1 = Me, R^2 = CH_2Ph \quad (24)$$

Similar reductive cyclization of 2-β-dinitrostyrene derivatives promoted by 5–10% Pd/C or Pt/C in alcohol, ethyl acetate or water produces indoles [87–90].

Reductive heterocyclization makes it possible to transform substituted α-nitro-β-phenylcrotonic esters into D,L-tryptophan and its 6-methyl and 6-chloro analogues [91].

So, there is a wide range of methods for the reduction of conjugated nitro compounds based on the application of metallic reducing agents or metallic (platinum, palladium, nickel) catalysts. Some other methods of reduction have also been devised to transform nitroalkenes into various types of compounds.

2-Phenylbenzimidazoline, formed *in situ* from *o*-phenylenediamine and benzaldehyde, is a mild, selective reducing agent which transforms arylnitroalkanes into the corresponding nitroalkanes [92, 93].

It has been found that the Hantzsch ester (**55**; equation 25) (HEH: 3,5-bis(ethoxycarbonyl)-2,6-dimethyl-1,4-dihydropyridine), an NAD(P)H analogue [94], is highly chemoselective on silica gel and reduces aliphatic, aromatic and heterocyclic nitroalkenes to give the corresponding nitroalkenes in 70–80% yield [95, 96].

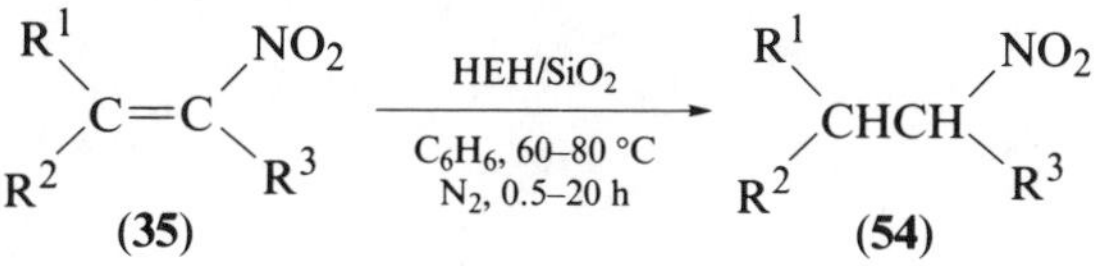

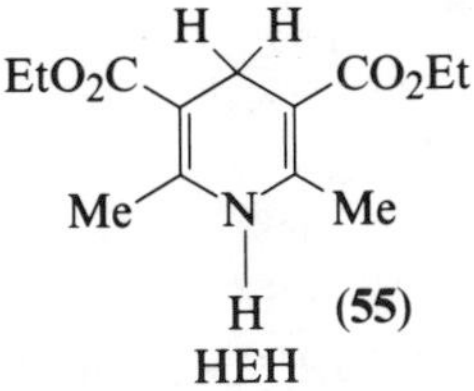

Nitroalkene reduction provides a route to enantioselective enzyme synthesis [97–99]. For example, 1-nitroalkanes (57; of equation 26) of high optical purity (83–98%) have been prepared under the action of baker's yeast [97, 98].

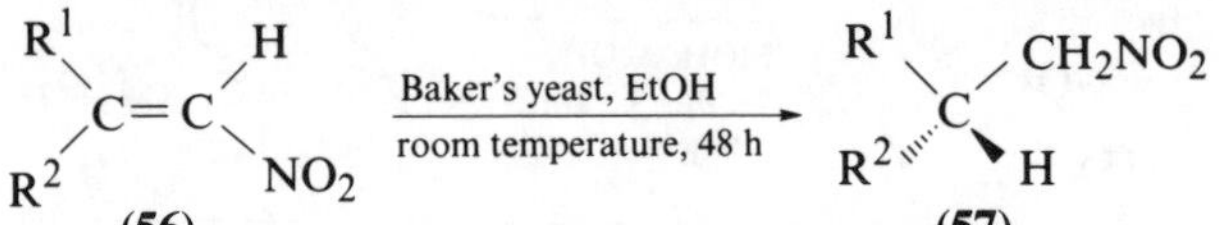

$$R^1 \diagdown \atop R^2 \diagup C = C \diagup^H \diagdown_{NO_2} \xrightarrow[\text{room temperature, 48 h}]{\text{Baker's yeast, EtOH}} R^1 \diagdown \atop R^2 \diagup C \diagup^{CH_2NO_2} \diagdown_H \qquad (26)$$

(56)   (57)

R^1 = Me, C_6H_{13}, Ph, Ph, Ph, Ph, Ph, p-ClC$_6$H$_4$, p-BrC$_6$H$_4$;
R^2 = C_6H_{13}, Me, H, Me, C_6H_{13}, Et, n-Pr, Me, Me

Sodium hypophosphite in combination with palladium reduces α,β-unsaturated nitroalkenes (8) to produce the corresponding oximes (9) in up to 72% yield [100]. Nitroalkanes and carbonyl compounds are also formed in small quantities (equation 27).

$$\text{ArCH}{=}\text{C(NO}_2)\text{R} \xrightarrow[\text{room temperature, 0.7–13 h}]{\text{NaH}_2\text{PO}_2/\text{Pd}} \text{ArCH}_2\text{C(R)}{=}\text{NOH} \qquad (27)$$

(8)   (9)

$$R = H, Me$$

Iodotrimethylsilane reduces nitroalkenes to oximes and ketones (equation 28). This method has been developed for steroid compounds [101].

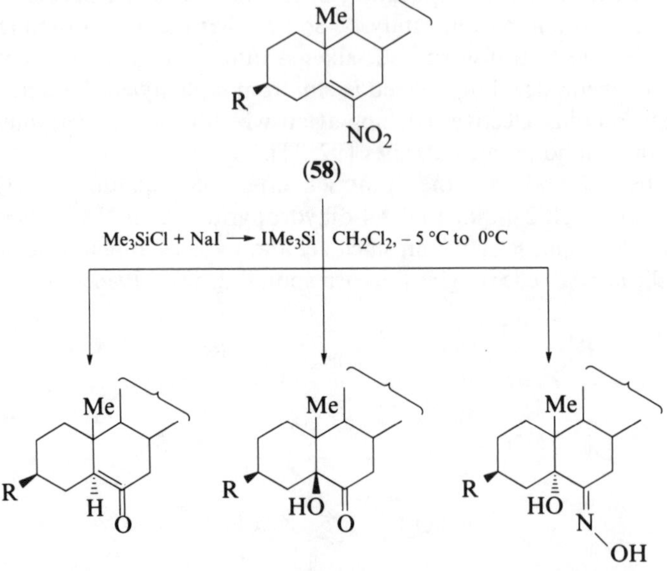

(58)

$$\text{Me}_3\text{SiCl} + \text{NaI} \longrightarrow \text{IMe}_3\text{Si} \mid \text{CH}_2\text{Cl}_2, -5\,°\text{C to } 0\,°\text{C} \qquad (28)$$

(59)	(60)	(61)
70–85%	2–8%	3–10%

R = H, OH, OMe, OAc, Cl

Nitroalkenes are transformed into carbonic acid amides (62; equation 29) under the action of ammonium hydrosulfide in pyridine at 180 °C and under pressure (sealed tube) [102].

$$RCH{=}CHNO_2 \xrightarrow[180\,°C,\,6\,h]{NH_4SH\,+\,S,\,C_5H_5N} RCH_2CONH_2 \qquad (29)$$

$$\textbf{(4)} \qquad\qquad\qquad \textbf{(62)}$$

$$R = Pr, Ph$$

Electrochemical reduction is a convenient method for the transformation of arylnitroalkenes into amines [103]. Several examples of the electrochemical reduction of nitroalkenes to oximes and ketones have been described [104–107]. According to Convert et al. [104] the first polarographic wave of reduction in a proton-containing medium corresponds to the process in equation (30).

$$R^1R^2C{=}C(R^3)NO_2 + 4e^- + 4H^+ \longrightarrow R^1R^2CHC(R^3){=}NOH + H_2O \quad (30)$$

$$\textbf{(35)} \qquad\qquad\qquad\qquad \textbf{(37)}$$

A similar result has been achieved for vicinal halonitroalkene reduction [104, 108]. However, the polarographic behaviour of geminal halonitroalkenes [107] and α-nitroacrylates [109, 110] is more complicated.

The reduction of aromatic nitroalkenes on the mercury or carbon cathode depends on the potential and is characterized by high selectivity (equation 31).

$$ArCH{=}CR(NO_2) \xrightarrow[i\text{-}PrOH/H_2O]{e^-,\,Hg\,or\,C\,cathode}$$

$$\xrightarrow{-0.3\,V\,to\,-0.5\,V} ArCH_2C(R){=}NOH$$

$$\textbf{(9)}\ 85\text{–}93\%$$

$$\textbf{(8)}$$

$$\xrightarrow{-1.1\,V\,to\,-1.3\,V} ArCH_2CH(R)NH_2 \qquad (31)$$

$$\textbf{(23)}\ 60\text{–}66\%$$

$$R = H, Me$$

In the case of nitroalkadienes such as 1-(3-cyclohexene-1-yl)-2-nitroethene the double bond attached to the nitro group is reduced selectively, while the isolated double bond remains intact [111].

The polarographic properties of α-nitrocinnamic and α-nitro-β-(3-indolyl) acrylic esters are discussed in detail in the literature [109–113]. Methyl α-nitro-β-(3-indolyl)acrylic ester was found to give a six-electron polarographic wave in acidic medium. The process is controlled by diffusion and corresponds to a reduction to the hydroxylamine derivative or imine via the intermediate oxime of the ester, i.e. (3-indolyl)pyruvic ester [110, 112, 113].

The electrochemical reductions of the methyl esters of α-nitrocinnamic acid and α-nitro-β-(3-indolyl)acrylic acid give D, L-phenylalanine and D, L-tryptophan in 70% and 60% yields, respectively (current efficiencies are 30% and 22%) [114–119]. The relatively low yield of tryptophan is attributed by the authors [116, 120] to the instability of the intermediate products of reduction, particularly the oxime of indolylpyruvic ester, in the acid medium.

Investigations of electrochemical reduction in polarographic and preparative electrolysis under potential-controlled conditions for some α,β-dinitroalkenes and 1,4-dinitrobutadienes over a wide range of pH have exposed specific features of these compounds [104, 108, 121]. In contrast to mononitroalkenes, these substances are characterized by a reversible, two-electron wave over the whole pH interval. Polarographic analysis in combination with UV data has made it possible to suggest the following mechanism (equation 32) for such a wave [108, 122].

$$O_2N(CR{=}CR)_n NO_2 + 2e^- + 2H^+ \xrightarrow{\ pH < 2.5\ } \rightleftharpoons$$

(63)

$n = 1, 2$

$$HOON{=}CR(CR{=}CR)_m CR{=}NOOH \quad (32)$$

(64)

$m = 0, 1$

$$O_2N(CR{=}CR)_n NO_2 + 2e^- \xrightarrow{\ pH > 6.6\ } \rightleftharpoons {}^- OON{=}CR(CR{=}CR)_m CR{=}NOO^-$$

(63)   (65)

$n = 1, 2$ $m = 0, 1$

The primary attack upon the molecules of the compound under discussion is not directed towards the nitro group like in nitroalkenes but towards the α-carbon. The reduction terminates at the monomeric or dimeric dinitronates in neutral or basic medium. 1,4-Dinitrodienes produce bis-nitronic acids under acidic conditions [104]. 2,3-Dinitrobutene undergoes further reduction to 2,3-diaminobutane and the minor products of the preparative electrolysis are diacetyl, its monooxime and dimethylfuroxan [108].

The isolation of the dianions of the corresponding bis(nitronic) acids correlates with their formation in the electron transfer reactions [123].

Studies of nitroalkenes in electron transfer reactions have shown that such mononitroalkenes as β-nitrostyrene and β-ferrocenylnitroethene (66; equation 33) under the action of the electron-donating cyclooctatetraene dianion (COTDA) form the products of dimerization (68) [124].

$$FcCH{=}CHNO_2 \xrightarrow{\ e^-\ } Fc\overset{\cdot}{C}HCH{=}NOO^- \longrightarrow 1/2(FcCHCH{=}NOO^-)_2 \quad (33)$$

(66)   (67)   (68)

Electronic reduction of conjugated dinitro compounds (63; equation 34) proceeds along a different pathway [123, 125]. Such compounds readily interact with both strong reducing agents (sodium amalgam, potassium naphthalenide) and moderate donors (COTDA, KI in THF). Treatment with two equivalents of the donor results in nearly quantitative yield of the products of two-electron reduction, i.e. bis(nitronic) acid dianions (65), which can be transformed back to the starting materials by means of a two-electron oxidation or to the products of carbon protonation (69) under the action of acid [125].

$$O_2NCR^1{=}CR^2{-}(CR^2{=}CR^1)_nNO_2$$
(63)

$$-2e^- \Big\Updownarrow +2e^-$$

$$^-O_2N{=}CR^1{-}(CR^2{=}CR^1)_n{-}CR^1{=}NO_2^-$$ (34)
(65)

$$\Big\downarrow 2H^+$$

$$O_2NCR^1{-}CR^2{=}CR^1)_n{-}CHR^1NO_2$$
(69)

So, the two-electron reactions of conjugated, unsaturated dinitro compounds, followed by protonation of the resultant dianions, lead to selective reduction of the C=C bond with no effect on the nitro group.

2.2 REACTIONS WITH OXYGEN NUCLEOPHILES

Nitroalkenes react with alcohols to form β-alkoxynitroalkanes (1; equation 1) under the action of basic catalysts (AlkONa) [126–129].

$$R^1CR^2{=}CR^3NO_2 + R^4OH \xrightarrow{\text{base}} R^4OCR^1R^2CHR^3NO_2$$ (1)
(1)

$$R^1, R^2, R^3, R^4 = H, Alk$$

The most violent reaction occurs with nitroarylalkenes and, moreover, the nitrostyrene alkoxylation product is able to add to the initial nitroalkene. Substituted nitrostilbenes and alcohols undergo interaction with difficulty and produce diastereoisomers [130]. Halogenoalcohols have also been used in such reactions. Reaction between 1-nitro-1-propene and ethene chlorohydrin, 2-chloro-1-propanol, 1-chloro-2-propanol or 1,3-dichloro-2-propanol leads to the nitrochloroalkyl ether [131].

A convenient way to alkoxyketones (2; equation 2) is through the alkoxylation of nitrocyclohexene with subsequent treatment of the intermediate salt under the conditions of the Nef reaction [15, 132].

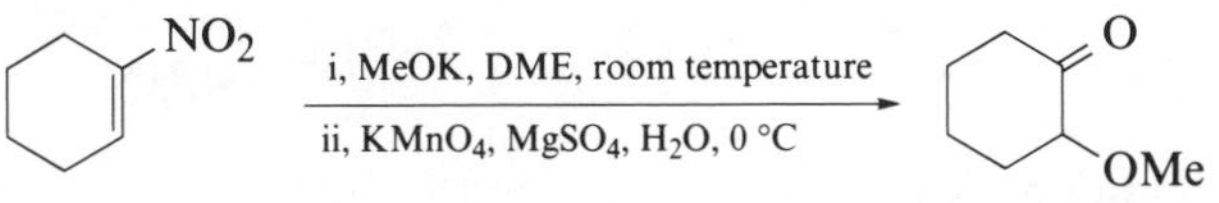

(2)

(2) 85%

The reduction of arylnitroalkenes with $SnCl_2$ in alcoholic medium results in the formation of oxime derivatives in high yield [15].

According to a kinetic study of alcohol additions to 1-nitro-1-propene, the alcohol reactivity decreases in the order MeOH > EtOH > PrOH > i-PrOH [133].

The mutagenic effects of 2-nitro-1-furylethene compared with the results of a kinetic study of its interactions with MeOH, EtOH and nucleic acids [134].

Hoz et al. [135] have prepared the addition product from nitrovinylfluorene and the methylate ion and have studied the kinetics of the reaction.

The stereochemical aspects of alcoholate additions to trans-nitroalkenes (equation 3) under protonation at −78 °C show the prevailing formation of anti-β-nitro ethers which can be transformed into anti-β-aminoalcohols of the type widely present in natural compounds [136, 137].

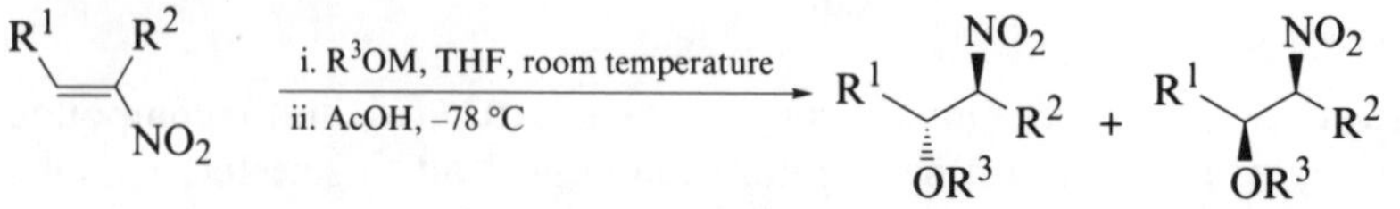

$$R^1, R^2 = Alk, Ph, PhCH_2CH_2; \quad R^3 = Alk, PhCH_2, Ph(CH_2)_n \ (n = 1\text{–}3); \quad M = Li, Na, K$$

The stereoselectivity of the process is reduced from $anti/syn = 88\text{–}96{:}12\text{–}4$ to 2:1 when the substituent $R^1 = Ph$; R^2 does not influence the process. If a diastereoisomeric mixture of 2-nitro-3-benzyloxyoctane (3a) containing mostly the anti form is treated with a catalytic amount of Et_3N in MeCN, the ratio anti/syn becomes 52:48 owing to thermodynamic control (equation 4). The mixture easily returns to the initial state upon deprotonation followed by protonation at low temperature [136].

$$anti\text{-}(3a) + syn\text{-}(3a) \ \rightleftharpoons \ anti\text{-}(3a) + syn\text{-}(3a) \quad (4)$$
$$\qquad 85\% \qquad\quad 15\% \qquad\qquad\qquad\qquad 52\% \qquad\quad 48\%$$

Hori et al. [138] have made a theoretical study of the stereoselectivity of alcohol addition to nitroalkenes.

Compound (4) is the product of anti-selective synthesis (equation 5) [137].

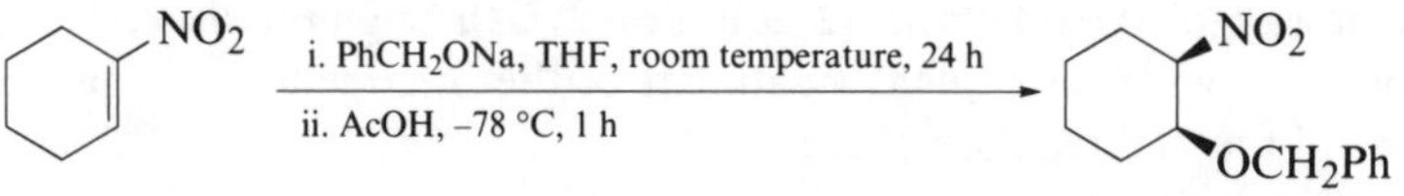

$$trans/cis = 4{:}96$$
$$(4)\ 71\%$$

Alcohol addition to some unsaturated nitrosugars (5; equation 6) is highly stereoselective [130]. For example, the methylate ion combines with sugar derivatives containing a terminal nitrovinyl fragment, as shown in equation (6). Alkoxylation of compound (5) followed by the Nef reaction leads to the corresponding 2-O-methylaldose (6).

$$
\begin{array}{ccc}
\text{CH}=\text{CHNO}_2 & & \text{MeOCHCH}=\text{O} \\
\text{AcO}-\!\!\!\!\!-\text{H} & & \text{HO}-\!\!\!\!\!-\text{H} \\
\text{H}-\!\!\!\!\!-\text{OAc} & \xrightarrow[\text{ii. H}^+,\text{H}_2\text{O}]{\text{i. MeONa}} & \text{H}-\!\!\!\!\!-\text{OH} \\
\text{H}-\!\!\!\!\!-\text{OAc} & & \text{H}-\!\!\!\!\!-\text{OH} \\
\text{CH}_2\text{OAc} & & \text{CH}_2\text{OH} \\
\textbf{(5)} & & \textbf{(6)}
\end{array}
\tag{6}
$$

The intramolecular oxygen addition of the alcohol hydroxy group in nitro-alkene carbohydrates, which are generated in the condensation of pentoses and hexoses with nitromethane, is one of the steps in the nitromethyl C-glucoside synthesis [135].

Certain acids can act as a catalyst in the reaction between alcohols and nitroalkenes. For example, the monomethyl ether of glycol condenses with nitroethene under the action of H_3PO_4 at 20 °C [126].

Alcohol interaction with 1,1,1-trichloro-3-nitro-2-propene [130, 139] is violent without a catalyst at 100–120 °C (equation 7).

$$
\text{Cl}_3\text{CCH}=\text{CHNO}_2 + \text{ROH} \longrightarrow \text{Cl}_3\text{CCH}(\text{OR})\text{CH}_2\text{NO}_2 \tag{7}
$$

$$\textbf{(7)}$$

The position of the nitro group defines the direction of nucleophilic attack in α-nitroacrylate and β-nitroacrylate alkoxylation by alcohols (equation 8) [140].

$$
\underset{\overset{|}{\text{NO}_2}}{\text{R}^1\text{CH}}=\text{CCO}_2\text{R}^2 \xrightarrow[\text{ii. H}^+]{\text{i. MeONa, MeOH, below}\,-10\,°\text{C}} \underset{\overset{|}{\text{MeO}}\ \overset{|}{\text{NO}_2}}{\text{R}^1\text{CHCHCO}_2\text{R}^2}
$$

$$\textbf{(8)}\ \ 65\text{--}78\%\tag{8}$$

$$
\underset{\overset{|}{\text{NO}_2}}{\text{R}^1\text{C}}=\text{CHCO}_2\text{R}^2 \xrightarrow[\text{ii. H}^+]{\text{i. MeONa, MeOH, }-10\,°\text{C to }15\,°\text{C}} \underset{\overset{|}{\text{O}_2\text{N}}\ \overset{|}{\text{OMe}}}{\text{R}^1\text{CHCHCO}_2\text{R}^2}
$$

$$\textbf{(9)}\ \ 56\text{--}66\%$$

$$\text{R}^1 = \text{Alk, Ph;}\quad \text{R}^2 = \text{Me, Et}$$

The addition of methanol to 3-nitroacrylic acid succeeds under reflux with p-toluenesulfonic acid [141].

Strong accepting groups, for example $C=O$, located in the β position to the nitrovinyl fragment activate the process and partially redirect the addition. The reaction takes place without a catalyst and is accompanied by a 'pushing out' of the nitro group. Subsequent addition of a second equivalent of alcohol gives the acetal (11; equation 9) [142].

$$PhCCH{=}CHNO_2$$

O | MeOH

(9)

$$PhCCHCH_2NO_2 \qquad\qquad PhCCH_2CH(OMe)_2$$

OOMe O

(10) 43%   **(11)** 13%

Dinitroalkenes are even more active; for instance, geminal dinitroalkenes (like 1,1-dinitro-2-phenylethene) combine with alcohols without a catalyst [143].

1,1-Dinitro-2,2-diphenylethene coupling with alcoholates leads preferentially to benzophenone (12; equation 10). In the case of sodium methylate, intermediate (13) was isolated from the reaction mixture at low temperature after acidification [144].

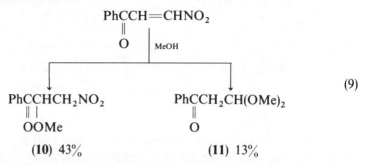

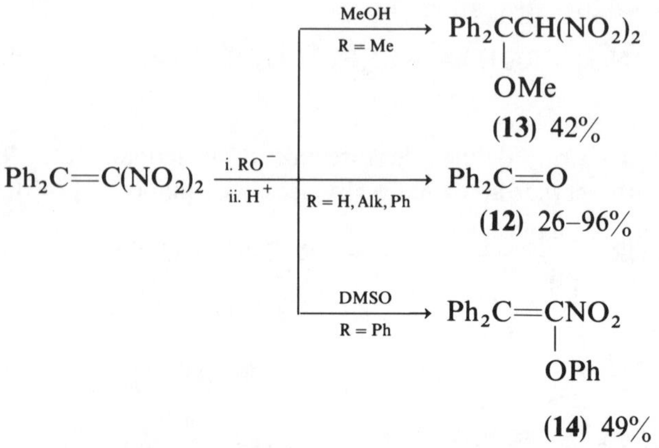

$$Ph_2C{=}C(NO_2)_2 \xrightarrow[\text{ii. H}^+]{\text{i. RO}^-}$$

MeOH
R = Me

$$Ph_2CCH(NO_2)_2$$
OMe
(13) 42%

R = H, Alk, Ph

$$Ph_2C{=}O$$
(12) 26–96%

(10)

DMSO
R = Ph

$$Ph_2C{=}CNO_2$$
OPh
(14) 49%

The product of nitro group substitution (14) was obtained in the reaction of phenolate ion with 1,1-dinitro-2,2-diphenylethene.

Vicinal dinitroalkenes react with alcoholates at low temperature (− 50 °C to −25 °C) to form the products of substitution (15; equation 11). At room temperature the addition of a second equivalent of alcohol to (15) gives the nitroacetal (16). A more reactive dinitroalkene such as 1,2-dinitro-2-phenylethene will transform into the nitroacetal even at − 50 °C [145].

The reaction of unsaturated nitro compounds with alcohols opens the way to various alkoxynitroalkanes and arylnitroalkanes [146] which can be easily

$$O_2NC(R^1){=}CR^1NO_2 \xrightarrow{R^2ONa,\ -50\,°C\ to\ -25\,°C} O_2NC(R^1){=}CR^1OR^2$$

(15)

$$\xrightarrow[\text{ii. H}^+]{\text{i, }R^2ONa,\ 18\text{--}20\,°C} O_2NCHR^1CR^1(OR^2)_2 \qquad (11)$$

(16)

$$R^1 = \text{Me, Ph}; \quad R^2 = \text{Me, i-Pr}$$

transformed into physiologically active amino ethers. Acid hydrolysis of β-alkoxynitroalkanes is the synthetic approach to α-hydroxyketones and α-hydroxy carboxylic acids substituted at the hydroxy group [147].

Various oxygen-containing heterocycles can be obtained in the reaction of nitroalkenes with unsaturated alcohols. The interaction between nitroalkenes and β,γ-unsaturated alcohols, which leads to THF derivatives, has been well studied by Ono et al. [148] and Ono and Kaji [149]. The process represented in equation (12) proceeds via the addition products (17). The latter were enriched by additional functional groups in Michael or hydroxymethylation reactions to give adducts (18). Subsequent reductive elimination of the NO₂ group and heterocyclization gave the tetrahydrofurans (19).

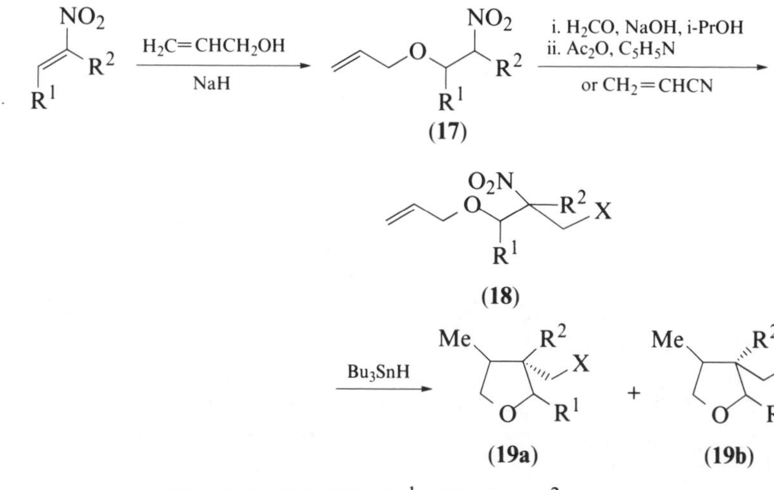

$$X = \text{OAc, CH}_2\text{CN}; \quad R^1 = \text{H, Ph}; \quad R^2 = \text{Me, Ph}$$

The 2-methylidenetetrahydrofuran derivative (20) was produced along such lines in the reaction of α-nitro-α-methylstyrene with propargyl alcohol followed by interaction with acrylonitrile [148, 149]. The cyclization usually proceeds stereoselectively. If the reaction is based on cyclohexene derivatives it leads to cis-fixed bicyclic compounds (21) (from 1-nitrocyclohexene) and tricyclic compounds (22) (from 1-nitrocyclohexene and cyclohexene-2-ol).

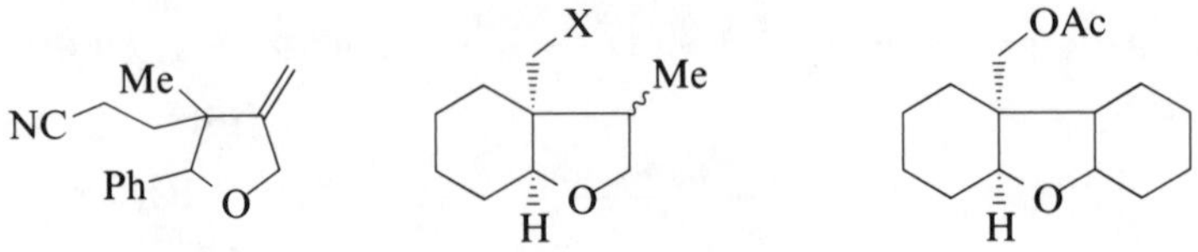

(20)   **(21)**   **(22)**

$$X = OAc, CH_2\underset{\underset{O}{\|}}{C}Me$$

This route to THF derivatives from nitroalkenes is preferable owing to the availability of the starting materials in comparison to other common methods.

The product of furyl alcohol addition to nitrostyrene is the starting material for tricyclic isoxazoline (**24**; equation 13) synthesis. The reaction proceeds via formation of the nitrile oxide (**23**) followed by its intramolecular cycloaddition [150].

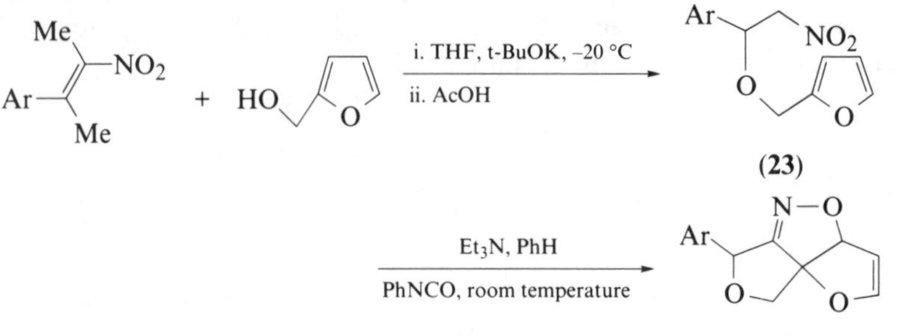

The participation of salicylic aldehyde and its derivatives in the reaction widens the scope for producing oxygen-containing heterocycles [151–154]. For example, nitrostyrenes react with salicylic aldehyde and its alkoxy derivatives (equation 14) in the presence of Et$_3$N [152] without a solvent to form compounds of type (**25**) in one step or on Al$_2$O$_3$ without a solvent under sonic agitation [151, 153]. Compounds of type (**25**) are the representatives of very important biologically active substances, namely the 2-nitrochromenes.

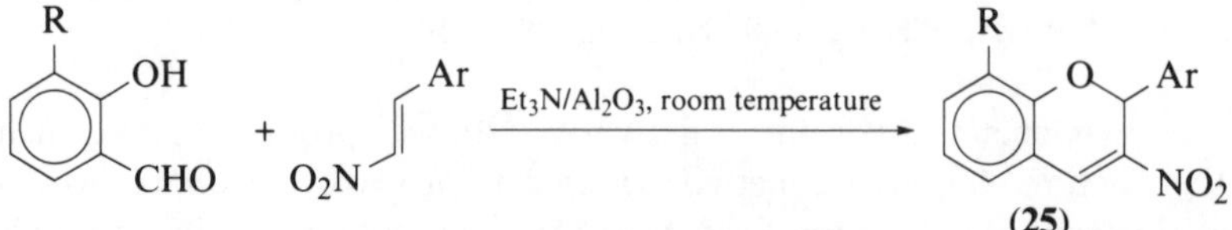

Under the catalysis of similar conditions CuCl$_2$/NEt$_3$ 1-nitro-2-naphthyl-ethenes form 2-naphthyl-3-nitrochromenes in 70–84% yield [153].

4,6-Dimethoxy-5-(2-nitrovinyl)pyrimidine condensation with salicylic alde-hyde is a synthetic method for the 3-nitrochromene-2-pyrimidine derivative (**26**; equation 15) [152]. The 4,6-dichloro analogue gives the products of mono-substitution and disubstitution.

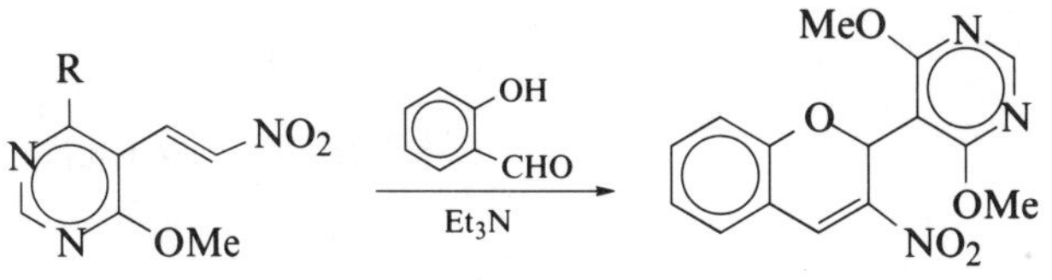

$$(15)$$

$$(\mathbf{26})$$

Recently, some 3-nitro-2*H*-chromenes (equation 16) were prepared on the basis of nitroethene or its precursors, e.g. nitroethyl alcohol [154]. 2-Hydroxy-naphthaldehyde was also used in this reaction.

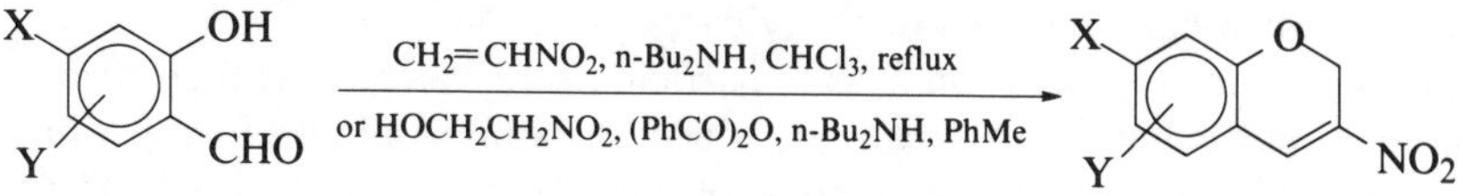

$$(16)$$

$$X = H, OMe$$
$$Y = H, Me, Br, OMe$$

(**27**)
49–95%

Oxidative addition of the hydroxide ion to nitroalkenes to give α-nitrooxiranes (**29**; equation 17) has been investigated [155a, 155b]. It proceeds via the one-electron oxidation of β-hydroxynitronate (**28**), which is the product of hydroxide ion addition, followed by intramolecular cyclization.

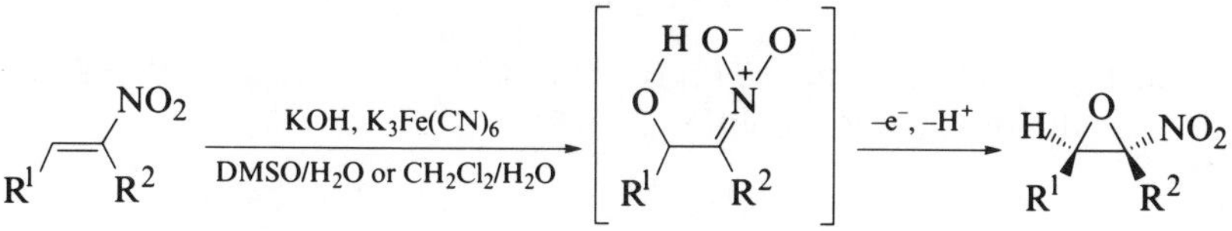

$$(17)$$

$$(\mathbf{28})$$

$$R^1 = H, Me, Et; \quad R^2 = Alk, Ar; \quad R^1, R^2 = (CH_2)_4$$

Best results are obtained under the action of H_2O_2 in alkaline solution [155a, 155b].

2.3 REACTIONS WITH SULFUR NUCLEOPHILES

Hydrogen sulfide easily adds to unsaturated nitro compounds without a catalyst to form 2-nitroalkane (arene)-1-thiols (**1**; equation 1) and, further, nitroalkyl(aryl) sulfides (**2**) [156].

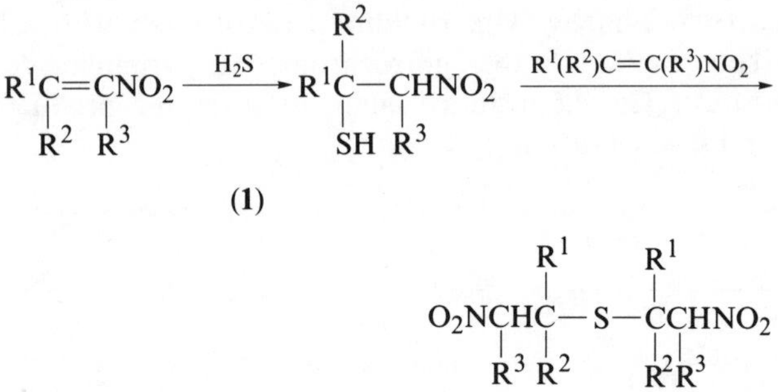

$$R^1C{=}CNO_2 \xrightarrow{H_2S} \overset{R^2}{\underset{SH}{\underset{|}{R^1C}}}{-}\underset{R^3}{\underset{|}{CHNO_2}} \xrightarrow{R^1(R^2)C{=}C(R^3)NO_2}$$

$$\underset{R^2\ R^3}{\underset{|\ \ |}{}}$$

(1)

$$O_2NCH\underset{R^3\ R^2}{\underset{|\ \ |}{\overset{R^1}{\overset{|}{C}}}}{-}S{-}\underset{R^2R^3}{\underset{|\ |}{\overset{R^1}{\overset{|}{C}}CHNO_2}}$$

(1)

(2)

$R^1{=}H, Ph;\quad R^2, R^3{=}H, Me$

The reaction between aliphatic or aromatic thiols and nitroalkenes depends upon the structures of the starting materials and leads to β-nitro sulfides. Thiol addition needs a basic catalyst in many cases [157, 158].

β-Nitro sulfides (3; equation 2) can participate in various synthetic transformations. For example, they can be reduced to amino sulfides [156], which are used as radioprotectors. Compounds of type (3) are oxidized by 30% H_2O_2 to give the corresponding sulfones [157].

$$R^1SCR^2HCH_2NO_2 \underset{(3)}{\overline{\qquad\qquad}} \begin{array}{l} \xrightarrow{[H]} R^1SCR^2HCH_2NH_2 \\ \qquad\qquad (4) \\ \\ \xrightarrow{[O]} R^1SO_2CR^2HCH_2NO_2 \\ \qquad\qquad (5) \end{array}$$

(2)

$R^1, R^2 = Alk, Ar$

The reduction of nitrostyrenes by tin(II) chloride in the presence of ethanethiol leads to α-ethylthiooximes (6; equation 3) [15].

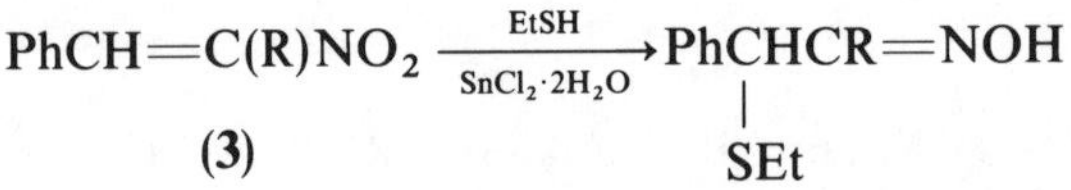

$$PhCH{=}C(R)NO_2 \xrightarrow[SnCl_2\cdot 2H_2O]{EtSH} \underset{SEt}{\underset{|}{PhCHCR}}{=}NOH$$

(3)

(3)

(6)

$R = H, Me$

Ono *et al.* [159, 160] have found that the β-thio group in nitro sulfides activates NO_2 groups towards stereospecific substitution by carbon nucleophiles (allyl, CN) and hydride ion. This allows the synthesis of compounds of type (8) and (9) from the products of interaction between nitroalkenes and thiophenol (7; equation 4). Under these conditions, compounds of type (10) that contain the

δ-aryl radical undergo Friedel–Crafts intramolecular reaction to give the substituted β-phenylthiotetrahydronaphthalenes (**11**; equation 5).

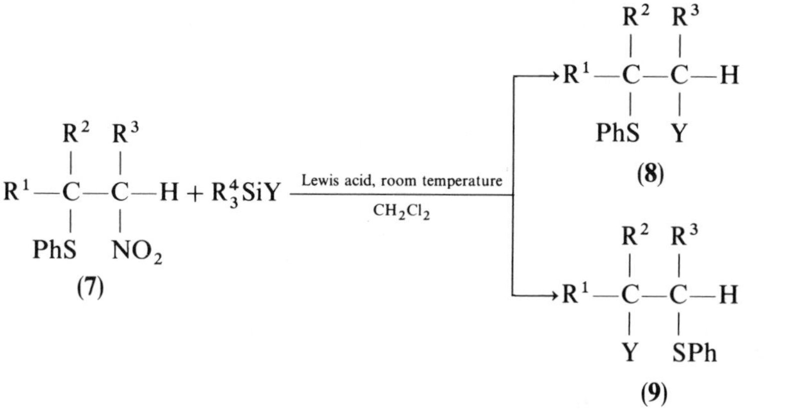

$$
\begin{array}{c}
R^2\ R^3 \\
|\quad| \\
R^1\!-\!C\!-\!C\!-\!H + R_3^4SiY \xrightarrow[\text{CH}_2\text{Cl}_2]{\text{Lewis acid, room temperature}} \\
|\quad| \\
PhS\ NO_2 \\
(\mathbf{7})
\end{array}
\qquad
\begin{array}{c}
R^2\ R^3 \\
|\quad| \\
\rightarrow R^1\!-\!C\!-\!C\!-\!H \\
|\quad| \\
PhS\ Y \\
(\mathbf{8}) \\[2mm]
R^2\ R^3 \\
|\quad| \\
\rightarrow R^1\!-\!C\!-\!C\!-\!H \\
|\quad| \\
Y\ SPh \\
(\mathbf{9})
\end{array}
\tag{4}
$$

$R^1, R^2 = $ H, Me, Et, Ph; $\quad R^3 = $ H, Me, C_6H_{13}; $\quad R^4 = $ Me, Et;
$R^1, R^2 = (CH_2)_4$; $\quad Y = $ H, $CH_2CH{=}CH_2$, CN

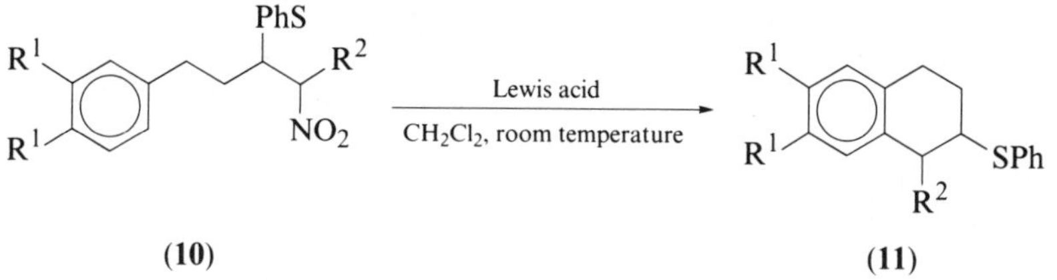

$$
(\mathbf{10}) \xrightarrow[\text{CH}_2\text{Cl}_2,\ \text{room temperature}]{\text{Lewis acid}} (\mathbf{11})
\tag{5}
$$

$R^1 = $ H, MeO; $\quad R^2 = $ Me, Ph, CH_2CH_2COMe

These reactions occur with preservation of configuration via the intermediate episulfonium ion with the participation of a Lewis acid ($SnCl_4$, $TiCl_4$, $AlCl_3$). The structure of the ion defines the direction of nucleophilic attack; when $R^3 = $ H, isomer (**9**) prevails; when $R^3 = $ Me or Et, the main product is isomer (**8**).

(Z)-2-Nitro-2-butene was obtained in the oxidation of the product of thiophenol addition to (E)-2-nitro-2-butene. This sequence can be used as a convenient way of transforming (E) isomers to (Z) isomers [160].

β-Nitro sulfides are probably the intermediates in the arylnitroalkene reductive elimination reactions that result in alkenes (**13**; equation 6) [161].

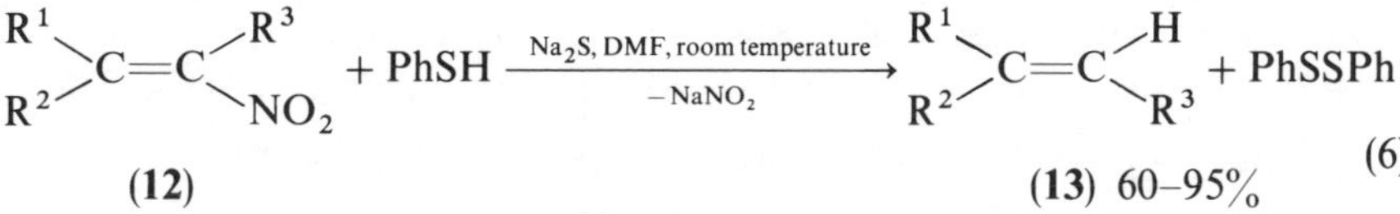

$$
\begin{array}{c}
R^1 \quad\quad R^3 \\
\diagdown C{=}C\diagup \\
R^2 \diagup \quad \diagdown NO_2 \\
(\mathbf{12})
\end{array}
+ PhSH \xrightarrow[-NaNO_2]{\text{Na}_2\text{S, DMF, room temperature}}
\begin{array}{c}
R^1 \quad\quad H \\
\diagdown C{=}C\diagup \\
R^2 \diagup \quad \diagdown R^3 \\
(\mathbf{13})\ 60\text{–}95\%
\end{array}
+ PhSSPh
\tag{6}
$$

$R^1 = $ Ar; $\quad R^2 = $ H, Ph; $\quad R^3 = $ H, Et, i-Pr, CO_2Et, Ar

Nitrocyclohexene can be thiolated by lithium isopropylthiolate to give the salt (15), which contains four reactive centres. The transformations of the salt (15) in accordance with equation (7) produce various functionalized thio derivatives of cyclohexane [132].

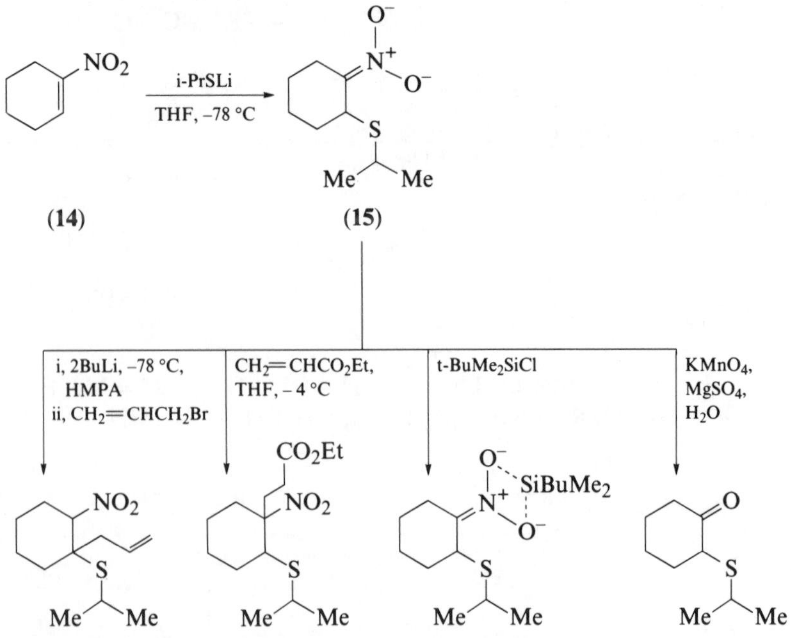

(7)

A kinetic study of thiolation by thiolate ions of α-nitrostilbenes in 50% DMSO [162, 163] has shown preferential thiolate ion addition over amines of similar basicity owing to 'soft acid–soft base' coupling. The same features were proven by a kinetic investigation of the same thiolate ion addition to substituted β-nitrostyrenes (20; equation 8) in water [164].

$$Ar^1CH{=}CNO_2 + RS^- \underset{k_{-1}}{\overset{k_1}{\rightleftharpoons}} Ar^1CHC{=}NO_2^- \underset{K_p}{\overset{H^+ (AcOH)}{\rightleftharpoons}} Ar^1CHCHNO_2 \quad (8)$$

with subscripts: under left term Ar^2; under middle term RS Ar^2; under right term RS Ar^2

(20)   (21)   (22)

$Ar^1 = C_6H_4X$; $X = H$, Me_2N, MeS, MeO, Cl, CN, NO_2; $Ar^2 = C_6H_4Y$; $Y = H$, Me, Br, NO_2; $R = Et$, n-BU, $HOCH_2CH_2$, $AcOCH_2$, $AcOCH_2CH_2$

Usually, the reaction of thiolation is not stereospecific [137, 165]. For example, (E)-2-nitro-2-butene coupling with thiophenol under the action of TEA leads to syn/anti-β-nitro sulfides in the ratio 60:40. The same nitroalkene combines with lithium thiophenolate at $-78\,°C$ to form only the *anti* isomer. Similar results have been described [137] for other nitroalkenes, including nitrocyclohexene. The literature also presents the reactions of nitroalkenes with phenylselenolate ion. The high *anti* selectivity is achieved by intermediate β-thionitronate ion protonation at low temperature and can be explained by the direction of proton attack to a specific conformation. Stereoselectivity decreases in the presence of bulky substituents ($R^1 = Ph$) and vanishes when R^1 or $R^2 = i$-Pr (equation 9).

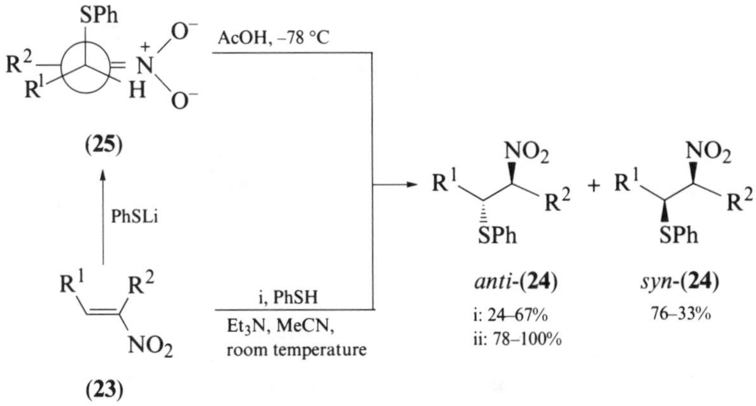

$$R^1 = Me, Et; \quad R^2 = Me, Et, Ph, cyclo\text{-}C_6H_{11}, cyclo\text{-}C_7H_{13}$$

Equation (10) presents a variation of such a method involving an aldehyde as the third reagent. This one-step procedure leads to γ-phenylthio-β-nitroalcohols (26) [165] with the same stereochemical regularity [166]. A diastereoisomeric mixture (about 1:1) is formed as a result of benzenethiol and aldehyde interaction in the presence of a catalytic amount of base. The introduction of phenylthiolate into the reaction followed by acidification at low temperature ($-78\,°C$) makes the process stereoselective (*anti/syn* ratio about 9:1) [166].

The products (26; equation 10) can be used as starting materials for the stereoselective synthesis of (E)-allyl alcohols (27) and their derivatives. To eliminate functional groups from β-nitro sulfides and β-nitro sulfones, tributyltin chloride was used [165–167].

The reaction of nitroalkenes with phenylselenolate is similar to that with thiophenolate. *Syn* elimination of benzeneselenic acid from the *anti* isomer (29; equation 11) gives (Z)-nitroalkenes (28) in 100% yield. This method was devised by Ono *et al.* [168] and is a general way of converting (E)-nitroalkenes to (Z)-nitroalkenes.

Alkaloids catalyse the asymmetric reactions of β-nitroarylethenes with

$$R^1 \diagup\diagdown_{NO_2} \xrightarrow[\substack{\text{ii. } CH_2O, \text{ 3 h} \\ \text{iii. } AcOH, -78\,°C, \text{ 1 h}}]{\text{i. PhSLi, THF, room temperature}} R^1 \underset{SPh}{\diagup\overset{NO_2}{\diagdown}} OH$$

$$(\textbf{26})$$

$$\xrightarrow[\substack{\text{ii. } Bu_3SnH, \\ PhMe, AIBN}]{\substack{\text{i. } Bz_2O \\ \text{or } Ac_2O}} R^1 \diagdown\diagup_{OR^2} \qquad (10)$$

$$(\textbf{27})$$

$$R^1 = Alk, Ph, PhCH_2CH_2; \quad R^2 = H, Ac, Bz$$

$$R^1 \underset{NO_2}{\overset{R^2}{\diagdown\diagup}} \xrightarrow[\text{ii. } AcOH, -78\,°C]{\text{i. PhSeNa, EtOH}, -78\,°C} R^1 \underset{SePh}{\diagup\overset{NO_2}{\underset{R^2}{\diagdown}}} \xrightarrow[-\, PhSeH]{H_2O_2, CH_2Cl_2, 0\,°C} R^1 \underset{R^2}{\overset{NO_2}{\diagdown\diagup}} \quad (11)$$

$$(E)\text{-}(\textbf{28}) \qquad\qquad \underset{68-82\%}{(\textbf{29})} \qquad\qquad (Z)\text{-}(\textbf{28})$$

$$R^1 = Alk, Ph; \quad R^2 = Me, Et$$

benzenethiol [169] and thioglycine [170], leading to enantiomeric addition products (**30**; equation 12).

$$R^1HC=C(R^2)NO_2 + HSCH_2R^3 \xrightarrow[PhMe, 20\,°C]{AN/CA} R^1(R^3CH_2S)\overset{*}{C}H\overset{*}{C}H(R^2)NO_2 \quad (12)$$

$$(\textbf{30})$$

$$AN/CA = \text{acrylonitrile/cinchona alkaloid}; \quad R^1 = Ar; \quad R^2 = H, Me; \quad R^3 = Ph, \\ CO_2H$$

The interaction between 2-nitro-2-propenyl pivaloate and lithium thiophenolate at low temperature does not lead to the addition product but proceeds via carboxylate group substitution to give 2-nitro-2-phenylthiopropene [171].

1-Nitrodienes easily undergo thiolation. Under the action of an equimolar quantity of MeONa, 2,3-disubstituted 1-nitro dienes combine with thiols to form the products of 1,4 addition [172]. The same conditions force thiols to add to 1-nitrobutadiene and 1-nitro-1-alken-3-ynes at the nitrovinyl fragment [172]. These reactions occur via radicals [173, 174]. Benzenethiol reacts with 1-nitrobutadiene to give the products of both 1,2 and 1,4 addition [175].

The reaction pathways of geminal and vicinal dinitroalkenes with CH compounds are different. The geminal compounds give the products of addition (equation 13). However, two aryl substituents in geminal dinitroalkenes direct the process towards substitution of one nitro group (equation 14). Vicinal dinitroalkenes (equation 15) also produce the products of a single nitro group substitution [144, 176].

$$PhCH{=}C(NO_2)_2 \xrightarrow{H_2S} PhC(SH)HCH(NO_2)_2 \qquad (13)$$
$$\quad\ \ \textbf{(31)} \qquad\qquad\qquad \textbf{(32)}$$

$$Ph_2C{=}C(NO_2)_2 \xrightarrow[\text{RS}^-,\,\text{DMSO}]{\text{RSH, room temperature}} Ph_2C{=}C(SR)NO_2 \qquad (14)$$
$$\quad\ \ \textbf{(33)} \qquad\qquad\qquad\qquad\qquad \textbf{(34)}$$

$$O_2NCR^1{=}CR^2NO_2 \xrightarrow[\text{or R}^3\text{S}^-,\,\text{MeOH, 0 °C}]{\text{R}^3\text{SH}} O_2NCR^1{=}CR^2SR^3 \qquad (15)$$
$$\qquad\ \ \textbf{(35)} \qquad\qquad\qquad\qquad\qquad \textbf{(36)}$$

In an excess of thiolate, reaction of the geminal dinitroalkene (33) proceeds via an addition process and results in the formation of α,α-diphenyl-β-thio-phenylvinyl sulfide and diphenyl sulfide [176]. The reactions between (E)-α,β-dinitroalkenes or (Z)-α,β-dinitroalkenes and arylthiolate ions give mainly the (Z)-2-α-nitro-β-arylthioalkenes (see Section 3.3). The reactions of 2,3-dinitro-2-butene and 3,4-dinitro-3-hexene with SCN⁻ lead entirely to the (Z) products [177]. These stereochemical results are interpreted as the strong interaction between the nucleophile (S atom) and the remaining nitro group (O atom) in the intermediate. This has been proven by X-ray analysis of the final products.

1,4-Dinitrodienes also form the products of single nitro group substitution with H_2S and ArSH (see Section 3.3). In the case of aliphatic thiols with lower acidity (e.g. EtSH, $K_a = 10^{-11}$) the products of bis addition (38; equation 16) are produced [178]. When $R^2 = Ph$, along with thiol addition at one nitro-vinyl group there occurs nitro group substitution at the second nitrovinyl group.

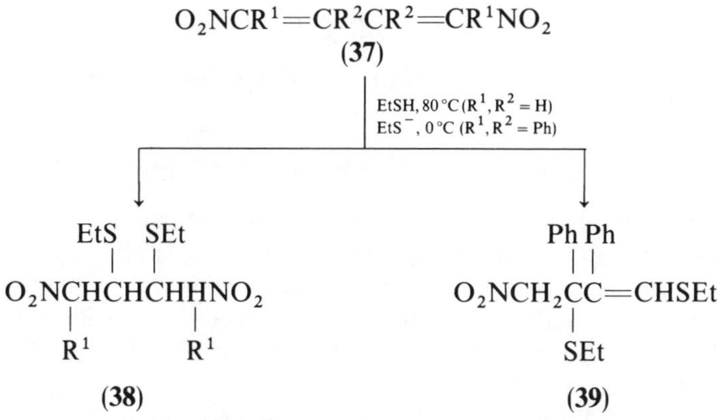

$$O_2NCR^1{=}CR^2CR^2{=}CR^1NO_2$$
$$\textbf{(37)}$$

EtSH, 80 °C (R^1, R^2 = H)
EtS⁻, 0 °C (R^1, R^2 = Ph)

(16)

EtS SEt	Ph Ph
O₂NCHCHCHHNO₂	O₂NCH₂CC=CHSEt
R¹ R¹	SEt
(38)	**(39)**

It is possible to synthesize β-nitro sulfones (40; equation 17) via the facile addition of arylsulfinic acids to the double bonds of arylnitroalkenes [157].

The products (**40**) can be reduced to the corresponding β-amino sulfones.

$$p\text{-}RC_6H_4SO_2H + PhCH\!=\!CHNO_2 \longrightarrow p\text{-}RC_6H_4SO_2CHPhCH_2NO_2 \quad (17)$$

$$(\textbf{40})$$

$$R = H, Me, NHCOMe$$

Tamura *et al.* [179, 180] have suggested a new synthetic method for allyl sulfones (equation 18). The method involves palladium (O)-catalysed reaction of phenylsulfinic acid with α,β-disubstituted aliphatic nitroalkenes.

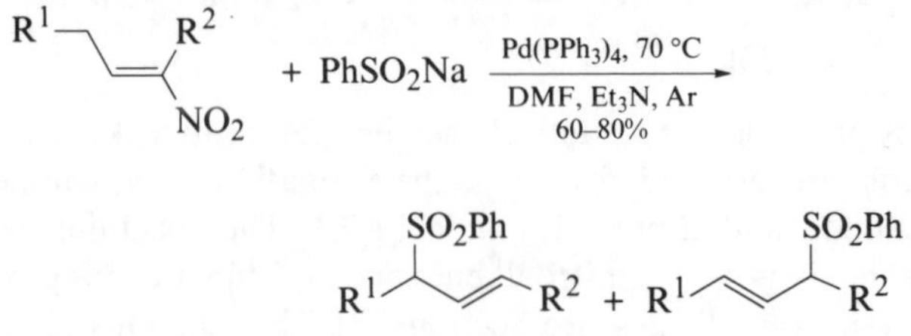

$$(18)$$

$$R^1 = H, Me, Et; \quad R^2 = Me, Et; \quad R^1, R^2 = -CH_2-$$

The first step is the isomerization of the double bond to the β,γ position under the action of Et_3N. Then the nitroallyl isomer transforms via a π-allylpalladium complex into the final products. The isomer (**41**) is either the only product or the major product. 3-(Phenylsulfonyl) cyclohexene is obtained from 1-nitro-cyclohexene in a similar way.

It is worth mentioning the 1,4-nucleophilic addition of sulfinic acids to 2-nitro-1,3-alkadienes (**43**; equation 19) [181]. Coupling of sulfinic acids with 1,4-dinitrodienes results in the products of substitution [182].

$$CH_2\!=\!C(NO_2)CMe\!=\!CHR^1 + p\text{-}R^2C_6H_4SO_2H \longrightarrow$$

$$(\textbf{43})$$

$$p\text{-}R^2C_6H_4SO_2CH_2C(NO_2)\!=\!CMeCH_2R^1 \quad (19)$$

$$(\textbf{44})$$

$$R^1 = H, Me; \quad R^2 = H, Me, NHCOMe$$

The structure of geminal halonitroethenes defines the pathway of the reaction with sulfinic acids (addition or dehydrohalogenation) [183].

The reactions of 1,1-dinitro-2,2-diphenylethene with arylsulfinic acids result in the formation of either benzophenone in dimethyl sulfoxide (DMSO) or p-arylsulfonylbenzophenone in DMF. The process occurs with C—C bond splitting in the intermediate [144].

The addition of $NaHSO_3$ to unsaturated nitro compounds (equation 20) is a rare case of organic compound sulfonation [184–187]. This method is of particular interest owing to the reduction of compounds of type (**45**) into the corresponding 1,2-aminosulfonic acids (46) [185].

$$O_2NCR^1{=}CR^2R^3 \xrightarrow{\text{NaHSO}_3} O_2NCHR^1CR^2R^3SO_3Na \xrightarrow{[H]}$$
$$\textbf{(45)}$$

$$H_2NCHR^1CR^2R^3SO_3Na \qquad (20)$$
$$\textbf{(46)}$$

A kinetic study of this reaction has been performed [186].

2.4 REACTIONS WITH NITROGEN NUCLEOPHILES

Those works devoted to the addition of ammonia, amines and other nitrogen-containing bases to nitroalkenes which were published before 1968 are discussed in a review by Baer and Urbas [130].

Ammonia reacts with nitroalkenes and leads to the products of addition, whereas β-nitrostyrene gives the products of bis addition. The same process with α-nitrostilbenes results in the formation of products of further transformations [130].

The majority of simple aliphatic amines initiate nitroethene polymerization. Mono-adducts and bis-adducts can be prepared from such amines as cyclohexylamine, benzylamine, hexamethylenediamine and aromatic amines [128], while L-cysteine [188], aminoacetonitrile and a series of α-amino acids [189] produce mono-adducts with nitroethene. Investigations with different kinds of amines have shown that the optimum basicity for the nitroethenation reaction is in the range $pK_a = 5-6$. The higher pK_a value promotes nitroethene polymerization [189]. It has been shown that nitroethenation proceeds via the zwitterion (equation 1), and the reaction is effective when the intermediate is able to rearrange quickly to the covalent form.

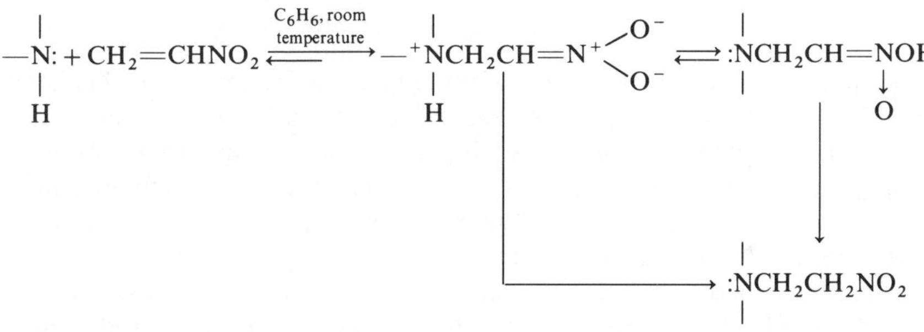

$$(1)$$

Equation (2) presents an example of the synthetic importance of nitro-ethenation [189].

To avoid polymerization in the synthesis of β-nitroethylamines (4) from amines with $pK_a > 8$ it was proposed to introduce 1-benzoyloxy-2-nitroethane

$$OH \quad \xrightarrow{CH_2=CHNO_2} \quad \cdots \quad \longrightarrow \quad (2)$$

(1)   **(2)**

(**3**; equation 3) into the reaction instead of nitroethene [190].

$$BzOCH_2CH{\overset{NO_2}{\underset{R^1}{\diagdown}}} \quad \xrightarrow[\substack{EtOH\ room \\ temperature}]{R^2R^3NH} \quad \left[CH_2{=}C{\overset{NO_2}{\underset{R}{\diagdown}}} \right] \quad \longrightarrow R^2R^3NCH_2CH{\overset{NO_2}{\underset{R^1}{\diagdown}}}$$

(3)   **(4)** 20–100% (3)

$R^1 = H$, Ph; $R^2R^3NH = c\text{-}C_5H_{10}NH$, $c\text{-}C_6H_{11}NH_2$, $BnNH_2$, Bu_2NH, 1,2,4-
triazole, imidazole

The first synthesis of the products of α-nitrostyrene amination was performed in a similar way. With primary amines a mixture of mono-adducts and bis-adducts was obtained.

Baer and Urbas [130] also describe the products of aniline addition to other α-nitroalkenes. Heterocyclic amine derivatives are prepared from 1-nitropropene [191] and 2-nitropropene [192].

The additions of aromatic amines and 2,4-dinitrophenylhydrazine have been described for α-nitro-γ-dihaloalkenes and α-nitro-γ-trihaloalkenes, whereas 1,1-bis(trifluoromethyl)-2-nitroethene (**5**; equation 4) combines with the same amines to form the products of substitution (**6**) owing to redirection of the nucleophilic attack [193].

$$(F_3C)_2C{=}CHNO_2 + H_2NR \longrightarrow (F_3C)_2C{=}CHNR \qquad (4)$$

        (5)   **(6)**

$$R = Ph,\ 2,4\text{-}(NO_2)_2C_6H_3NH$$

The amination of β-nitrostyrene and its homologues has been studied in detail [130, 194, 195]. Among the products described are those from the addition of primary and secondary aliphatic and alicyclic amines and nitrogen-containing heterocycles. A wide range of adducts is obtained from aromatic amines. Lough and Cuine [194] have investigated the influence of ring substituents upon the reaction rate. α-Nitrostilbenes give rise to the corresponding 1-arylamino-2-nitro-1,2-diphenylethanes, but the similar products from aliphatic amines are not isolated [130]. However, the product of cyclohexylamine addition to (Z)-α-nitrostilbene has been prepared. From the reactions of the latter with morpholine, piperidine and N,N-dimethylaniline, the charge transfer complexes have been isolated [196].

Detailed quantitative studies of the reversible additions of butylamine or aniline to β-nitrostyrenes [197] and piperidine or morpholine to substituted α-nitrostilbenes [198] in water/DMSO have made it possible to discuss the two-step reaction mechanism, which proceeds via the zwitterionic adduct (equation 5). The latter exists in equilibrium with the related base [163].

$$ArCH=CR^1NO_2 + R_2^2NH \underset{k_{-1}}{\overset{k_1}{\rightleftharpoons}} ArCH(R_2^2N^+H)CR^1=NO_2^-$$

$$\underset{H^+}{\overset{K_a}{\rightleftharpoons}} ArCH(R_2^2N)CR^1=NO_2^- \qquad (5)$$

$$R = H, Ph$$

The interactions between nitrostyrenes (7; equation 6) and p-fluoroaniline are accompanied by nitromethane elimination from the product of addition and lead to azomethines (8) [199].

$$ArCH=CHNO_2 + p\text{-}FC_6H_4NH_2 \xrightarrow[-MeNO_2]{C_6H_6} ArCH=NC_6H_4F\text{-}p \qquad (6)$$
$$\quad (7) \qquad\qquad\qquad\qquad\qquad\qquad\qquad\qquad (8)$$

The reactions of o-phenylenediamine (9; equation 7) with nitrostyrenes proceed via the azomethine and its cyclization. Further aromatization takes place with the participation of a second molecule of nitrostyrene which acts as a hydrogen acceptor, resulting in the formation of 1-nitro-2-arylethanes (10) and 2-arylbenzimidazoles (11) [200].

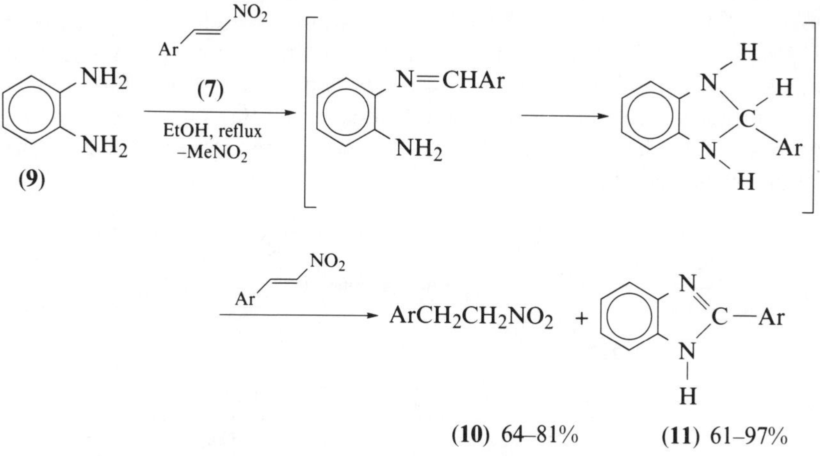

$$\text{(10) 64–81\%} \qquad \text{(11) 61–97\%}$$

$$Ar = Ph, p\text{-}MeC_6H_4, p\text{-}MeOC_6H_4, p\text{-}ClC_6H_4$$

The reaction of 1-methyl-2-(β-nitro-β-phenylvinyl)benzimidazole (12; equation 8) with benzylamine results in the azomethine (13). Aromatic amines lead to the products of addition (14) [201].

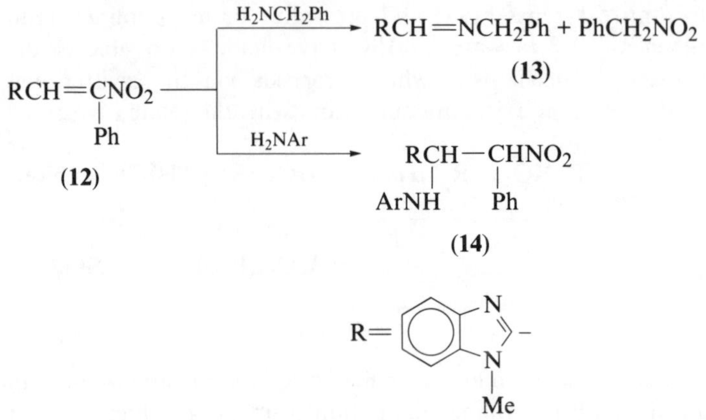

$$RCH=\underset{\underset{\textbf{(12)}}{Ph}}{\overset{}{C}}NO_2 \quad
\begin{cases}
\xrightarrow{H_2NCH_2Ph} & RCH=NCH_2Ph + PhCH_2NO_2 \quad \textbf{(13)} \\[2em]
\xrightarrow{H_2NAr} & \underset{\underset{\textbf{(14)}}{ArNH \quad Ph}}{RCH-CHNO_2}
\end{cases} \qquad (8)$$

$$R= \text{[benzimidazole ring]} -$$
$$Me$$

Addition products from the reactions of 5-substituted 2-(β-nitrovinyl)furans with morpholine, piperidine and pyrrolidine can be obtained in 80–90% yield. The reaction rate increases with increasing amine nucleophilicity [202]. Rosenberg *et al.* [203] have performed a wide-ranging kinetic study of the reactions of 1-(5-nitrofuryl-2)-2-nitroethene with butylamine, aniline, various types of biologically active amines, amino acids, and some amino-containing and hydroxy-containing biopolymers. A kinetic study [204] of the reaction of (*E*)-furylnitroethene with piperidine in different solvents at 25 °C has revealed the zwitterionic character of the intermediate.

Nitroalkene amination by γ-aminocrotonic esters or unsaturated δ-aminoketones (**15**) is the basis of the novel synthesis of pyrrolidone and piperidine cyclic systems [205–207]. These nucleophiles contain a double bond activated by an electron-withdrawing group, owing to which five-membered or six-membered ring closure in the primary adduct occurs spontaneously (equation 9). Application of reagents containing the *N*-benzylamino group prevents the formation of the bis-adducts.

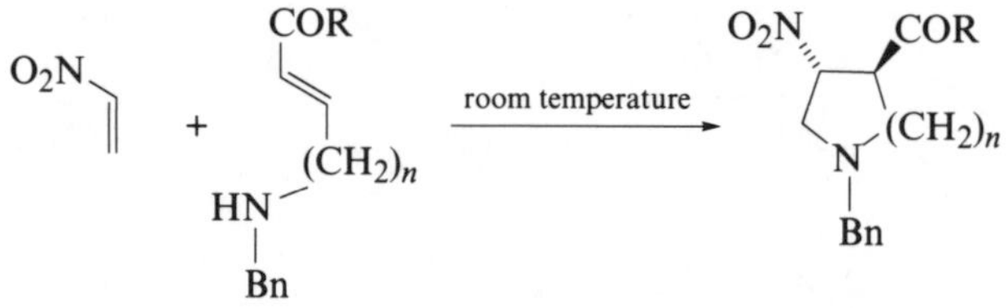

$$\underset{\textbf{(15)}}{O_2N \diagdown\|} \quad + \quad \underset{\underset{Bn}{HN}}{\overset{COR}{\diagup}}(CH_2)_n \quad \xrightarrow{\text{room temperature}} \quad \underset{\underset{Bn}{N}}{\overset{O_2N \quad COR}{\diagup\diagdown}}(CH_2)_n \qquad (9)$$

$$\textbf{(16)}$$

$$R = OEt, n = 1; \quad R = Me, n = 2$$

The reaction is stereospecific and leads to the product with *trans*-oriented substituents in the heterocycle in good yield [205]. A similar interaction is the key step in the synthesis of the kainoids – a group of naturally occurring, non-proteinogenic amino acids. In their synthesis, optically active esters of type (**18**) are used (equation 10). The corresponding pyrrolidines (**19**) are synthesized from nitroalkenes with the 2-nitroethene fragment, which are in turn prepared *in situ* from 1-benzoyloxy-2-nitroalkanes (**17**) and nucleophiles (**18**).

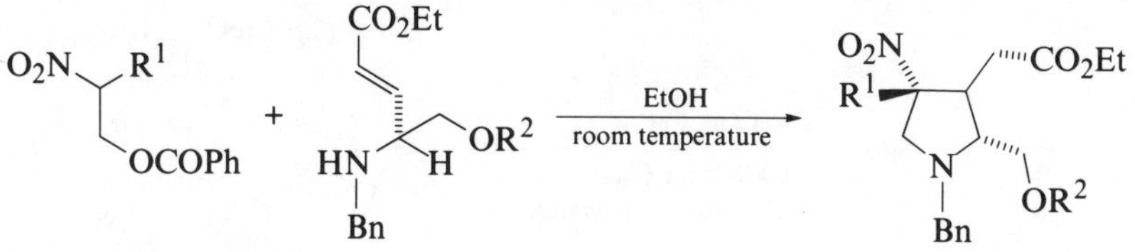

$$
\begin{array}{c}
\text{(17)} \quad + \quad \text{(18)} \xrightarrow[\text{room temperature}]{\text{EtOH}} \text{(19) 75–90\%}
\end{array}
\tag{10}
$$

2-Nitro-3-methyl-1,3-butadiene obtained *in situ* was used in the synthesis of the precursor of α-kainic acid, which has an isopropenyl fragment on the pyrrolidone ring (equation 11) [206, 207].

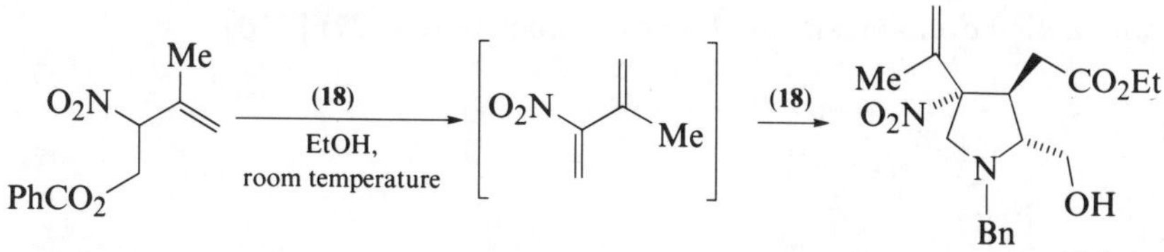

$$
\xrightarrow[\text{room temperature}]{\substack{\text{(18)} \\ \text{EtOH,}}} \left[\text{O}_2\text{N} \diagup \diagdown \text{Me} \right] \xrightarrow{\text{(18)}} \text{(20) 88\%}
\tag{11}
$$

The presence of the nitro group in the starting alkene defines the stereochemical and regiochemical control of the cyclization process.

Nitroalkene sugars react with ammonia to produce aminosugars and diaminosugars [130]. For example, N-acylamino-D-mannose (**23**) was prepared in accordance with equation (12).

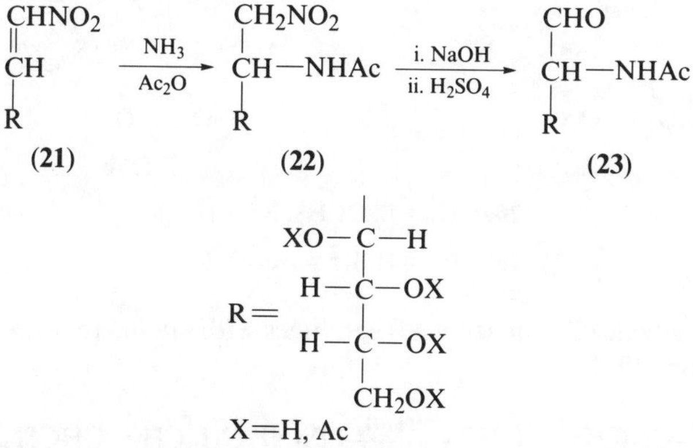

$$
\begin{array}{ccc}
\text{CHNO}_2 & \text{CH}_2\text{NO}_2 & \text{CHO} \\
\| & | & | \\
\text{CH} & \text{CH—NHAc} & \text{CH—NHAc} \\
| & | & | \\
\text{R} & \text{R} & \text{R} \\
\text{(21)} & \text{(22)} & \text{(23)}
\end{array}
$$

$$
\begin{array}{c}
\text{CHNO}_2 \xrightarrow[\text{Ac}_2\text{O}]{\text{NH}_3} \text{CH}_2\text{NO}_2 \xrightarrow[\text{ii. H}_2\text{SO}_4]{\text{i. NaOH}} \text{CHO}
\end{array}
\tag{12}
$$

$$
\text{R} = \begin{array}{c}
\text{XO—C—H} \\
\text{H—C—OX} \\
\text{H—C—OX} \\
\text{CH}_2\text{OX}
\end{array}
$$

X=H, Ac

Baumberger *et al.* [208] describe the products of stereoselective addition of ammonia and primary amines to 1-*C*-nitroglycal (**24**; equation 13). The adduct of (24) with ammonia was transformed into the azomethine (25) to protect the amino group. The product was used in the synthesis of 6-amino-6-deoxysialoic acid [209], which is a sialidase inhibitor. The azomethine was isolated in the form of an anomeric mixture containing mostly the α anomer.

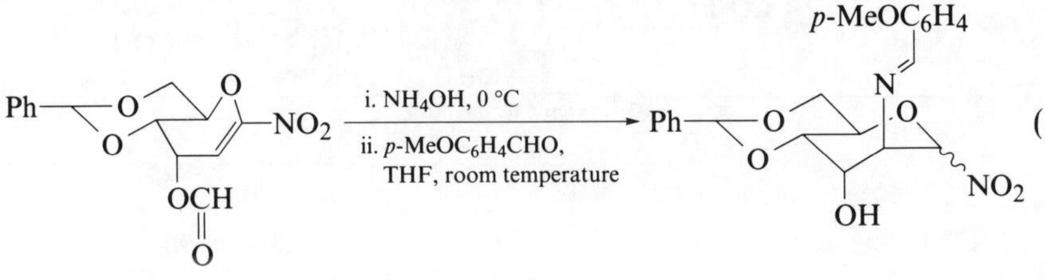

(24)   (25) 79%

Stereoselective addition of adenine to anomers of nitrocyclohexene alkoxy derivatives (26; equation 14) is the key step in the new synthetic approach to optically active cyclohexane analogues of nucleosides (27) [210].

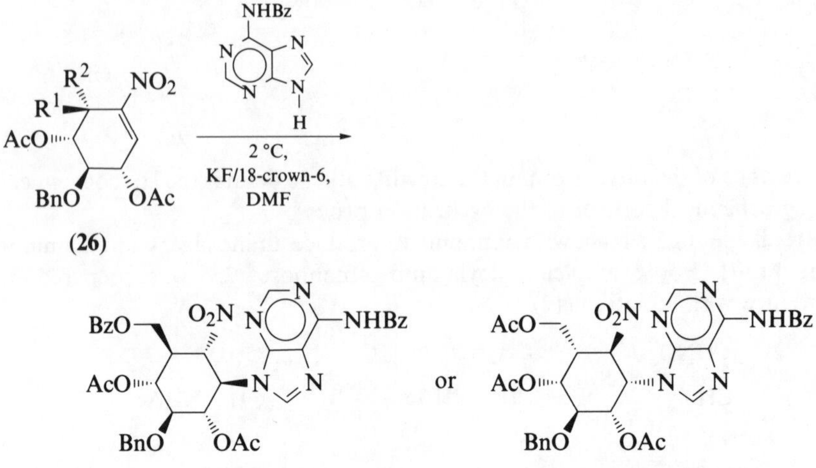

(26)

(27a) 72%   (27b) 71%

(26a): R^1 = BzOCH$_2$, R^2 = H

(26b): R^1 = H, R^2 = AcOCH$_2$

1-Nitrobutadiene (28; equation 15) combines with aniline to form the product of 1,4 addition (29) [175].

$$CH_2{=}CHCH{=}CHNO_2 \xrightarrow[Et_2O]{PhNH_2} PhNHCH_2CH{=}CHCH_2NO_2 \quad (15)$$

(28)   (29) 57%

1-Nitro-1-alken-3-ynes and 1-nitro-4-trimethylsilyl-1-buten-3-yne combine with aniline with the participation of their nitrovinyl groups [211].

1,1-Dinitro-2,2-diphenylethene reacts with primary and secondary amines to give the addition products, but these quickly and irreversibly decompose to benzophenone and dinitromethane anion [212]. A kinetic study has made it possible to work out the mechanism of this process.

α,β-Dinitroalkenes (**30**; equation 16) react with ammonia and amines to form the products of substitution (**31**) (nitroenamines) [177, 213, 214].

$$O_2NCR^1{=}CR^2NO_2 \xrightarrow{\text{NHR}^3\text{R}^4} \underset{O_2N}{\overset{R^1}{\diagdown}}C{=}C\underset{NR^3R^4}{\overset{R^2}{\diagup}} \tag{16}$$

$$\underset{(Z)\,\text{or}\,(E)}{(\mathbf{30})} \qquad\qquad\qquad (\mathbf{31})$$

The products of the reactions between α,β-dinitrostilbene and secondary amines [213] or morpholine [177] also have the (Z) structure, which was attributed by the authors to the strong coupling between the oxygen atoms of the nitro group and the nitrogen atom of the amino group in the zwitterionic reaction intermediate.

A spectrophotometric study of the reactions between α,β-dinitrostilbene and morpholine, aniline and its N-alkylated derivatives has established the formation of a charge transfer complex at the preliminary reaction step [196]. The results of the kinetic study of the reactions of (E)-α,β-dinitrostilbene with piperidine, morpholine and aniline have been discussed also [215] in terms of a multistep process that proceeds via the zwitterion.

Non-conjugated dinitrodienes (**32**) [216] and conjugated 1,4-dinitrodienes without substituents at the middle carbon atoms [214] form the products of bis addition (**33**; equation 17) with amines.

$$CH_2{=}C(NO_2)(CH_2)_nC(NO_2){=}CH_2 \xrightarrow[\text{EtOH, 60°C}]{2p\text{-MeC}_6\text{H}_4\text{NH}_2} \tag{17}$$

$$(\mathbf{32})$$

$$p\text{-MeC}_6\text{H}_4\text{NHCH}_2\text{CH(NO}_2)(CH_2)_n\text{CH(NO}_2)\text{CH}_2\text{NHC}_6\text{H}_4\text{Me-}p$$

$$(\mathbf{33})$$

$$n = 1\text{--}4$$

The alkyl groups of the 1,4-dinitrobutadiene derivative (**34**) at C(2) and C(3) depress the reactions that lead to the addition products (**35, 36**; equation 18). The mono-adduct transforms into the allylic isomer, which is passive to further addition.

Phenyl groups at C(2) and C(3) shield the corresponding carbon atoms and direct the attack to the terminal atom of the diene system to give the product of substitution of one nitro group (**38**; equation 19).

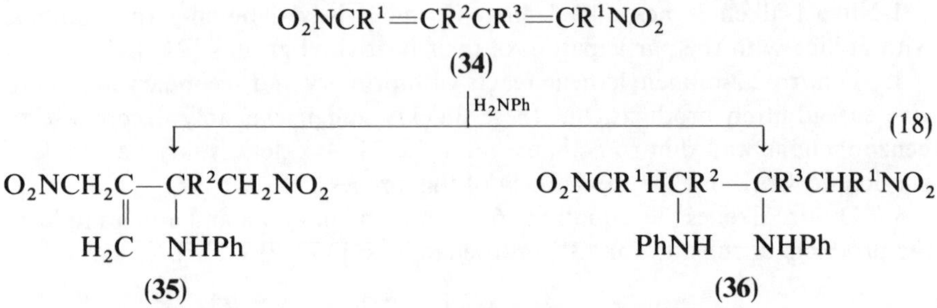

$$O_2NCR^1=CR^2CR^3=CR^1NO_2$$

(34)

$$\downarrow \overset{|\,H_2NPh}{}$$ (18)

$$O_2NCH_2C-CR^2CH_2NO_2 \qquad\qquad O_2NCR^1HCR^2-CR^3CHR^1NO_2$$
$$\overset{\|}{\;}\quad\overset{|}{\;}$$
$$H_2C\;\;NHPh \qquad\qquad\qquad\qquad\quad PhNH\quad NHPh$$

(35) (36)

$$R^2 = H, Me$$

$$R^1 = R^2 = R^3 = H$$
$$R^1 = R^3 = H, R^2 = Me$$
$$R^1 = H, R^2 = R^3 = Me$$
$$R^1 = Ph, R^2 = R^3 = H$$

$$O_2NCH=CPhCPh=CHNO_2 \xrightarrow{PhNH_2} O_2NCH=CPhCPh=CHNHPh \quad (19)$$

(37) (38)

Apart from amines, other nitrogen nucleophiles (hydroxylamine, arylhydrazines, semicarbazides) have been studied in the reaction with nitroalkenes [217, 218]. Hydroxylamine forms addition products with 1-nitroalkenes, β-(2-furyl)nitroethene and nitrostyrene, the stability of the product increases in that order. Phenylhydrazine easily combines with nitrostyrenes to produce arylhydrazones. The reaction of various nitrogen nucleophiles with 3-(β-nitrovinyl)indole (39) is illustrated in equation (20) [219].

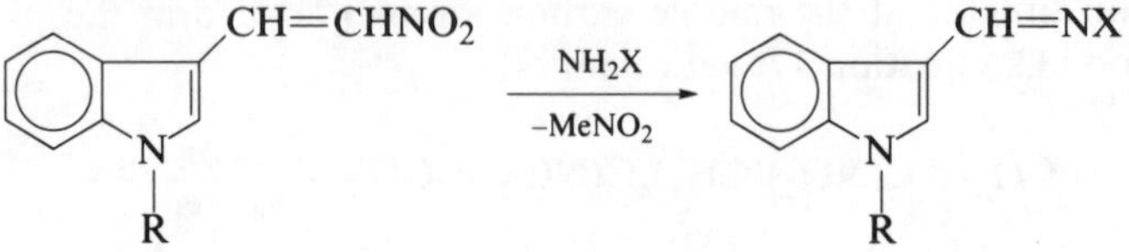

(39) (40)
R=H, Me;   X=OH, NH₂, NHCONH₂, NHPh

The products from the addition of a wide range of aliphatic aldehyde hydrazones (41; equation 21) to 1-nitropropene and nitroalkene sugar (42) are the intermediates in the synthesis of substituted pyrazoles (43). Adducts (42) were isolated in only two cases [220a].

The mechanism of the process has been proven by the synthesis of the pyrazole derivative from 1-nitropropene and benzaldehyde methylhydrazone. Substituted isomeric pyrazoles have also been prepared in 9–59% yields from D-galactose phenylhydrazone and several nitroalkenes [220b].

The intramolecular addition to the nitroalkene fragment in the azetidinone (44) gives an epimeric mixture (2:1) of bicyclic nitro compounds (45; equation 22).

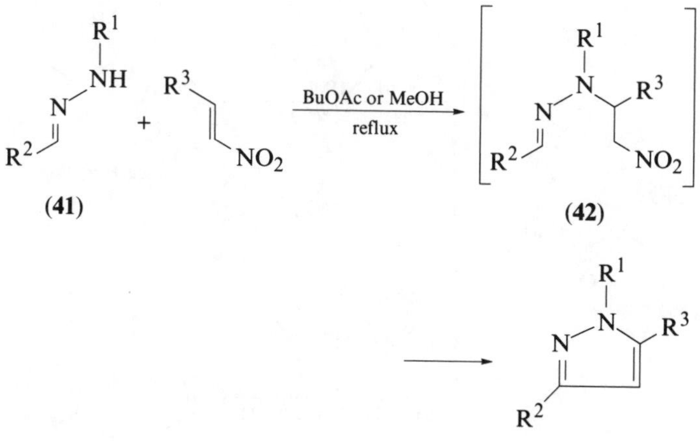

$$(21)$$

(43) 37–81%

$$R^1 = Me, Ph; \quad R^2 = Alk, Ar; \quad R^3 = Me,$$

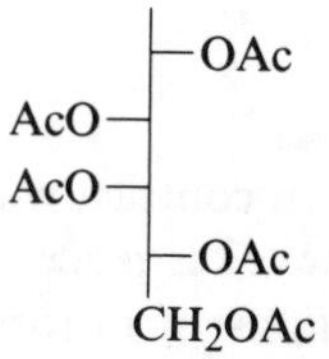

Compound **(45)** hydrolyses to a 2,2-dimethyl-1-carbapename derivative which is a β-lactam antibiotic [221].

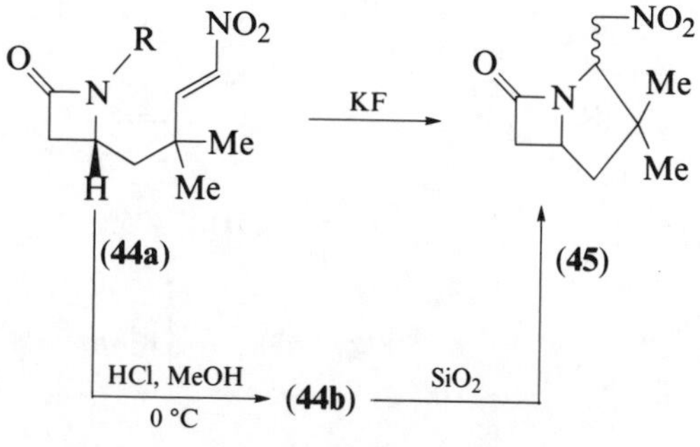

$$(22)$$

(44a): R = SiMe$_2$Bu-t
(44b): R = H

Apart from the reactions of addition (and substitution in the presence of a leaving group in the β position), some other types of amine reactions with nitroalkenes have been reported. 2-Nitropropenyl pivaloate (**46**; equation 23) makes it possible to nitroallylate N-methylaniline and diphenothiazine [171].

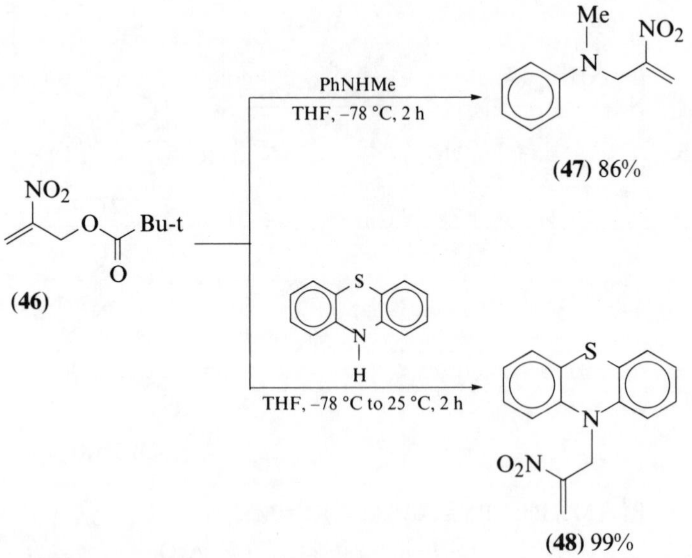

(23)

In this case, in contrast with the thiolation, substitution of the pivaloate group by the amine takes place.

The reactions of nitrostyrene and nitrocyclohexene with 2-vinyl-2-phenyl-aziridine lead to (Z)-nitroenamines [222].

The new pathway of the reaction of α,β-disubstituted nitroalkenes (49a) with secondary amines occurs under the action of a palladium (0) catalyst. The tertiary allylic amines (51; equation 24) are formed instead of the products of addition

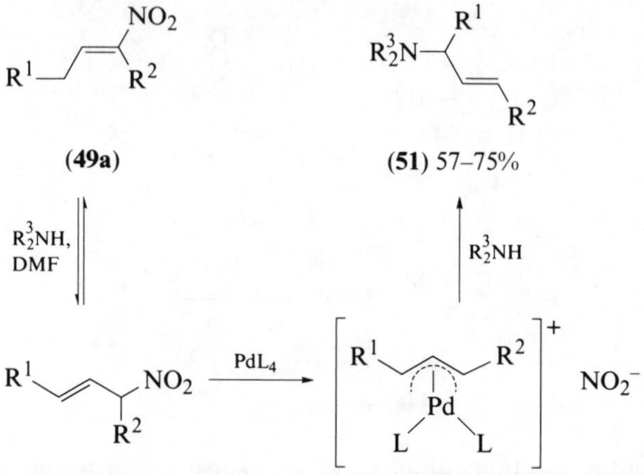

(24)

$R^1, R^3 = (CH_2)_n, n = 4, 5; \quad R^3 = Alk$
$R^1 = Alk; \quad R^2 = H, Alk; \quad L = PPh_3$

[179, 223]. The first step of the process is probably the amine-induced isomerization of (49a) to the nitroallylic isomer (49b); this then transforms to the π-allylpalladium complex (50) after NO_2^- elimination. Amine attack on the least bulky fragment results in the formation of the products of regioselective and stereoselective substitution, i.e. only the (E) isomer is formed.

2.5 REACTIONS OF UNSATURATED NITRO COMPOUNDS WITH CH ACIDS

The reaction of unsaturated nitro compounds with CH acids (Koller, Engelbrecht, 1919) is a further development of the Michael reaction. It occurs in the presence of basic catalysts, in accordance with the common scheme of nucleophilic addition to compounds with activated double bonds, and gives products in high yields [130, 163]. The process is illustrated in equation (1).

$$O_2NCH{=}CHR^1 + R^2CHXY \longrightarrow O_2NCH_2CHR^1CR^2XY \qquad (1)$$

$R^1 = H$, Alk, Ar, Het; $R^2 = H$, Alk, Ar;   X, Y = electron-withdrawing substituents

2.5.1 REACTIONS OF MONONITROETHENES WITH CH ACIDS

Unsaturated mononitro compounds of aliphatic, aromatic and heterocyclic series have been used successfully in reactions with various methyl, methylene and methine CH acids.

The significance of the reaction is underlined by the application of its products and the possibility to transform these compounds easily into new groups of organic substances such as α-amino acids, γ-amino acids, γ-aminoketones, amino derivatives of cyclic β-diketones, furans, pyrans, pyrroles, pyrrolidones and other heterocycles [2a, 130, 180, 224, 225].

The addition of nitroalkanes and arylnitroalkanes to mononitroalkenes is a common synthetic method for various polynitro compounds. For example, the reactions of nitroalkenes with nitroalkanes, dinitroalkanes and trinitroalkanes (equation 2) lead to the corresponding products (1–3) shown in equation (2) [225–234]. The process can be carried out with the potassium or sodium salt of the nitroalkane in the presence of MeONa or an organic base (Trilone B, diethylamine, triethylamine, piperidine and others) [226, 227, 229–231, 233, 234].

Mononitroalkanes can be added to nitroalkenes under the action of TEA at room temperature [225].

The strong CH acids, for example trinitromethane, react with nitroethenes without a catalyst at 0 °C [232].

The reaction of an equivalent amount of 1,1-dinitroethane, 1,1-dinitropropane or trinitromethane with β-nitrostyrene (and its derivatives) takes place in ethanol

$$R^1CH\!\!=\!\!CR^2NO_2$$

$\begin{array}{c}R^3\\ \diagdown\\ R^4\end{array}\!\!\!CHNO_2$

$$R^3C\!-\!CH\!-\!CHNO_2 \quad (1)$$
with R^4, R^1, R^2 substituents and NO_2

$R^3CH(NO_2)_2$

$$R^3C\!-\!CH\!-\!CHNO_2 \quad (2)$$
with NO_2, R^1, R^2 substituents and NO_2

(2)

$CH(NO_2)_3$

$$(NO_2)_3CCH\!-\!CHNO_2$$
with R^1, R^2 substituents

(3)

$$R^1, R^2, R^3, R^4 = H, Alk, Ar$$

at 78 °C (3–7 h) and results in the formation of the products of nucleophilic addition [233]. Under similar conditions in the presence of base or without a catalyst, nitroethenes combine with trinitrotoluene and other compounds with activated CH_2 groups such as mesomethylacridine, iodomethylates of quinaldine, and methylbenzothiazole [235].

Nitroalkane addition to fuctionalized nitroalkenes has been widely investigated. For example, the addition to 1-nitroheptafluoro-1-pentenes is described by McBee et al. [236]. The same process with 2-nitro-3-acetoxy-1-propenes is reported elsewhere [233, 237, 238].

The hydroxy group in 2-nitro-1-propen-3-ol makes the reactions with trinitro fluoromethane and dinitrofluoromethane more complicated. The products obtained are those from nucleophilic bis addition, i.e. pentanitroalkanes [234].

The reaction of 2-nitropropane or 2-nitrobutane (4; equation 3) with α-nitroacrylate derivatives or their analogues (5) in the presence of KF/Al_2O_3 produces heterocyclization products (6) [239].

The result of the reaction between phenylnitromethane and cis-nitrostilbene (7; equation 4a) is the formation of 1,3-dinitro-1,2,3-triphenylpropane (8) which is further transformed into triphenylisoxazoline N-oxide [240].

Studies of the kinetics of nitromethide ion addition to substituted nitrostyrenes [163, 197] have shown that the intrinsic rate constant for the first step increases with increasing electron-withdrawing capacity of the substitutents on the benzene ring. The resulting adducts are transformed into isoxazoles (equation 4b), and the rate of transformation is much faster than the rate of adduct decay to the starting materials.

$$
\begin{array}{c}
R^1 \\
\diagdown \\
\quad CHNO_2 \\
\diagup \\
R^2
\end{array}
+
\begin{array}{c}
R^3 \quad Y \\
\diagdown \diagup \\
C = C \\
\diagup \diagdown \\
H \quad NO_2
\end{array}
\xrightarrow{\text{MeCN, KF/Al}_2O_3}
\begin{array}{c}
R^3 \;\; Y \\
H \diagdown \diagup \\
R^1 \diagdown \diagup \diagdown N \\
R^2 \; O \; O \diagdown O
\end{array}
\qquad (3)
$$

$$
\textbf{(4)} \qquad\qquad \textbf{(5)} \qquad\qquad\qquad\qquad \textbf{(6)}
$$

$$
42\text{--}74\%
$$

$$R^1 = R^2 = Me; \quad R^1 = Me, R^2 = Et$$

$$R^3 = Ph, \text{ } p\text{-MeOC}_6H_4, \text{ } p\text{-Me}_2NC_6H_4, \text{ 2-furyl, 4-piperonyl}$$

$$Y = CO_2Me, COPh$$

$$
PhCH_2NO_2 + PhCH = C \begin{array}{c} \diagup Ph \\ \diagdown NO_2 \end{array}
\xrightarrow[\text{reflux}]{\text{MeONa, MeOH}}
\begin{array}{c}
PhCH-CH-CHNO_2 \\
\;\;\; | \qquad | \quad\;\; | \\
\;\; NO_2 \; Ph \;\; Ph
\end{array}
$$

$$
cis\text{-}\textbf{(7)} \qquad\qquad\qquad\qquad \textbf{(8)}
$$

$$
\xrightarrow{-H^+, -NO_2^-}
\begin{array}{c}
Ph \quad\;\; Ph \\
\diagdown \diagup \\
N \diagdown \diagdown \\
O \diagup \; O \diagdown Ph
\end{array}
\qquad (4a)
$$

$$
\textbf{(9)}
$$

$$
ArCH = CHNO_2 + CH_2 = NO_2^-
\underset{K_{-1}}{\overset{K_1}{\rightleftharpoons}}
\begin{array}{c}
ArCH-CH=NO_2^- \\
\;\;\;\; | \\
\;\; CH_2NO_2
\end{array}
\longrightarrow
\begin{array}{c}
Ar \\
\diagdown \\
\diagup \diagdown \\
N \;\; O
\end{array}
\qquad (4b)
$$

α-Nitroketones and their derivatives have been used successfully as starting materials in the synthetic approach to various pharmacologically active substances, such as adrenaline, quinacrine, chloromycetine, myosmine and other analogues.

Owing to their strong CH acidity, α-nitroketones react readily with aromatic and heterocyclic unsaturated nitrocompounds to form α,γ-dinitroketone derivatives [241].

Nitroacetic ester is a potential α-amino acid component. It is used in reactions with β-nitrostyrene, nitrocinnamic esters and furylnitroethene under the action of an equivalent amount of diethylamine (equation 5). The reaction results in the formation of salts of the nucleophilic addition products (10) [227, 242].

The dimethyl ester of α,α'-dinitro-β-methylglutaric acid is prepared by the addition of nitroacetic ester to methyl α-nitrocrotonic ester in the presence of AcONa in MeOH [80]. Under the action of an excess of BuNH₂, α,β-unsaturated α-nitro esters (11; equation 6) are transformed into the isoxazole-3,5-dicarbonic acid derivatives (12) upon refluxing in alcohol. The process probably proceeds via the fission of the intermediate product of amine addition to the nitroalkene followed by nitroacetic ester addition to (11) [243].

$$O_2NCHR^1{=}CHR^2 + O_2NCH_2CO_2Et$$

$$\xrightarrow[\text{Ligroin, room temperature}]{\text{EtNH}} \left[O_2NCHR^1CHR^2C{\overset{\nearrow NOO^-}{\underset{\searrow CO_2Et}{}}} \right] Et_2NH_2^+ \quad (5)$$

$$(10)$$

$$R^1 = H, CO_2Et; \quad R^2 = H, Ar, Het$$

$$RCH{=}\underset{\underset{NO_2}{|}}{C}{-}CO_2Et \xrightarrow[\text{reflux, 3 h}]{\text{BuNH}_2,\ \text{EtOH}} \underset{BuNHOC{-}\underset{\underset{N}{\overset{\|}{}}}{C}}{\overset{R}{\underset{O}{}}}\overset{\diagdown}{\underset{}{C}}{=}\overset{CONHBu}{\underset{}{C}}$$

$$(11) \hspace{5cm} (12)$$

$$R = Alk, Ar$$

$$\xrightarrow[\text{60 °C, 2 h}]{10\% \text{ NaOH}} \underset{HO_2C{-}\underset{\underset{N}{\overset{\|}{}}}{C}}{\overset{R}{\underset{O}{}}}\overset{\diagdown}{\underset{}{C}}{=}\overset{}{\underset{}{C}}{-}CO_2H \quad (6)$$

$$(13)$$

    Isoxazole derivatives are also formed via spontaneous denitration as the products of ω-nitrocarbonic ester addition to nitroalkencarbonic esters upon heating in xylene in the presence of diethylamine [244].

    Hydrogen cyanide in combination with β-nitrostyrene (Holleman, 1904) forms two stereoisomeric 1,4-dinitro-2-cyano-2,3-diphenylbutanes.

    The reactions of alkali metal cyanides with unsaturated nitro compounds (equation 7) usually lead to the synthesis of β-nitro cyanides (14) [245].

$$\underset{p\text{-}YC_6H_4}{\overset{p\text{-}XC_6H_4}{}}\hspace{-0.3cm}\overset{\diagdown}{\underset{\diagup}{C}}{=}\overset{NO_2}{\underset{H}{C}}\hspace{-0.1cm}\overset{\diagdown}{\diagup} \xrightarrow[\text{DMSO}]{\text{KCN}} \underset{p\text{-}YC_6H_4}{\overset{p\text{-}XC_6H_4}{}}\hspace{-0.3cm}\overset{\diagdown}{\underset{\diagup}{C}}{\underset{\underset{CN\ (14)}{|}}{}}{-}CH_2NO_2 \quad (7)$$

$$X = OMe, Me, H, H, H, Cl, F, H, H;$$
$$Y = OMe, Me, OMe, H, Cl, H, F, NO_2, CN$$

    However, in the case of 1-nitro-2,2-bis(p-nitrophenyl)ethene (15; equation 8), the reaction results in the formation of the 'normal' product (16) in 24% yield and

1-cyano-2,2-bis(p-nitrophenyl)ethene (**17**) (10%) as the result of nucleophilic vinyl substitution [245].

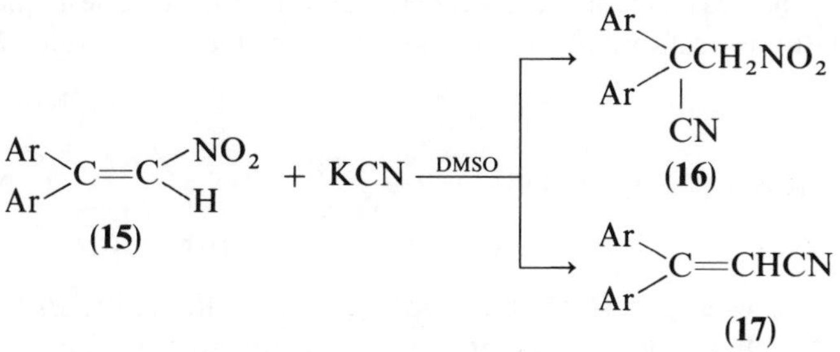

$$\text{Ar} = p\text{-nitrophenyl}$$

In water, compound (**15**) transforms only into (**16**) [245].

The regioselectivity of the process for the series of 9-(nitromethyl)fluorenes (equation 9) depends upon the solvent: in water, the cyanide anion attacks the C(9) position, while in DMSO and DMF it attacks the α position to give the product of nitro group substitution (**20**) [135a, 163].

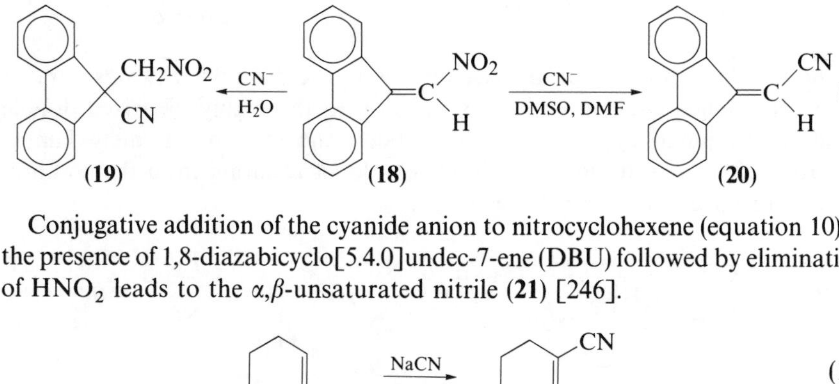

Conjugative addition of the cyanide anion to nitrocyclohexene (equation 10) in the presence of 1,8-diazabicyclo[5.4.0]undec-7-ene (DBU) followed by elimination of HNO_2 leads to the α,β-unsaturated nitrile (**21**) [246].

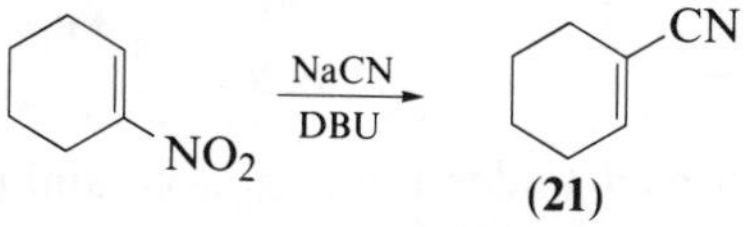

The reaction of nitroalkene sugars (**22**; equation 11) with HCN in the presence of Et_3N [247] leads to a similar result.

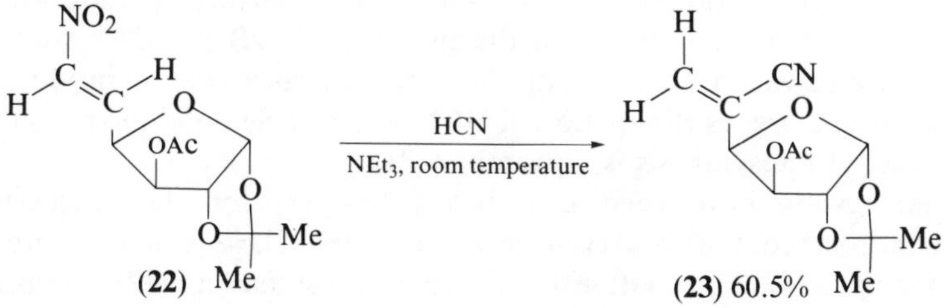

Condensation of substituted nitriles, for example cyanoacetic ester, nitroaceto-nitrile and others, with nitroalkenes gives nitroalkyl(aryl) cyanide derivatives [248, 249]. The reaction of tetracyanoethane with nitroethene (equation 12a) proceeds without a catalyst at 30–35 °C and gives the bicyclic product (24) [250].

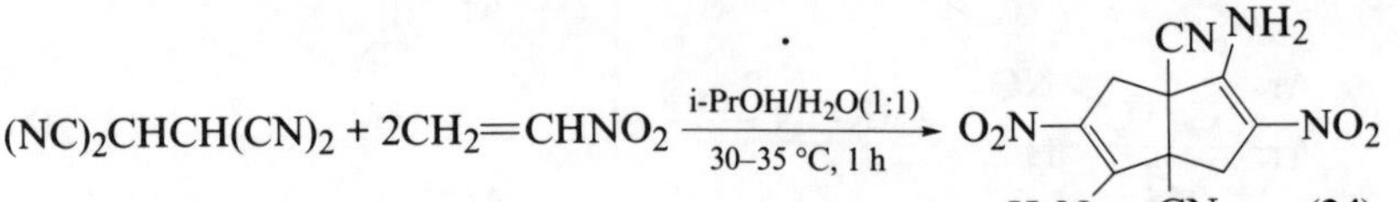

$$(NC)_2CHCH(CN)_2 + 2CH_2{=}CHNO_2 \xrightarrow[\text{30–35 °C, 1 h}]{\text{i-PrOH/H}_2O(1:1)}$$

$$(12a)$$

Cyanoacetic ester forms Michael-type adducts with α-nitroalkenes in the presence of sodium alcoholates or a catalytic quantity of TEA [251, 252]. Reaction of β-nitrostyrene with methylcyanoacetic ester potassium salt in DMSO results in the cyclopropane derivative (25; equation 12b) in good yield and high diastereoselectivity [180].

$$(12b)$$

The specific behaviour of nitroacetonitrile in the reaction with unsaturated nitro compounds is illustrated in its addition to the highly polarized double bond of geminal cyanonitrostyrene under the action of diethylamine (equation 13). The salt (26) is the Michael adduct resulting from the addition of two nitro acetonitrile molecules [253].

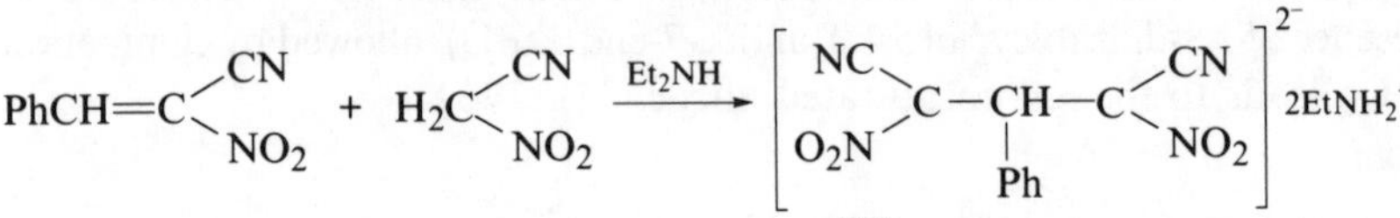

$$(13)$$

The salt (26) is also formed under the action of amines on geminal cyano-nitrostyrene [254].

More complicated substituents at the cyanomethyl group direct the process to the products of heterocyclization. Such a result has been achieved by Tsuge et al. [255] for the reaction of the N-(1-cyanoalkyl)imine (27; equation 14) with nitrostyrene. The process occurs in the presence of DBU and converts the starting materials via the intermediate (28) to the heterocyclization product (29), which further undergoes elimination of HCN and double bond migration. The final product of the synthesis is 1-pyrroline (31).

Unusual results have been described [256–259] for the reactions of alkyl(aryl)nitroethenes with alkyl isocyanoacetates. These reactions produce substituted pyrroles (35; equation 15) via the Michael adduct (32). Cyclization of (32) through the isocyano group followed by denitration of (33) gives the pyrrole.

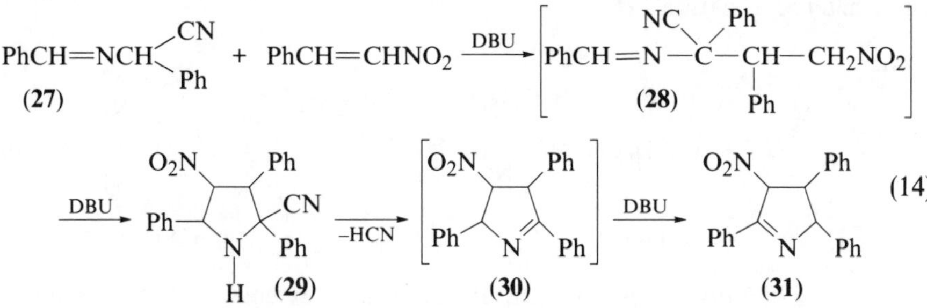

DBU = $(Me_2N)_2C=NBu$-t

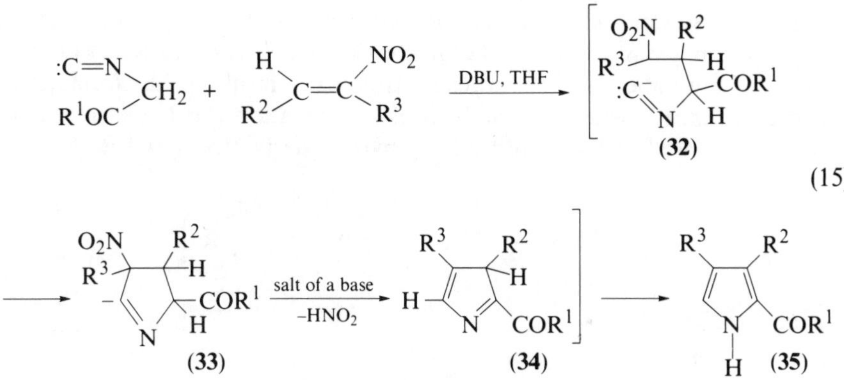

R^1 = EtO, t-BuO; R^2 = H, Me, Ar; R^3 = H, Me, Et

Nitroalkenes that contain galactose, fructose or ribose fragments react with ethyl isocyanoacetate in a similar way. The resulting pyrrole derivatives are converted to porphyrins (36), which show high antiviral activity [260].

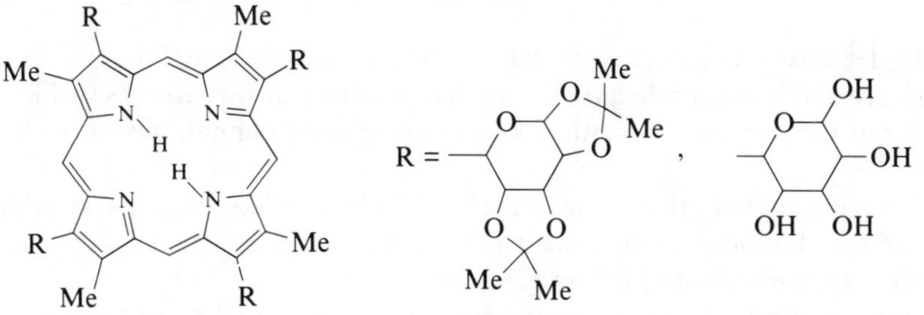

The CH acid 1-phenyl-3-methyl-5-pyrazolone (37; equation 16) forms with nitroalkenes cyclic nitro adducts [252, 261]. The structures of these compounds are close to those of drugs with a wide range of activity (analgesics, tranquillizers). The addition of (37) to geminal cyanonitrostyrene results in intramolecular heterocyclization [262].

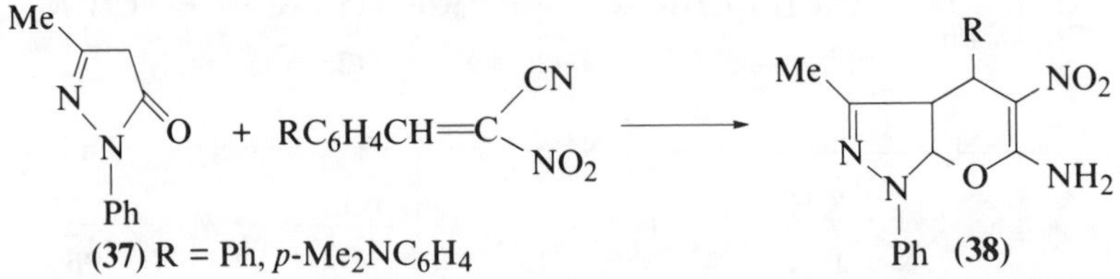

$$(16)$$

(37) R = Ph, p-Me$_2$NC$_6$H$_4$ Ph (38)

Coupling of β-diketones with nitroalkenes is used extensively in the synthesis of oxygen-containing and nitrogen-containing heterocycles such as furans, pyrroles, pyrans and pyrazoles.

The first description of the reaction between unsaturated nitro compounds and β-diketones was given in 1948 for acetyl(benzoyl) acetone. Reaction of the latter with aromatic and heterocyclic nitroalkenes results in the formation of dihydrofuran methyleneamino derivatives (39; equation 17). These compounds can be hydrolysed to the corresponding pyrrole derivatives (40) [263].

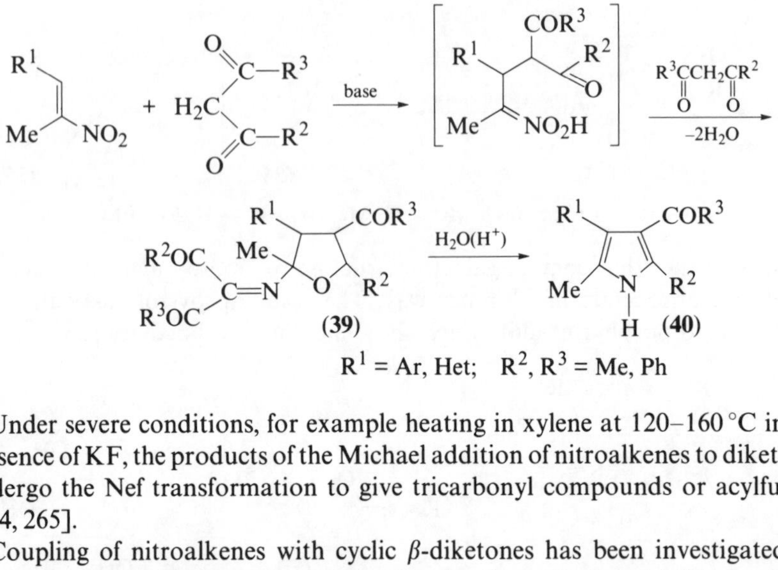

$$(17)$$

R^1 = Ar, Het;   R^2, R^3 = Me, Ph

Under severe conditions, for example heating in xylene at 120–160 °C in the presence of KF, the products of the Michael addition of nitroalkenes to diketones undergo the Nef transformation to give tricarbonyl compounds or acylfurans [264, 265].

Coupling of nitroalkenes with cyclic β-diketones has been investigated for dimedone, 1,3-indandione, 2-methyl-1,3-indandione, 2-phenyl-1,3-indandione, 4-hydroxycoumarin and others [266–268].

The reactions between dimedone and some aliphatic, aromatic and heterocyclic unsaturated nitro compounds (equation 18) involve the initial formation of the Michael adduct (41). Under the action of base (KF, MeONa, TEA), the adduct undergoes intramolecular heterocyclization to form either a mixture of *syn* and *anti* stereoisomers (42), furan derivatives (43) or a diastereoisomeric mixture (44). The nature of the mixture depends upon the nature of the substituents around the nitro group [264, 269–273].

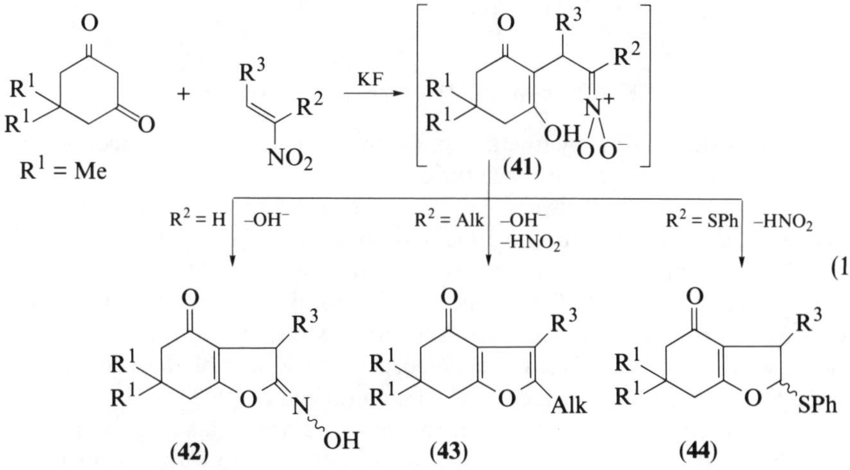

$$R^3 = Me, Ph, p\text{-}BrC_6H_4$$

(18)

A nitrile group in the geminal position to the NO_2 group of nitroalkenes participates in heterocyclization (equation 19) in the reaction with dimedone. The process leads to the formation of the tetrahydrochromenone (46) [262].

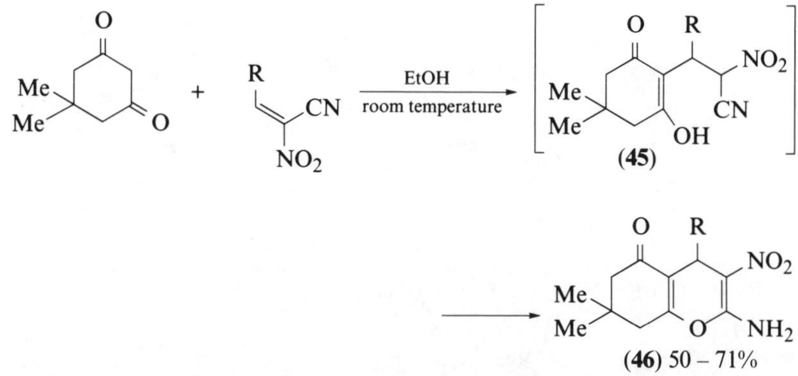

(19)

(46) 50 – 71%

R = Ph, $p\text{-}Me_2NC_6H_4$, 3-methoxy-4-oxyphenyl, 2-furyl, 2-thienyl, *N*-benzyl-3-indolyl

Barbituric acid is a rather active heterocyclic β-dicarbonyl compound owing to which it easily condenses with nitroalkenes [274, 275]. Its condensation with geminal cyanonitrostyrenes (equation 20) does not end at the addition step but proceeds further and results in the product of intramolecular heterocyclization (47) [253]. The nitrile group participates in this heterocyclization.

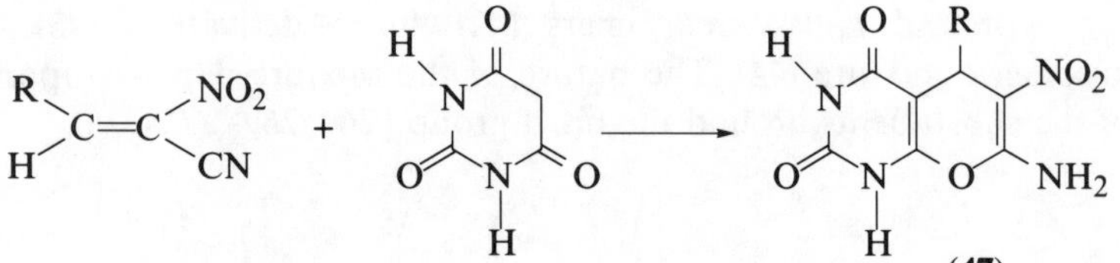

$$R = Ph, p\text{-}MeOC_6H_4, 3\text{-methoxy-4-oxyphenyl}$$

This is a wide-ranging synthetic approach to various pharmacologically important nitro derivatives of barbituric acid.

Some heterocyclic β-diketones, e.g. 1,2-diphenyl-3,5-dioxo-4-alkyl(aryl)pyra-zolidines [276] and (2R, 3S)-3,4-dimethyl-2-phenylperhydro-1,4-oxazepin-5,7-dione [277], combine with nitroalkenes to give nucleophilic addition products.

The regularities of cyclic β-diketone addition to nitroalkenes have been used successfully for their reactions with unsaturated nitrosugars [278]. Gomez-Sanchez et al. have established the stereospecific features of the process and defined the configuration of the novel C(3) chiral centre [278].

Acetoacetic ester reacts with nitroalkenes, arylnitroalkenes and their derivatives in the presence of basic catalysts [279–283] to form the products of condensation (48; equation 21). These compounds eliminate HNO_2 under the action of base and transform into furan carbonic esters (49) [284, 285]. The reduction of (48) leads to Δ^1-pyrrolines (50a) [281, 285] and Δ^2-pyrrolines (50b) via ring closure [282].

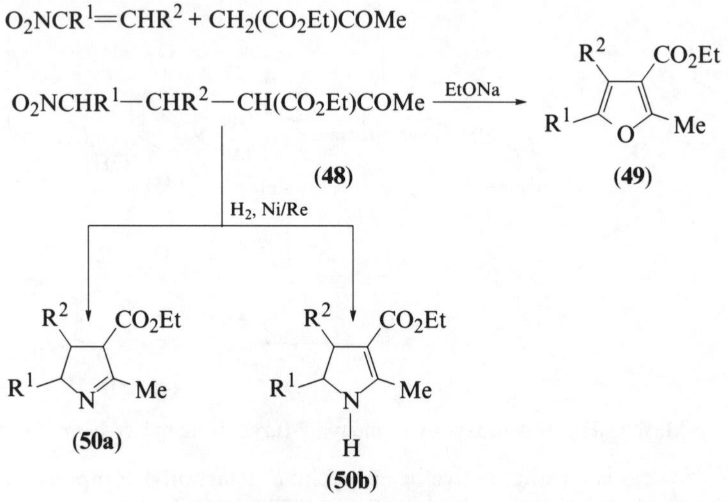

$R^1, R^2 = H, Alk, Ar$

Gomez-Sanchez et al. [286] have found that the structures of the products of reaction between β-nitrostyrene and acetoacetic ester in the presence of NaOH depend upon the nature of the solvent. The common Michael adduct is isolated from benzene, but in methanol solution bis(dihydrofuryl)hydroxylamine (51) is formed. Under the action of amines (51) transforms into the pyrrole derivative [286].

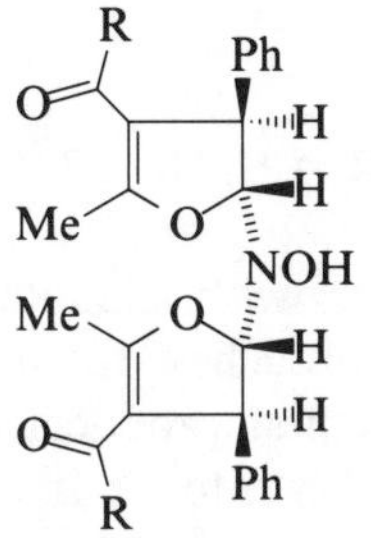

(51)

R = OMe, OEt, Me

Studies [224, 287–289] of the reactions between acetoacetic ester and substituted aromatic and heterocyclic nitroethenes (with adduct hydrolysis) and the same processes under the action of amines have revealed facile synthetic approaches to oxygen-containing and nitrogen-containing heterocycles such as substituted pyrroles and dihydrofurans.

Highly electron-deficient geminal cyanostyrene makes it possible to deliver the synthesis in equation (22) without a catalyst. The participation of the nitrile group in the heterocyclization results in the formation of the pyran derivative (52) [253].

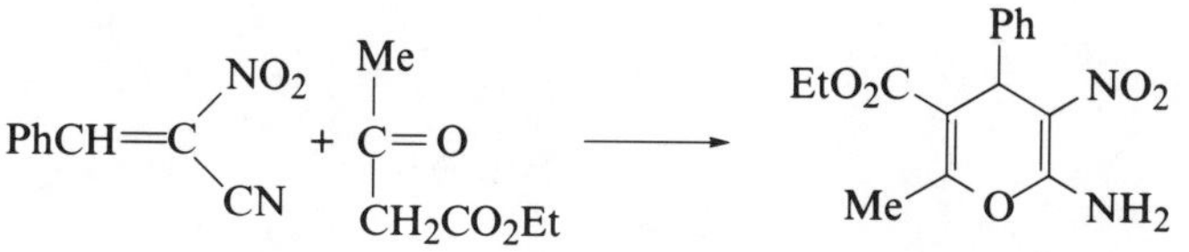

$$\tag{22}$$

(52)

Malonic ester reacts with aliphatic, aromatic and heterocyclic unsaturated nitro compounds (equation 23) [290].

$$O_2NCR^1{=}CR^2H + R^3CH(CO_2R^4)_2 \longrightarrow O_2NCHR^1CHR^2CR^3(CO_2R^4)_2$$

(53)

$$R^1, R^2, R^3 = H, Alk, Ar, Het; \quad R^4 = Me, Et \tag{23}$$

The reactions of dialkyl malonates with 3-oxy-1-nitro-1-cyclohexene and its derivatives (54; equation 24) result in the stereoselective formation of the products of intramolecular heterocyclization, i.e. the γ-butyrolactones (55) [291].

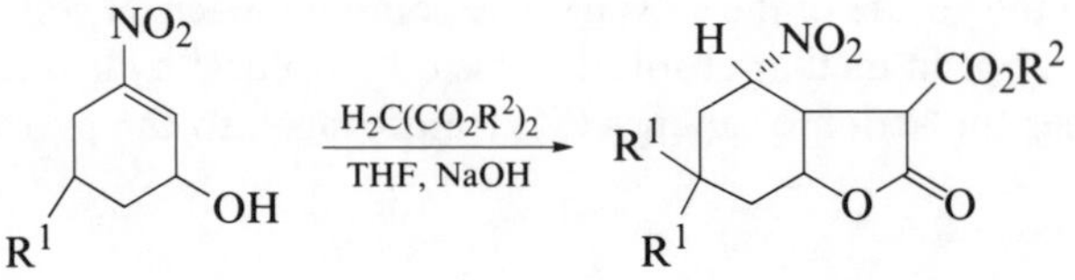

NO2
OH
R1
H2C(CO2R2)2
THF, NaOH
H
NO2
CO2R2
R1
O
O
R1

(24)

(54) (55)

R^1 = H, Me; R^2 = Me, Et

The reactions of malonic esters with aliphatic and heterocyclic nitroethenes provide general synthetic approaches to substituted γ-amino acids and α-pyrrolidones [292–295]. The first products in the Michael reactions with γ-nitro esters are converted further into α-pyrrolidones. Subsequent hydrolysis results in the formation of γ-amino acids. A typical example of such a process is the reduction of (**56**) on a skeletal nickel catalyst (equation 25). The resultant pyrrolidone (**57**) can be hydrolysed in acid or base to produce either 4-amino-3-phenylbutyric acid (**58**) or 4-phenyl-2-pyrrolidone (**59**), respectively.

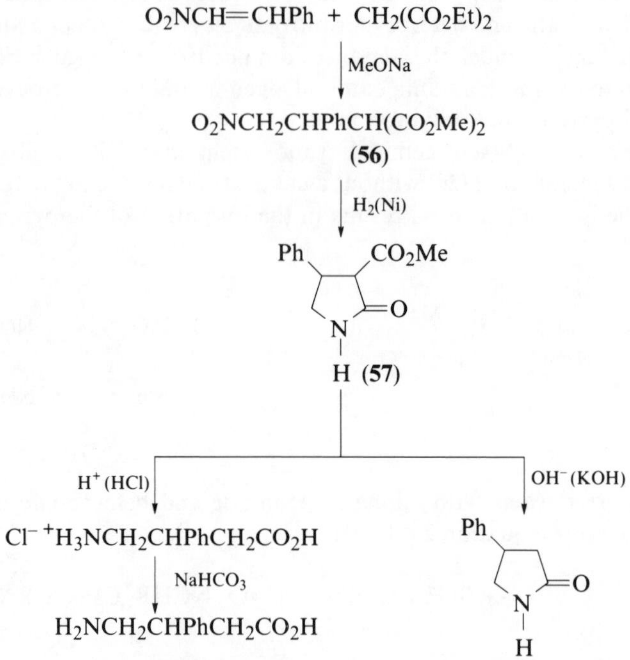

O2NCH=CHPh + CH2(CO2Et)2
MeONa
O2NCH2CHPhCH(CO2Me)2
(56)
H2(Ni)
Ph
CO2Me
O
N
H (57)
H+ (HCl)
OH− (KOH)
Cl− +H3NCH2CHPhCH2CO2H
NaHCO3
H2NCH2CHPhCH2CO2H
Ph
O
N
H

(25)

(58) (59)

The interest of biologists and pharmacologists in the closely related γ-aminobutyric acid and α-pyrrolidones was heightened after the detection of a significant amount of γ-aminobutyric acid in brain tissues and the discovery of its important functions in central nervous system activities (participation in the retarding process).

β-Phenyl-γ-aminobutyric acid is the parmaceutical drug 'Phenybut', the very first representative of a novel group of drugs which are analogues of the products of natural metabolism. This drug has no side effects and is practically non-toxic. Phenybut is a tranquilizer with a nootropic component, it relieves stress, and regulates the normal sleep pattern [296]. It is applied successfully to neuroses and logoneuroses and is an effective antisickness drug.

A series of various CH acids, such as chloromethyl phenyl sulfone, chloro-acetophenone and α-chloropropionitrile, can be used in the reaction with nitro-styrene in the presence of t-BuONa. The presence of a group which can eliminate chloride ion provides a route to either common adducts (60) or the unsaturated nitro compounds (61; equation 26), depending on the reaction conditions [297].

$$PhCH=CHNO_2 + PhSO_2CH(Et)Cl \xrightarrow[\text{t-BuONa}]{\text{THF}}$$

$$\xrightarrow{-50\,^{\circ}C} PhSO_2C(Et)CH(Ph)CH_2NO_2$$
$$\underset{\displaystyle \overset{|}{Cl}}{} $$
$$\textbf{(60)} \ 50\%$$

$$\xrightarrow{20\,^{\circ}C} PhSO_2CH(Et)C(Ph)=CHNO_2$$
$$\textbf{(61)} \ 35\% \qquad\qquad (26)$$

Under the action of NaH in a THF/t-BuOH mixture, nitroalkenes of type (62) undergo intramolecular Michael addition to form products of type (63), which are prospective biologically active substances (equation 27) [298].

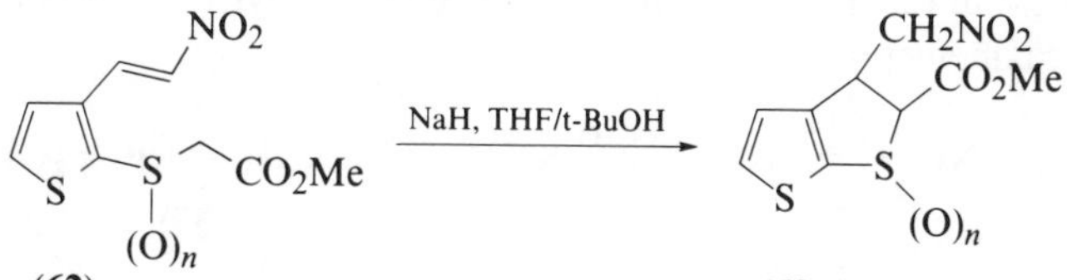

$$\xrightarrow{\text{NaH, THF/t-BuOH}} \qquad\qquad (27)$$

(62)   (63) $n = 0, 2$

2.5.2 REACTIONS OF MONONITROALKADIENES AND MONONITROALKENYNES WITH CH ACIDS

The main features of CH acid addition to 1-nitro-1,3-dienes and 1-nitro-1-en-3-ynes are defined by the nature of the nitrodiene and nitroenyne systems. 1-Nitro-1,3-butadienes undergo nucleophilic reactions with CH acids under the action of a catalyst (equation 28) to give mainly the products of 1,4 addition (64) [299–301].

$$O_2NCH{=}CR^1CR^2{=}CH_2 + O_2NCHR^3CO_2Et \xrightarrow{Et_3N}$$

$$O_2NCH_2CR^1{=}CR^2CH_2CR^3(NO_2)CO_2Et \qquad (28)$$

$$\textbf{(64)} \ 23\text{--}54\%$$

$$R^1, R^2 = H, Me; \quad R^3 = H, CO_2Et$$

Unsubstituted 1-nitro-1,3-butadiene combines with nitroacetic and nitro-malonic esters in 32% and 35% yields, respectively [301]. The same reactions can be achieved with such compounds as 1-nitro-4-iodo-2-butene [302] or 1-nitrobutenyl nitrates, which transform into nitrodiene under the reaction conditions [300]. The resulting compounds can be converted into D,L-lysine dichlorohydrate in about 70% yield by hydration on a Pd/C catalyst in acidic alcohol solution followed by hydrolysis [301].

The interaction between 2-nitrofurans (**65**; equation 29) and nitroalkanes takes place at the nitroethene group and involves the elimination of HNO_2 [303].

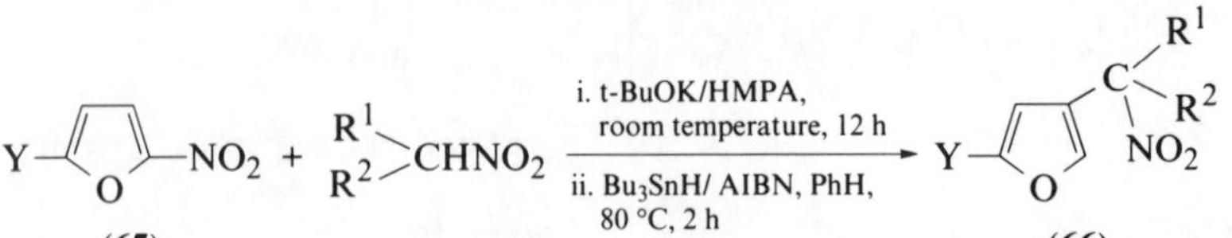

$$(29)$$

$$\textbf{(65)} \qquad \qquad \textbf{(66)}$$

AIBN = 2,2′-azobisisobutyronitrile
$Y = CO_2Me$, $R^1 = Me$, n-C_6H_{13}, n-C_8H_{17}, $(CH_2)_2CO_2Me$, Ph,
$R^2 = Me$; $\quad Y = COMe$, $R^1 = Me$, $(CH_2)_2CO_2Me$, $R^2 = Me$

2-Nitro-1,3-butadienes (equation 30) react with methylene components at the nitrovinyl fragment [304].

$$RCH{=}CMeC(NO_2){=}CH_2 \xrightarrow{XCH_2CO_2Et} RCH{=}CMeCH(NO_2)CH_2CHXCO_2Et$$

$$\textbf{(67)}$$

$$R = H, Me; \quad X = NO_2, CO_2Et \qquad (30)$$

Sterically overloaded CH acids, for instance 2-phenyl-1,3-indandione, give 1,4-addition products with 2-nitrobutadienes [305].

CH acids combine with 1-nitroalkenynes (equation 31) at the nitrovinyl fragment when catalysed by TEA (for nitroacetic and cyanoacetic esters) or MeONa (nitromethane and malonic ester) [306].

$$RC{\equiv}CCH{=}CHNO_2 \xrightarrow{XCH_2CO_2Et, \ base} RC{\equiv}CCH(CH_2NO_2)CHXCO_2Et$$

$$\textbf{(68)}$$

$$R = Me, Et, Me_3Si; \quad X = CN, NO_2, CO_2Et \qquad (31)$$

2.5.3 REACTIONS OF DINITROETHENES, DINITRODIENES AND DINITROTRIENES WITH CH ACIDS

1,2-Dinitroalkenes give the products of nitro group substitution with strong CH acids (nitromethane, acetylacetone). This is typical for all nitroalkenes with a substituent in the β position which can be eliminated in the form of a stable anion (Hal, CN, NO_2). The reaction takes place without base at room temperature (1,2-dinitroethene) or with heating (1-phenyl-1,2-dinitroethene) [307]. The process (equation 32) results in the formation of products of type (69) (R = H) or (70) (R = Ph) [307].

$$RC\!\!=\!\!CHNO_2 + MeNO_2 \xrightarrow[-HNO_2]{18-20\,°C,\,5h} RC\!\!=\!\!CHCH_2NO_2 \xrightarrow[80-90\,°C,\,2h]{MeNO_2}$$

$$\underset{NO_2}{|} \qquad\qquad\qquad \underset{\underset{(69)\ R=H}{NO_2}}{|}$$

$$\left[\underset{NO_2}{\overset{RCHCH(CH_2NO_2)_2}{|}} \right] \xrightarrow[-HNO_2]{} RCH\!\!=\!\!C(CH_2NO_2)_2$$

$$(70) \quad R = Ph \tag{32}$$

2,3-Dinitro-2-butene and 1,2-dinitrostilbene, in contrast with 1,2-dinitroethene, 1,2-dinitrostyrene [307] and β-nitroacrylic ester [308], do not react with CH acids.

1,1-Dinitroethene, unlike 1,2-dinitroethene, forms the products of addition with methylene components (equation 33). For example, polynitroalkanol (72) was synthesized from 1,1-dinitroethene (71) formed in situ [309].

$$K^+(NO_2)_2\bar{C}CH_2OH \xrightarrow{H^+} [(NO_2)_2CHCH_2OH] \xrightarrow[-H_2O]{}$$

$$[(NO_2)_2C\!\!=\!\!CH_2] \xrightarrow[ii.\,H^+]{i.\,K^+(NO_2)_2\bar{C}CH_2OH} (NO_2)_2CHCH_2C(NO_2)_2CH_2OH \tag{33}$$

$$(71) \qquad\qquad\qquad\qquad\qquad (72)\ 55\%$$

Owing to a weak cross-influence between the two nitrovinyl groups in unconjugated dinitrodienes, CH acids add to both nitrovinyl fragments simultaneously [310].

2,5-Dinitro-1,3,5-trienes contain the formally conjugated dinitrotriene system. However, owing to steric hindrance from the nitro groups, various substituents add the distortion of the planar structure of the conjugated system, these compounds react in such a way as if their nitrovinyl fragments were unconjugated. For example, 3,4-dimethyl-1,6-diphenyl-2,5-dinitro-1,3,5-hexatriene (73; equation 34) adds two equivalents of malonic ester in the 1,2 and 5,6 positions [310].

$$PhCH\!\!=\!\!C(NO_2)CMe\!\!=\!\!CMeC(NO_2)\!\!=\!\!CHPh + 2CH_2(CO_2Me)_2 \longrightarrow$$
$$(73)$$

$$(MeO_2C)_2CHCHPhCH(NO_2)CMe\!\!=\!\!CMeCH(NO_2)CHPhCH(CO_2Me)_2$$

$$(74) \tag{34}$$

Lipina and Perekalin [311, 312] have found that the reactivity of conjugated 1,4-dinitrodienes in the process of nucleophilic addition–substitution depends upon the radicals location and their nature. For example, the reaction between 1,4-dinitro-1,3-butadiene (75; equation 35) and dimedone proceeds in a similar manner to the reaction of mononitroalkenes and leads to 1,2 addition followed by vinyl–allyl isomerization to give the most stable system (76).

$$O_2NCH{=}CHCH{=}CHNO_2 \xrightarrow[\text{base}]{\text{NuH}} O_2NCH_2CNu{=}CHCH_2NO_2 \quad (35)$$
$$(75) \hspace{5cm} (76)$$

$$NuH = \text{dimedone}$$

Owing to steric hindrance the less electrophilic 2,3-dimethyl-1,4-dinitrobutadiene (77; equation 36) does not participate in the Michael reaction. Under the influence of a base it undergoes vinyl–allyl isomerization and stabilizes in the form of (78) [312].

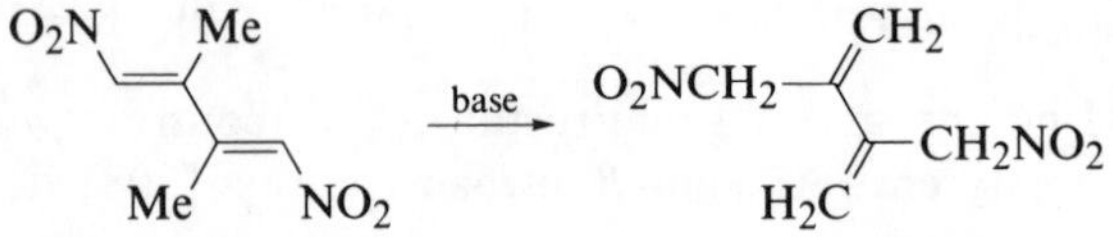

$$(36)$$

1,4-Dinitro-2,3-diphenyl-1,3-butadiene (79; equation 37) has a non-planar structure which results in behaviour similar to mononitroalkenes. However, the shielding effect of the phenyl substituents at C(2) and C(3) causes the nucleophilic attack to be directed to the terminal carbon atoms of the diene chain, leading to the products of substitution (80) [312].

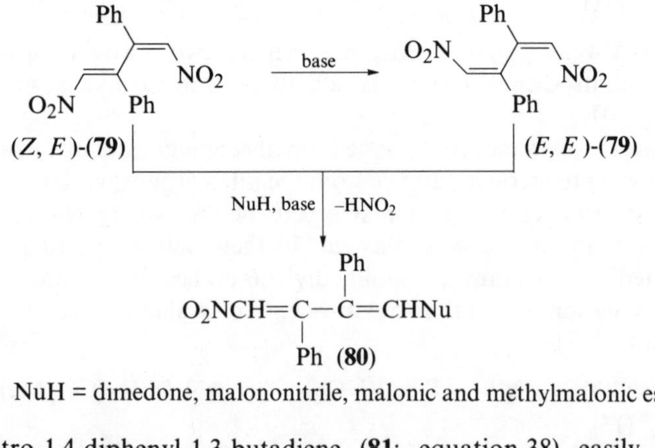

$$(37)$$

$$O_2NCH{=}C{-}C{=}CHNu$$
with Ph substituents, (80)

$$NuH = \text{dimedone, malononitrile, malonic and methylmalonic esters}$$

1,4-Dinitro-1,4-diphenyl-1,3-butadiene (81; equation 38) easily gives the product of 1,2 addition with one equivalent of CH acid. The resulting adduct undergoes vinyl–allyl isomerization to form the more stable 1,4-dinitro-2-butene (82).

$$PhC(NO_2){=}CHCH{=}C(NO_2)Ph \xrightarrow[\text{base}]{\text{NuH}} PhCH(NO_2)CH{=}CCH(NO_2)Ph$$

(81)   (82) $\overset{|}{Nu}$ (38)

NuH = malonic and methylmalonic esters, malononitrile

2.5.4 INTERACTIONS BETWEEN NITROALKENES AND WEAK CH ACIDS WITH THE PARTICIPATION OF ORGANOLITHIUM COMPOUNDS

The reaction of nitroalkenes with organolithium compounds presents various synthetic possibilities for the well-studied Michael reaction. Weak CH acids that do not react under common, basic conditions can be used in the form of monocarboanions.

Weak CH acids such as ketones, ethers, sulfur-containing compounds (1,3-dithianes,1,3-dithiacarbonic acid derivatives) and amides can be converted to highly reactive anions with the help of organolithium compounds and used in Michael reactions with nitroalkenes at -78 to $90\,°C$ [313–317].

The application of lithium diisopropylamide (LDA) in the reactions of ketones with nitroalkenes (equation 39) effects the convenient synthesis of polyalkylpyrroles (85) [318].

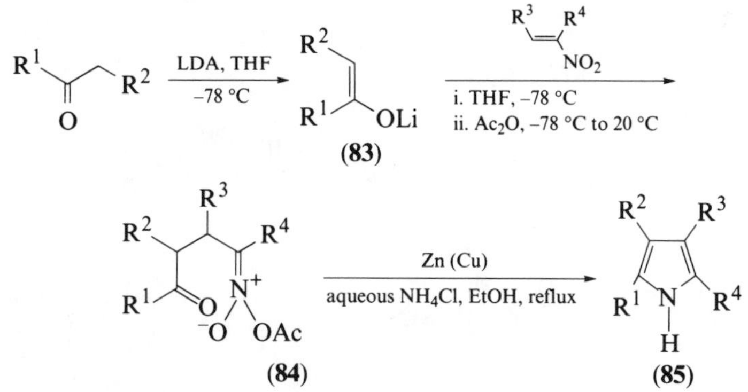

$$R^1, R^2, R^3, R^4, = H, Me, Et, Bu$$

Acetonitronic anhydrides (84) hydrolyse under the action of BF_3 to form 1,4-diketones [319].

Some substituted 1,4-diketones (87; equation 40) are also obtained in the hydrochloric acid hydrolysis of Li salts of type (86) (not isolated). The salts are formed from aliphatic ketones and nitroalkenes [320].

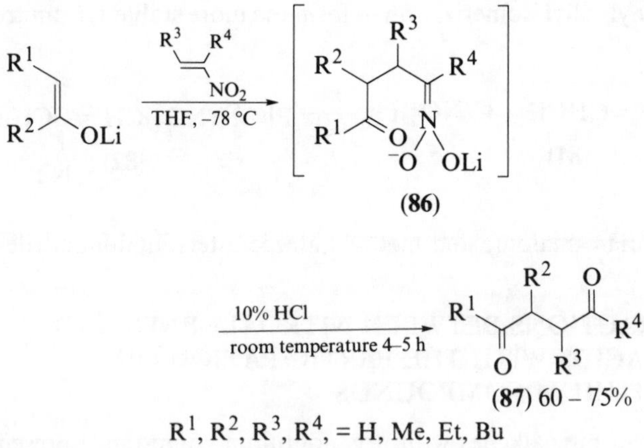

$$(40)$$

(87) 60 – 75%

$$R^1, R^2, R^3, R^4, = H, Me, Et, Bu$$

The reaction with cyclic ketones proceeds in a similar way.

Cyclohexenone metallation by lithium diisopropylamide in THF at $-78\,^{\circ}\mathrm{C}$ followed by addition of (E)-1-nitropropene and neutralization with AcOH (equation 41) results in a mixture of *erythro* and *threo* diastereoisomers (90). The reaction mixture under reflux in HMPA, instead of neutralization of the intermediate (89), transforms into the tricyclooctanones (91) via intramolecular Michael condensation and denitration [321, 322].

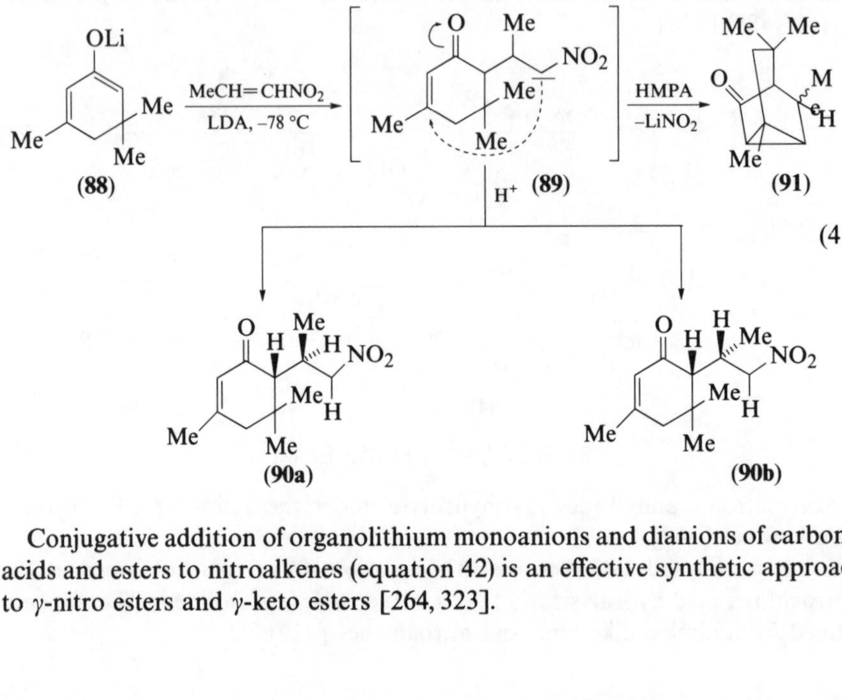

$$(41)$$

Conjugative addition of organolithium monoanions and dianions of carbonic acids and esters to nitroalkenes (equation 42) is an effective synthetic approach to γ-nitro esters and γ-keto esters [264, 323].

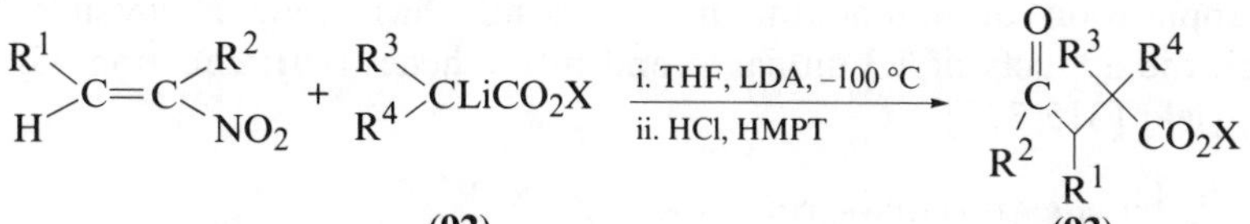

$$\underset{(92)}{\overset{R^1}{\underset{H}{>}}C=C\overset{R^2}{\underset{NO_2}{<}} + \overset{R^3}{\underset{R^4}{>}}CLiCO_2X} \xrightarrow[\text{ii. HCl, HMPT}]{\text{i. THF, LDA, }-100\,°C} \underset{(93)}{\overset{O}{\underset{R^2}{\overset{\|}{C}}}\underset{R^1}{\overset{R^3}{\underset{|}{C}}R^4}CO_2X} \quad (42)$$

$R^1 = R^3 = H$, $R^2 = Me$, $R^4 = (CH_2)_5$, Ph; $R^1, R^2 = (CH_2)_4$, $R^3 = H$,
$R^4 = (CH_2)_5$, Ph;   X = Li, Me;   HMPT = hexamethyl phosphortriamide

This method has been extended to cyclic enolates [317, 324].

One-pot reactions of nitroalkenes with ester enolates and formaldehyde or acrylates result in the nitroalkane intermediates (94). These are further transformed into 2-pyrrolidones (95; equation 43) by means of catalytic hydration [325].

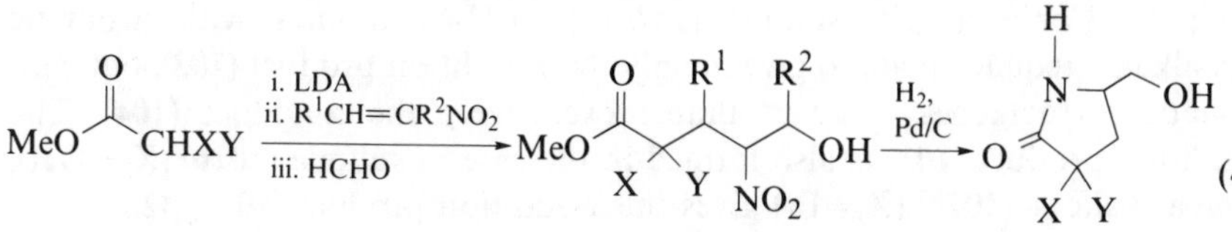

$$\underset{MeO}{\overset{O}{\|}}\underset{CHXY}{C} \xrightarrow[\substack{\text{ii. }R^1CH=CR^2NO_2 \\ \text{iii. HCHO}}]{\text{i. LDA}} \underset{X\ Y\ NO_2}{\overset{O\ R^1\ R^2}{MeO}}OH \xrightarrow[\text{Pd/C}]{H_2,} \underset{(95)\ 40\text{--}80\%}{O=\underset{X\ Y}{N}\underset{}{OH}} \quad (43)$$

X, Y = H, Me, Ph, SPh;   (94) 48–77%   (95) 40–80%
R^1 = H, Ph; R^2 = H, Me

This method has been used in the synthesis of the pyrrolizidine alkaloid (96).

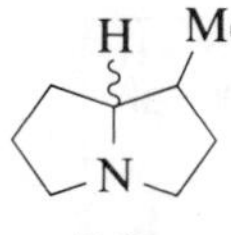

(96) 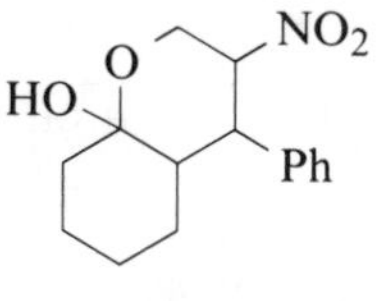   (97)

Under similar conditions cyclohexanone enolate reacts with β-nitrostyrene and formaldehyde to give the oxygen-containing heterocyclic compound (97) in 74% yield [325].

Miyashita *et al.* [324, 326] have prepared γ-keto acids and their lactones from sulfur-containing carbonic acid lithium salts and nitroalkenes (equation 44).

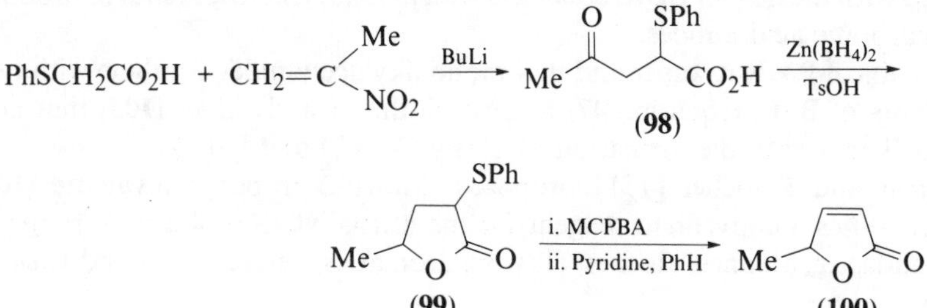

$$PhSCH_2CO_2H + CH_2=C\overset{Me}{\underset{NO_2}{<}} \xrightarrow{BuLi} \underset{(98)}{Me}\overset{O\quad SPh}{\underset{}{CO_2H}} \xrightarrow[\text{TsOH}]{Zn(BH_4)_2}$$

$$\underset{(99)}{Me}\overset{SPh}{\underset{O}{}}O \xrightarrow[\text{ii. Pyridine, PhH}]{\text{i. MCPBA}} \underset{(100)}{Me}\overset{}{\underset{O}{}}O \quad (44)$$

The application of organolithium compounds has made it possible to approach the adducts of 1,3-dithianes and nitroethenes (**101**; equation 45) in 65–90% yield [315, 327].

$$\begin{array}{c} \text{(structure with S, Li, } R^1) \\ \xrightarrow[\text{ii. AcOH}]{\text{i. } R^2CH{=}CR^3NO_2, \text{ THF, } -78\,°C} \end{array} \quad \begin{array}{c} \text{(structure with S, } R^1, CHR^3CHR^3NO_2) \\ \textbf{(101)} \end{array} \tag{45}$$

R^1 = H, Me, ⟨cyclohexyl⟩, Ph, ⟨O-O benzodioxole-Me⟩ ;

R^2 = Me, Ph, p-MeOC$_6$H$_4$, 3,4-CH$_2$O$_2$C$_6$H$_3$, 3,4-(MeO)$_2$C$_6$H$_3$, 2,4,5-(MeO)$_3$C$_6$H$_2$;
R^3 = H, Me

The Michael additions of allyl sulfone anions, obtained under the action of LDA, to nitroethenes depend on the nature of the substituents in the starting materials. For example, sulfone (**102a**) (X = H) combines with aromatic nitroalkenes (equation 46) to give mainly the γ-addition product (**103**), but with aliphatic nitroalkenes it gives almost exclusively the α-adduct (**104**). The γ-addition product (**103**) is also formed in the case of sulfone (**102b**) (X = OH), whereas sulfone (**102c**) (X = Br) gives the α-addition product (**104**) [328].

$$RCH{=}CHNO_2 \quad + \quad \begin{array}{c} \text{(allyl sulfone, } CH_2X) \\ SO_2Ph \\ \textbf{(102a – 102c)} \end{array}$$

| R = Ar; X = H, OH | R = Alk, X = H;
R = Ar, X = Br |

$$\begin{array}{cc} \begin{array}{c} SO_2Ph \\ OH \\ R \quad NO_2 \\ \textbf{(103)} \end{array} & \begin{array}{c} R \\ O_2N \qquad X \\ SO_2Ph \\ \textbf{(104)} \end{array} \end{array} \tag{46}$$

Lithium salts of carbonic acid dialkylamides in the form of their enolates, obtained with the help of LDA, condense with alkyl(aryl)nitroethenes to produce γ-nitrocarbonic acid amides.

By means of PMR it was found that monoalkylacetamides combine with two equivalents of BuLi (equation 47) to give dilithium derivatives (**105**) that add to nitroalkenes with the formation of a new C—C bond [314].

Seebach and Knochel [171] proposed 2-nitro-2-propenyl pivaloate (**107**; equation 48) as a convenient reagent for the nitroallylation of a wide range of compounds, e.g. alkanes, alkenes, alkynes, aromatics, heterocyclics and various ketones.

$$MeC(O)NHR^1 \xrightarrow{\text{BuLi, THF, 0 °C}} \left[CH_2C(O)NR^1 \right]^{2-} 2Li^+ \tag{47}$$

$$\textbf{(105)}$$

$$\xrightarrow[\text{ii. H}_2\text{O, AcOH (100\%)}]{\text{i. R}^2\text{CH}=\text{CHNO}_2\text{, THF, } -90\text{ °C, 1.5–3 h}} O_2NCH_2CHR^2CH_2C(O)NHR^1$$

$$\textbf{(106)} \quad 75\% \ (R^1 = Me)$$
$$32\% \ (R^1 = t\text{-Bu})$$

$$R^1 = Me, t\text{-Bu}; \ R^2 = \ \text{(benzodioxole group)}$$

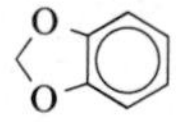

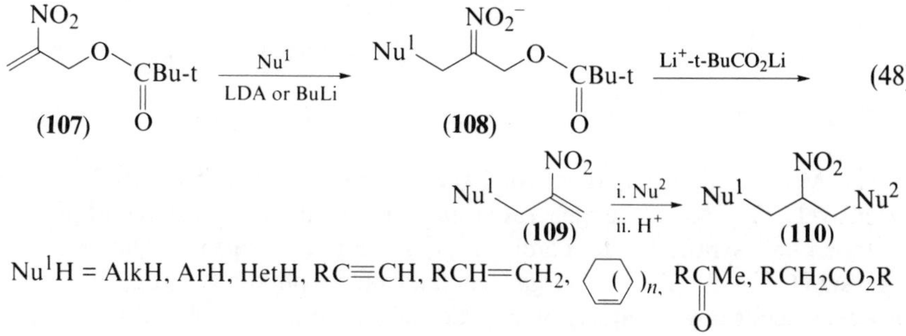

$$\text{(107)} \xrightarrow[\text{LDA or BuLi}]{\text{Nu}^1} \text{(108)} \xrightarrow{\text{Li}^+\text{-t-BuCO}_2\text{Li}} \tag{48}$$

$$\text{(109)} \xrightarrow[\text{ii. H}^+]{\text{i. Nu}^2} \text{(110)}$$

$$Nu^1H = AlkH, ArH, HetH, RC\equiv CH, RCH=CH_2, \ \text{(cyclohexene)}_n, \ RCMe, RCH_2CO_2R$$
$$\qquad \qquad \qquad \qquad \qquad \qquad \qquad \qquad \qquad \qquad \qquad \qquad \qquad \qquad \qquad \| \atop O$$

A CH_2CO_2Bu-t fragment (i.e. $OCOBu$-t) in the α position of nitroethene behaves as a strong nucleofuge. That is why stabilization of the anionic form of adduct (108) proceeds via the corresponding carbonic acid elimination. Addition of a second nucleophilic molecule to (109) extends the synthesis to pivaloyloxynitropropenes.

The use of lithium or titanium derivatives of cyclo(L-Val-Gly)III bislactam ester (111; equation 49) in the reaction with (E)-nitroalkenes provides a route to the asymmetric synthesis of some methyl α-amino-γ-nitrocarbonic esters (112) with high stereoselectivity [329].

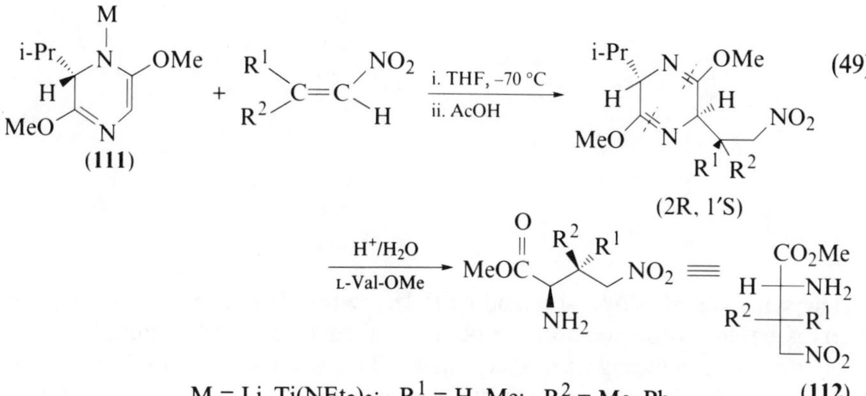

$$\xrightarrow[\text{L-Val-OMe}]{\text{H}^+/\text{H}_2\text{O}}$$

$$M = Li, Ti(NEt_2)_3; \ R^1 = H, Me; \ R^2 = Me, Ph \tag{112}$$

An enantioselective synthesis of $(3R, 4R)$-3-amino-1-hydroxy-4-phenyl-2-pyrrolidinone (**114**; equation 50) has been devised with the aim of producing drugs for the treatment of cerebral ischaemia. The synthesis includes the addition of substituted glycine to nitrostyrene [330].

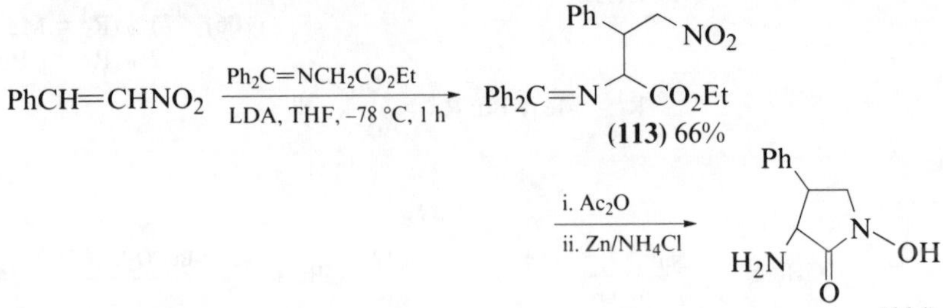

$$
\text{PhCH}{=}\text{CHNO}_2 \xrightarrow[\text{LDA, THF, }-78\,^\circ\text{C, 1 h}]{\text{Ph}_2\text{C}{=}\text{NCH}_2\text{CO}_2\text{Et}} \quad \underset{(\textbf{113})\ 66\%}{\text{Ph}_2\text{C}{=}\text{N}}\overset{\overset{\text{Ph}}{\diagup}\diagdown\text{NO}_2}{\diagdown\text{CO}_2\text{Et}} \tag{50}
$$

$$
\xrightarrow[\text{ii. Zn/NH}_4\text{Cl}]{\text{i. Ac}_2\text{O}} \quad (\textbf{114})
$$

The Michael addition of nitroalkenes to enolates has been studied [331] extensively. Enolates are generated *in situ* in the conjugative addition of cyclopentene synthones to complex organocopper reagents. The latter can consist of CuI, t-Bu$_3$P and an organometallic nucleophile, e.g. BuLi. The reaction proceeds according to equation (51). Column chromatography of the resulting mixture separates the diastereoisomeric 3-butyl-2-(2'-nitropropyl)cyclopentanones (**116**). The hydrolysis of the intermediate (**115**) results in the formation of the diketone (**117**).

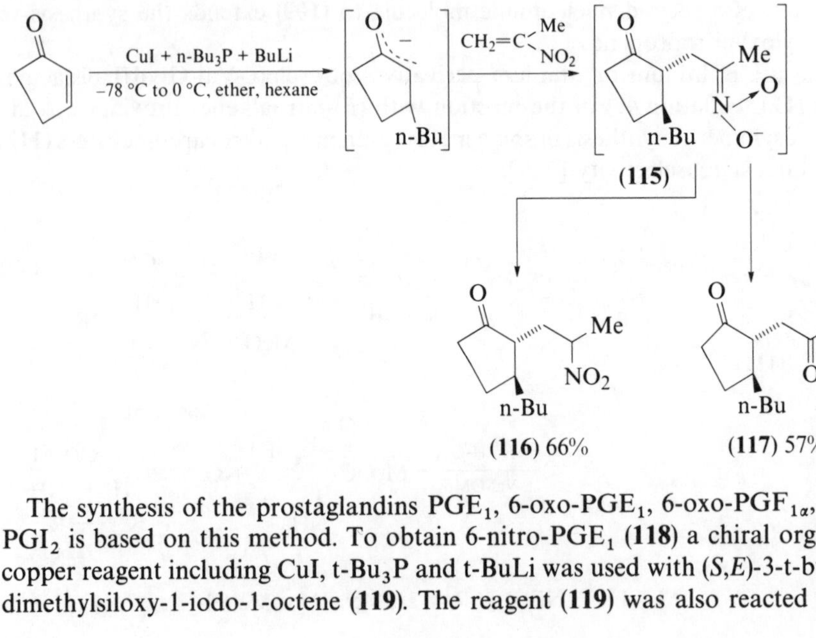

$$
\xrightarrow[-78\,^\circ\text{C to 0 }^\circ\text{C, ether, hexane}]{\text{CuI} + \text{n-Bu}_3\text{P} + \text{BuLi}} \quad \underset{\text{n-Bu}}{[\quad]} \xrightarrow{\text{CH}_2{=}\text{C}\overset{\text{Me}}{\underset{\text{NO}_2}{\diagup}}} \quad \underset{(\textbf{115})}{[\quad]} \tag{51}
$$

$$
\underset{(\textbf{116})\ 66\%}{\text{n-Bu}} \qquad \underset{(\textbf{117})\ 57\%}{\text{n-Bu}}
$$

The synthesis of the prostaglandins PGE$_1$, 6-oxo-PGE$_1$, 6-oxo-PGF$_{1\alpha}$, and PGI$_2$ is based on this method. To obtain 6-nitro-PGE$_1$ (**118**) a chiral organocopper reagent including CuI, t-Bu$_3$P and t-BuLi was used with (S,E)-3-t-butyldimethylsiloxy-1-iodo-1-octene (**119**). The reagent (**119**) was also reacted with the chiral enone (120) and methyl 6-nitro-6-heptenoate (121). 6-Nitro-PGE$_1$ (118) was isolated as an epimeric (6R)/(6S) mixture in a total yield of 71%.

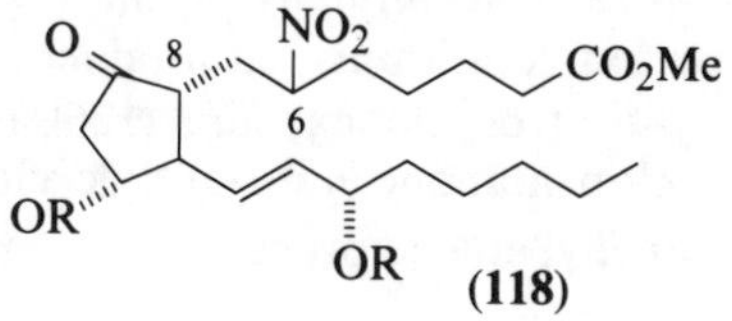

(118)

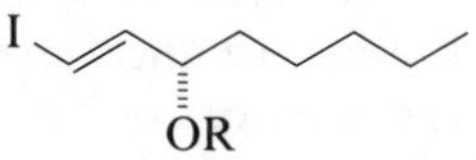

(119)

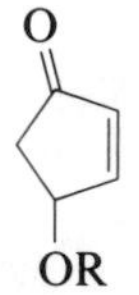

(120)

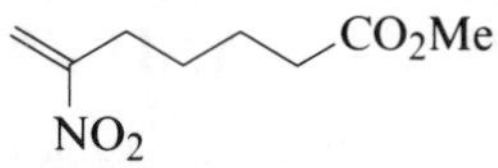

(121)

R = SiMe$_2$Bu-t

Apart from the weak CH acids, strong ones such as β-arylnitroalkanes and β-dicarbonyl compounds have been used in the reactions with nitroethenes with participation of organolithium reagents. The reaction occurs at the more basic reaction centre of the CH acid dianion, in contrast to the classical behaviour shown in this type of reaction. For example, β-nitrophenylethane couples with 1-nitro-2-phenylethene (equation 52) to give 1,4-dinitro-2,3-diphenylbutane (123) [332].

$$PhCH_2CH_2NO_2 \xrightarrow{\text{BuLi, THF, } -90\,°C} \left[Ph\overset{-}{C}HCHNO_2 \right] 2Li^+ \xrightarrow{PhCH=CHNO_2}$$

(122)

$$O_2NCH_2CHPhCHPhCH_2NO_2 \tag{52}$$

(123)

(124)

$$O_2NCH_2CHR^2CH_2CCH_2CR^1 \tag{53}$$

(125)

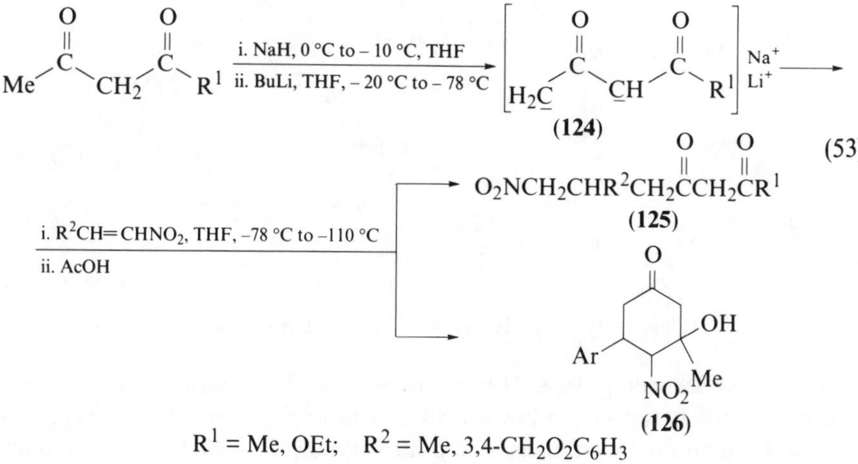

(126)

R^1 = Me, OEt;   R^2 = Me, 3,4-CH$_2$O$_2$C$_6$H$_3$

Dianions of β-dicarbonyl compounds (acetylacetone, acetoacetic esters) (124; equation 53), resulting from reaction with lithium diisopropylamide or multistage reaction with NaH and BuLi, react initially at the deprotonated methyl group. Depending on the reaction conditions, the product of addition (125) or cyclization, i.e. the 3-hydroxy-4-nitrocyclohexanone (126), is obtained [313, 314]. The additions of 2-acetylcyclopentanone and 2-acetylcyclohexanone to 1-nitro-2-(3,4-methylenedihydroxyphenyl)ethene proceed in a similar way.

2.6 REACTIONS OF NITROALKENES WITH ENOLSILANES AND ENAMINES

Nitroalkene reactions with enolsilanes and enamines can be discussed along with those of Li enolates as Michael additions of weak CH acids, because the final result of the transformation is nitroalkylation of the ketone in the β position [333]. This interesting tendency of nitroalkenes provides routes to simple syntheses of 1,4-nitrocarbonyl and 1,4-dicarbonyl compounds. It also develops the stereochemical aspects of nitroalkene chemistry, making it possible to carry out highly stereoselective syntheses and to obtain various heterocyclic systems that are the fragments of natural compounds.

Yoshikoshi et al. [264, 334, 335] have characterized the main pathways of nitroalkene reactions with enol silyl ethers. These elegant transformations, induced at low temperature by Lewis acids, are stereospecific. The initially formed silyl nitronates (2; equation 1) or the products of bipolar cycloaddition (3) [334] hydrolyse under mild conditions to give 1,4-diketones (4). These are easily transformed into cyclopentanones (5).

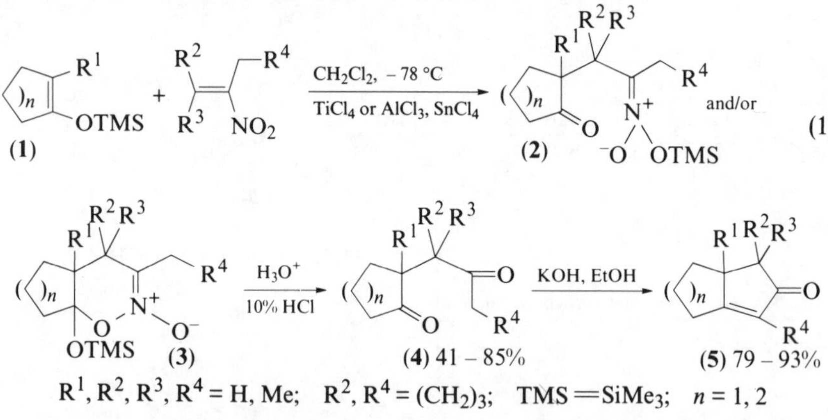

$$R^1, R^2, R^3, R^4 = H, Me; \quad R^2, R^4 = (CH_2)_3; \quad TMS = SiMe_3; \quad n = 1, 2$$

Brook and Seebach [336, 337] have shown that the structure of the product depends on the nature of the Lewis acid. For example, 1-trimethyl siloxycyclohexene (6; equation 2) combines with β-nitrostyrene at $-90\,^\circ$C in the presence of TiCl$_4$ to form a diastereoisomeric mixture of Michael adducts (**7a, 7b**). The participation of (i-PrO)$_2$TiCl$_2$ in the reaction leads to a stereoisomeric mixture of cyclonitronates (**8a–8c**) in a total yield of 81% [336].

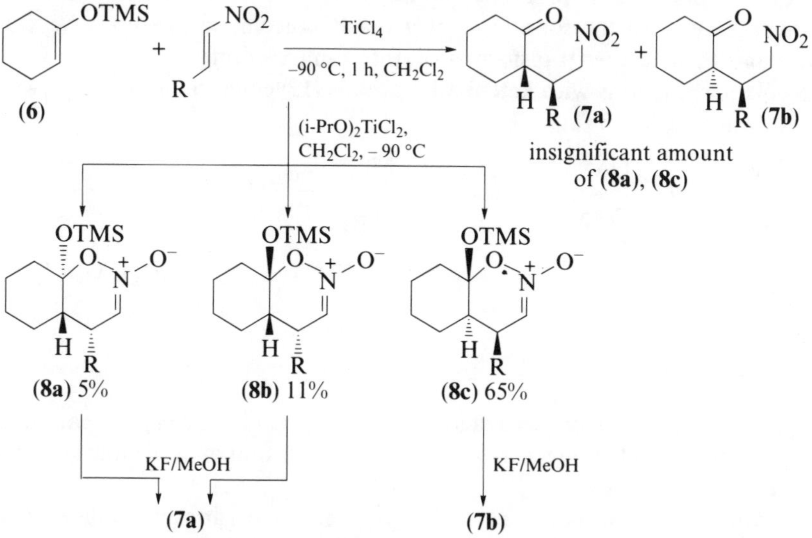

(2)

R = Ph, p-MeC$_6$H$_4$, p-MeOC$_6$H$_4$, p-NCC$_6$H$_4$

Under the action of KF in methanol the cycloadducts (**8**) are converted into a diastereoisomeric mixture of (**7a**) and (**7b**). This fact and the diastereoselectivity dependence upon the nitroalkene geometry do not exclude the possibility for the reaction to take place with TiCl$_4$ initiation via cycloaddition. Such a

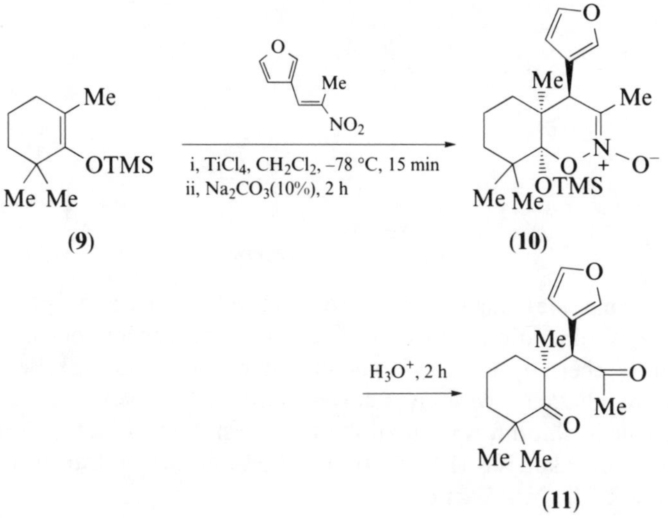

(3)

procedure can be illustrated by the reaction between the enolsilane of 2,2,6-trimethylcyclohexanone (**9**; equation 3) and (*E*-1-(3-furyl)-2-nitro-1-propene [338].

When initiated by $(i\text{-}PrO)_2TiCl_2$, the reaction of compound (**9**) with the heterocyclic nitroalkene proceeds similarly to the reaction described in equation (2), i.e. with the formation of a diastereoisomeric mixture of cycloadducts and their further transformation into γ-nitroketones.

Nitroalkene reactions with ketene silyl acetals (**12**; equation 4) open the way to γ-keto esters (**13**) [334].

$$\underset{\textbf{(12)}}{\text{Me(H)C}=\text{C(OMe)(OTMS)}} + CH_2=C(Me)NO_2 \xrightarrow[\text{ii. reflux, 3 h}]{\text{i. TiCl}_4\text{, Ti(i-PrO)}_4\text{, CH}_2\text{Cl}_2\text{, }-78\,^\circ\text{C}} \underset{\textbf{(13)}}{\text{MeC(=O)CH}_2\text{CH(Me)CO}_2\text{Me}} \qquad (4)$$

The synthetic potential of the reaction can be extended [334] by variation of the substituents in the linear fragments and application of cyclic nitroalkenes and ketene silyl acetals.

The reactions of nitroalkenes with cycloketone organotin enolates (**14**; equation 5) proceed similarly to the transformations described above. The application of organotin derivatives in the coupling reaction between β-nitrostyrene and cyclohexanone or cyclopentanone leads to a diastereoisomeric mixture of γ-nitroketones (**15**) containing mostly the *anti* adduct [339].

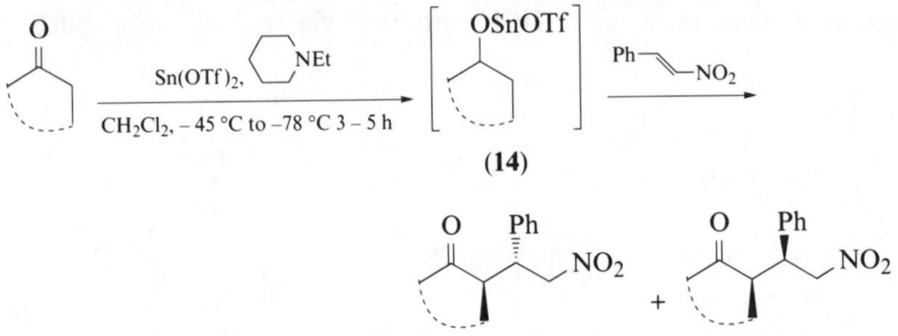

(5)

anti-(**15**) *syn*-(**15**)

Cyclic enamines, such as *N*-pyrrolidinocyclopentenes, *N*-morpholinocyclopentenes, *N*-piperidinocyclopentenes and their hexene and octene homologues, as well as their open-chain analogues, react with nitroalkenes under mild conditions (0–25 °C) without a solvent or in ether, petroleum ether, hexane, acetonitrile or chloroform (equation 6). The resultant adducts (**17**) can be hydrolysed to γ-nitroketones (**18**) [340], and these are easily transformed into 1,4-diketones in the Nef reaction.

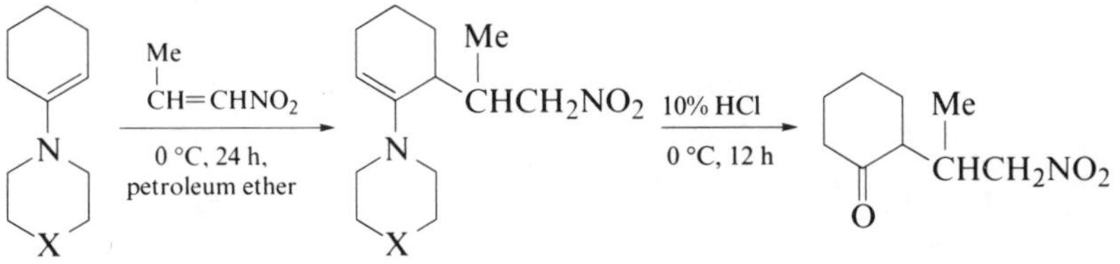

(16)   (17) 85%   (18) 80%

$$X = O, CH_2$$

(6)

The same process with β-nitrostyrene [341] proceeds stereospecifically: the resultant γ-nitroketone exists in the *erythro* configuration, as shown by X-ray analysis.

The ability of enamines (19) to form mono-adducts and bis-adducts in the reactions with nitroalkenes is noteworthy (equation 7). The bis-adducts can be transformed further into the products of carbocyclization (23) [342].

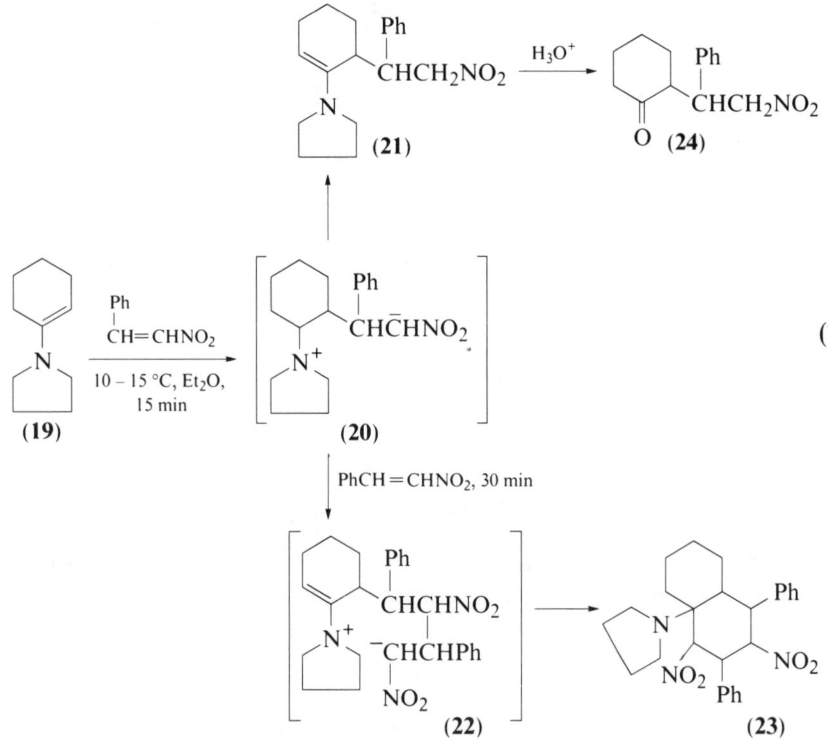

(7)

The preparation of another type of bis-adduct (25; equation 8) has been described [343].

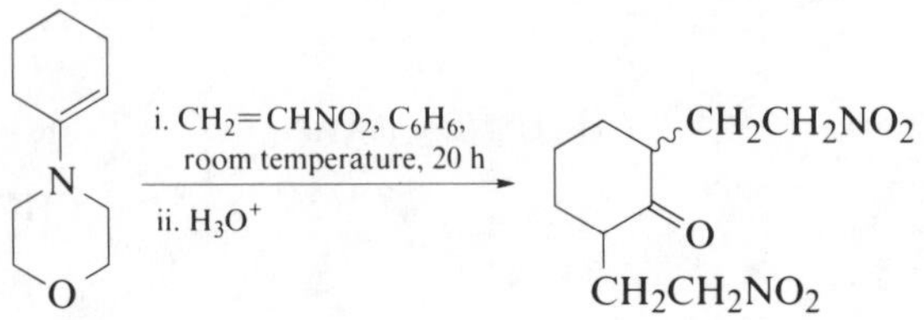

$$(8)$$

The equilibrium in enamines produced from asymmetric cycloketones promotes the formation of two regioisomers [343–345]. For example, β-nitro-styrene reacts with β-tetralone enamine (26; equation 9) to produce adducts (27) and (28) in the ratio 4:1. Acidic hydrolysis leads to the corresponding γ-nitroketones (29, 30). Although there are two asymmetric carbon atoms in each of these molecules, only the *erythro* isomer was isolated in all cases [345].

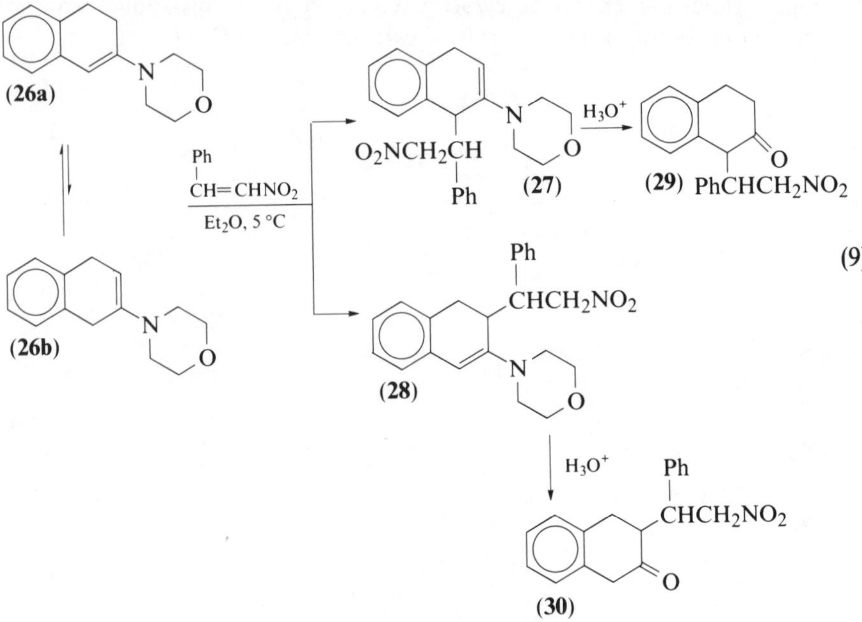

$$(9)$$

Nitroalkylation of the equilibrium mixture of N-morpholino-3-alkyl(aryl) cyclohexene enamines gives a combination of adducts (31, 32) [344, 346, 347].

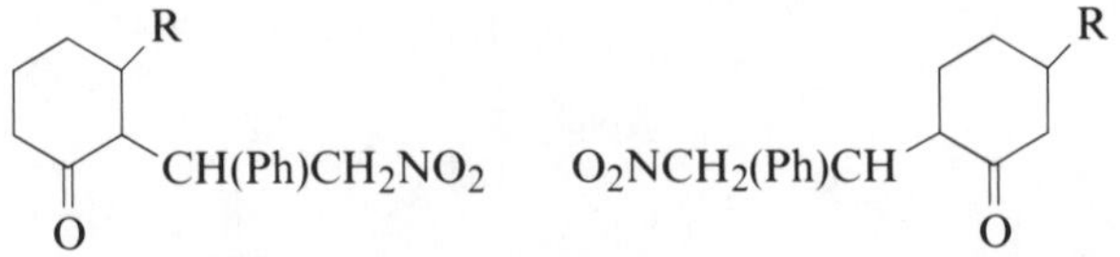

R = Me, Ph, t-Bu

The relative configurations and conformations of the products were determined by NMR methods and from the thermodynamic stabilities. It is concluded that a high stereoselectivity for nitroalkylation will be obtained with β-nitrostyrene.

X-ray analysis [348] of 2-(α-phenyl-β-nitro)ethyl-4-t-butylcyclohexanone (33; equation 10), produced by the method of Valentin et al. [347], shows that the compound crystallizes as an equilibrium mixture of two isomeric forms, namely trans-erythro and cis-threo.

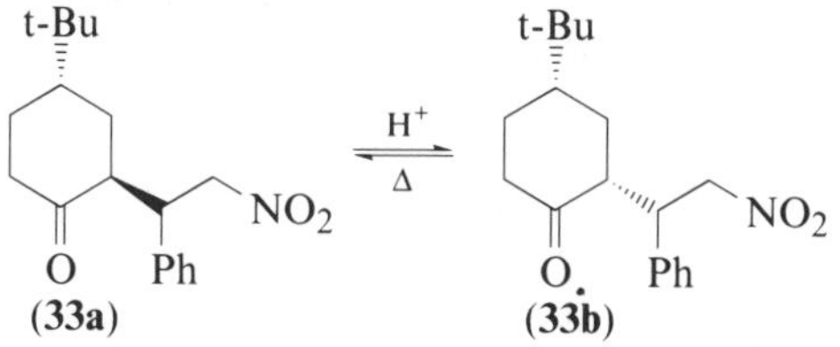

$$\text{(10)}$$

(33a) (33b)

By varying the nitroalkene and the reaction medium it has been shown that enamines, like enolsilanes, can react with nitroalkenes to form [4+2]-cycloaddition products, i.e. 1,2-oxazine-N-oxide derivatives [340, 349–352]. Equation (11) illustrates an example of such a process [350].

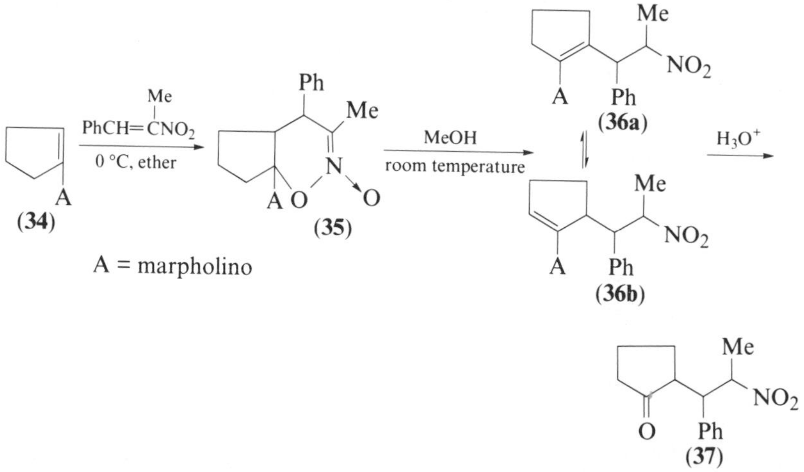

$$\text{(11)}$$

A = marpholino

(34) (35) (36a) (36b) (37)

Cyclonitronate (35) is stable only below $-15\,^\circ$C, and at room temperature it undergoes ring cleavage to form an equilibrium mixture of nitroalkylated enamines (36a, 36b). These enamines hydrolyse in a water/alcohol solution of acetic acid. Similarly, β-nitrostyrene reacts with conformationally fixed 1-(N-morpholino)-4-t-butyl-1-cyclohexene [353].

Reaction of a mixture of (E)-1-nitro-1-phenylpropene and (Z)-1-nitro-1-phenylpropene with aminocycloalkenes leads to extraordinarily unstable, substituted 1,2-oxazine N-oxides that upon treatment with CDCl$_3$ or MeOH undergo nucleophilic cleavage of the ring to give a mixture of diastereoisomeric nitroenamines. The latter hydrolyse to a mixture of (2S, αS, βS) and (2S, αS, βR) isomers of γ-nitroketones [354].

The formation of the 1,2-oxazine N-oxide in the O-alkylation of the initially formed bipolar intermediate (20; equation 7) is not the only result of the reaction between nitroalkenes and enamines. Also produced are cyclobutane derivatives [344, 346, 349, 355, 356] through intramolecular C-alkylation. For example, nitroethene reacts with cyclopentanone, cyclohexanone and 2-methylcyclo-hexanone enamines (38; equation 12) to give aminocyclobutanes (40) and nitroalkylated enamines (41) [356].

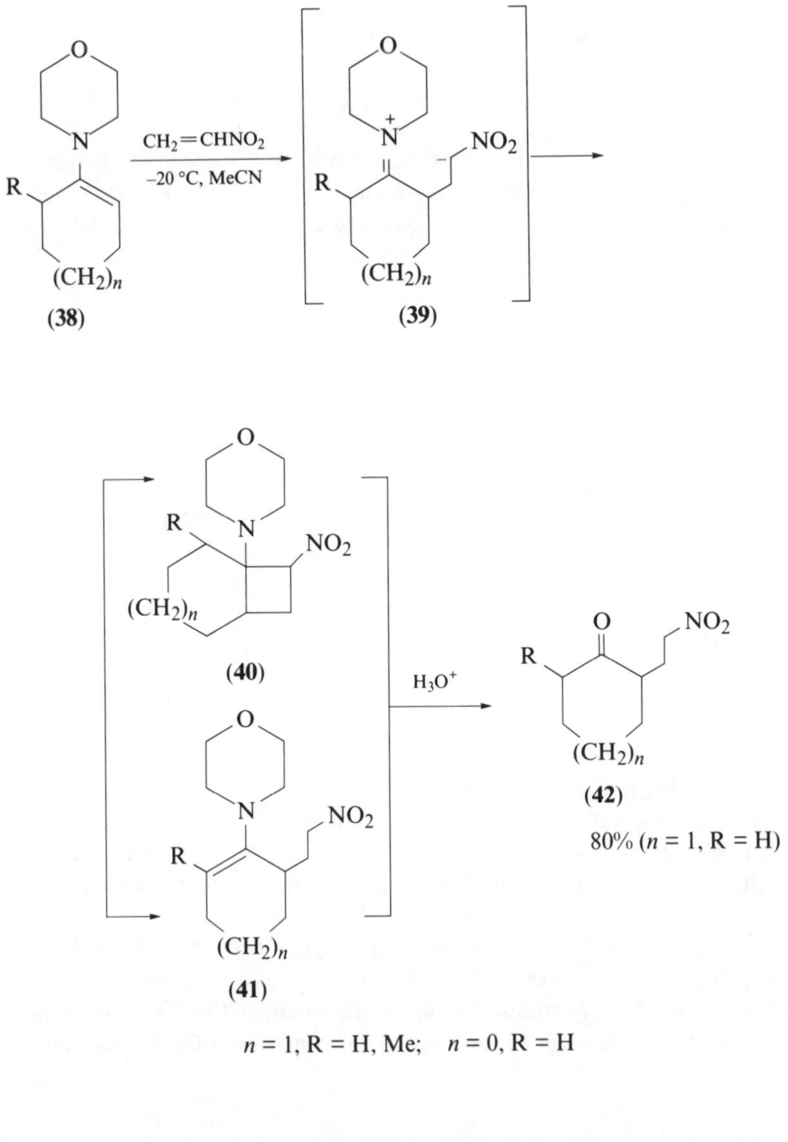

(12)

$n = 1$, R = H, Me; $n = 0$, R = H

This is a multivariable process because the enamines can both isomerize and form the products of carbocyclization and heterocyclization.

The typical transformation pathways for the reactions of nitroalkenes with enamine (43) have been summarized by Felluga et al. [357]. The pathways are shown in equation (13).

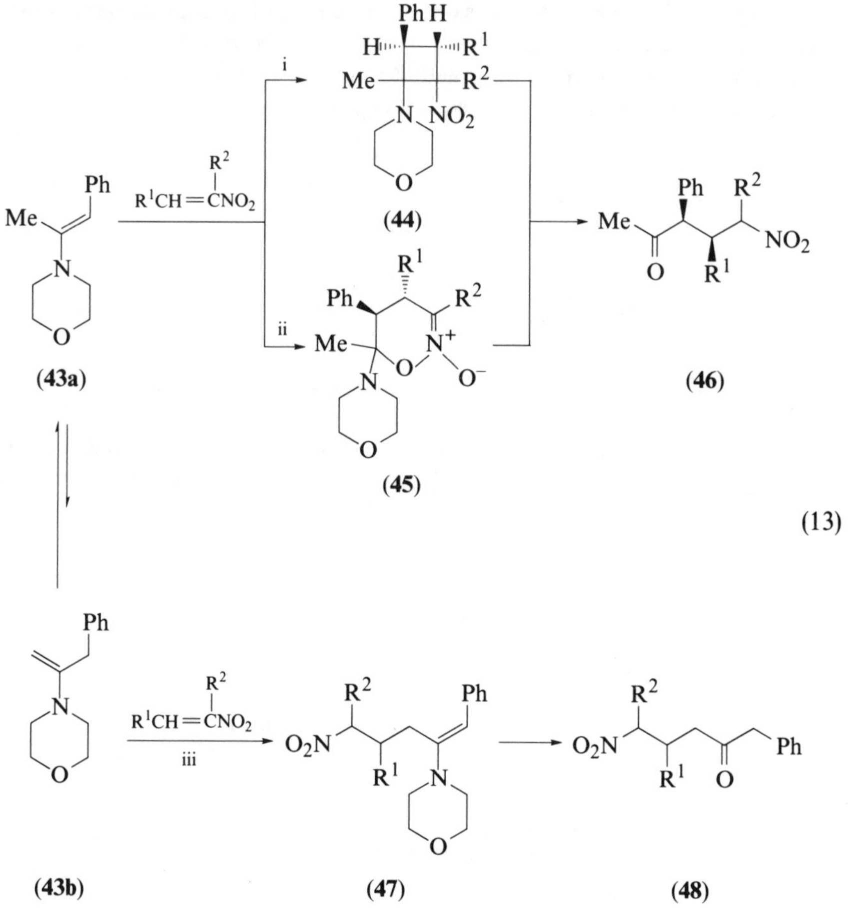

$$R^1 = Me, Ph; \quad R^2 = H, Me, Ph; \quad R^1, R^2 = (CH_2)_n, n = 3, 4$$

i: equimolar amounts, 10 min, absence of solvent
ii: equimolar amounts, 0 °C, 1 h
iii: equimolar amounts, absence of solvent; then CHCl$_3$, 24 h

This type of reaction has become the basis for the diastereoselective synthesis of some cyclic nitrone esters [358]. These nitrone esters react to give the Michael adducts in methanol. Hydrolysis of these systems is also highly stereoselective and leads to γ-nitroketones and γ-diketones.

The correlations established thus far facilitate the regioselective, diastereo selective and enantioselective nitroalkylations of enamines.

The stereochemistry of the reaction of nitroalkenes with enamines has been investigated in detail by Seebach *et al.* [359–362]. The authors have suggested a topological rule [359] that allows one to anticipate the prevailing configuration of the resulting diastereoisomer. Donors and acceptors with the (*E*) configuration are oriented in the reaction intermediate in such a way that their π systems have a synclinal relationship (**49**; equation 14).

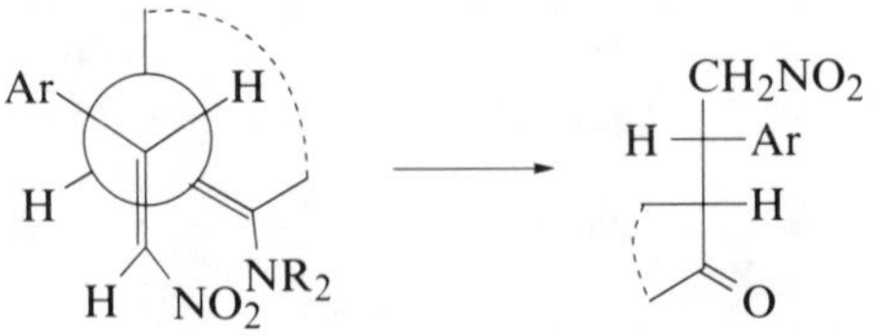

$$\tag{14}$$

(**49**)

re*re* approach

(**50**)

(*RS, SR*) or n-diastereoisomer

The reactions of β-nitrostyrenes with (2*S*)-*N*-(1-cyclohexenyl)-2-methoxy-methylpyrrolidine (**51a**) and (2*R*)-*N*-(1-cyclohexenyl)-2-methoxymethylpyrrolidine

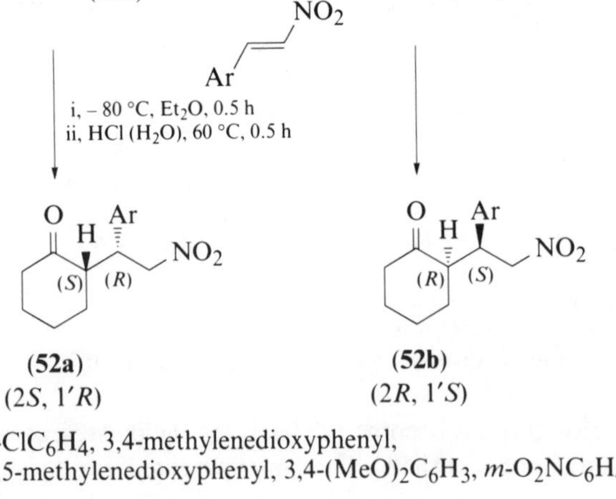

$$\tag{15}$$

(**51a**)

(**51b**)

i, – 80 °C, Et₂O, 0.5 h
ii, HCl (H₂O), 60 °C, 0.5 h

(**52a**)

(2*S*, 1′*R*)

(**52b**)

(2*R*, 1′*S*)

Ar = Ph, *p*-ClC₆H₄, 3,4-methylenedioxyphenyl,
2-bromo-4,5-methylenedioxyphenyl, 3,4-(MeO)₂C₆H₃, *m*-O₂NC₆H₄, naphthyl

(**51b**; equation 15) are examples of the asymmetric additions of optically pure enamines to nitroalkenes (stereoselective synthesis of the γ-nitroketone). Blarer *et al.* [360] showed that, in both cases, similar reaction conditions lead to only one of the four possible diastereoisomers with high enantiomeric purity. The absolute configurations of the stereoisomers were proven by chemical correlation methods and X-ray analysis.

Blarer *et al.* [360] also discuss the mechanism of the stereospecific reaction in accordance with the asymmetric synthesis classification system devised by Seebach and Prelog [363]. Blarer *et al.* [360] consider that the trigonal centre formed by combination of the components is characterized by an *lk* topology, and the chiral centre promotes *ul*-1,4 induction; the whole reaction can be described as an *lk, ul*-1,4 process (**53**).

(**53**)

It should be pointed out that a similar approach has been used to analyse the stereoselective addition of enolsilanes to nitroalkenes [338].

The enrichment of enamine structure extends the synthetic potential of such compounds in the reactions with nitroalkenes. It is worth mentioning the nitroalkene reaction with cross-conjugated enamines (2,3-diaminobutadiene and α-ketoenamines).

The combination of two enamino groups in the 2,3-diamino-1,3-butadienes (**54**; equation 16) defines the unusual result of their reaction with nitroalkenes. In this case the nitrocyclopentene derivatives are isolated as one stereoisomer or a mixture of diastereoisomers (**55, 56**). Hydrolysis of (**55**) and (**56**) with hydrochloric acid gives aminonitrocyclopentanones (**57, 58**) [364]. The reaction is likely to proceed via the bipolar intermediate. The structures of the products have been confirmed by X-ray analysis.

The carbonyl group in enamine molecules also directs their reaction with nitroalkenes towards the formation of five-membered rings [365]. Linear α-ketoenamines (**59**; equation 17) react with nitroalkenes without a solvent to give one stereoisomer (**62**) in 90–95% yield. Felluga *et al.* [365] suggest that the reaction proceeds via bipolar intermediates (**60, 61**) followed by [3 + 2] carbocyclization. Aminonitrocyclopentanol (**62**) hydrolysis results in the 2-hydroxy-3-nitrocyclopentanone (**63**).

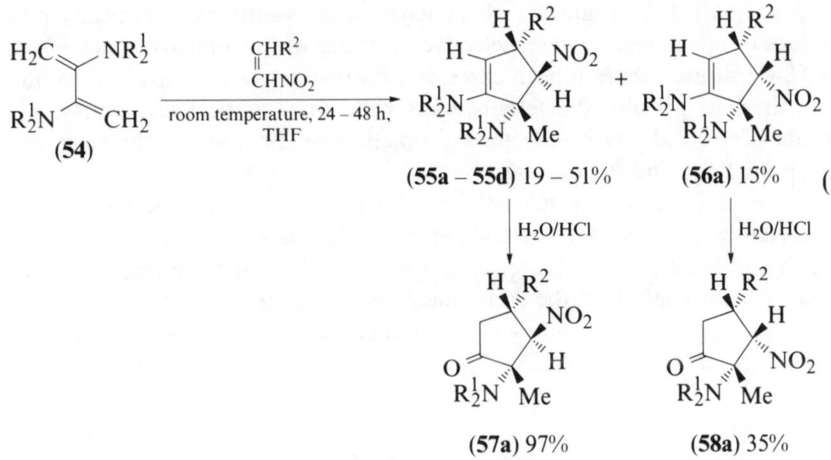

$$(16)$$

(55a – 58a)	(55b)	(55c)	(55d)
NR$_2^1$ = morpholino,	piperidino,	diethylamino,	morpholino
R^2 = Ph,	Ph,	Ph,	H

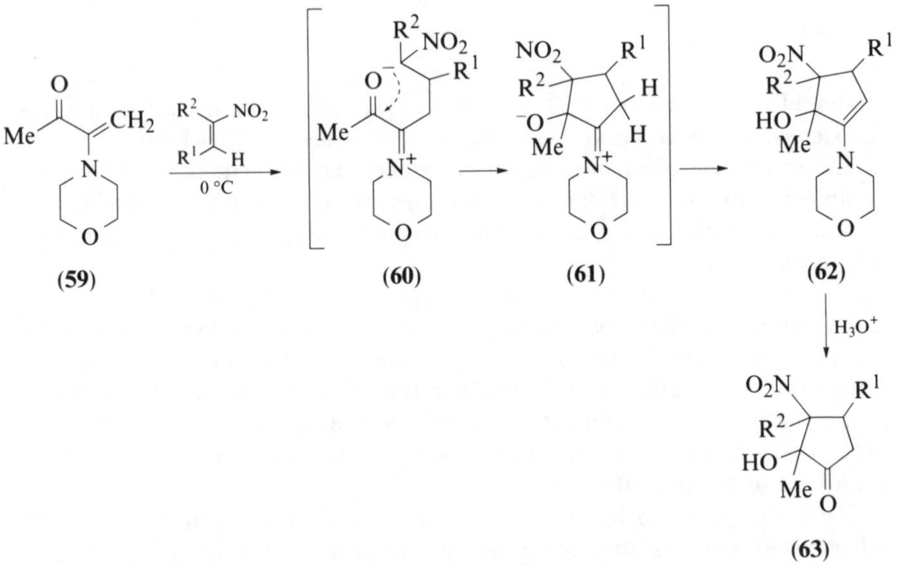

R^1 = Me, Ph; R^2 = H, Me, Ph

$$(17)$$

The reaction of (59) with cyclic nitroalkenes has different characteristic features. For example, the process with nitrocyclohexene (equation 18) results in the formation of the [4 + 2]-heterocyclization product (64). This compound is stable in the pure state but undergoes nucleophilic degradation under the action of solvents. The further transformations are analogous to the reaction discussed above and lead to the indene nitroalcohol (66) [366a].

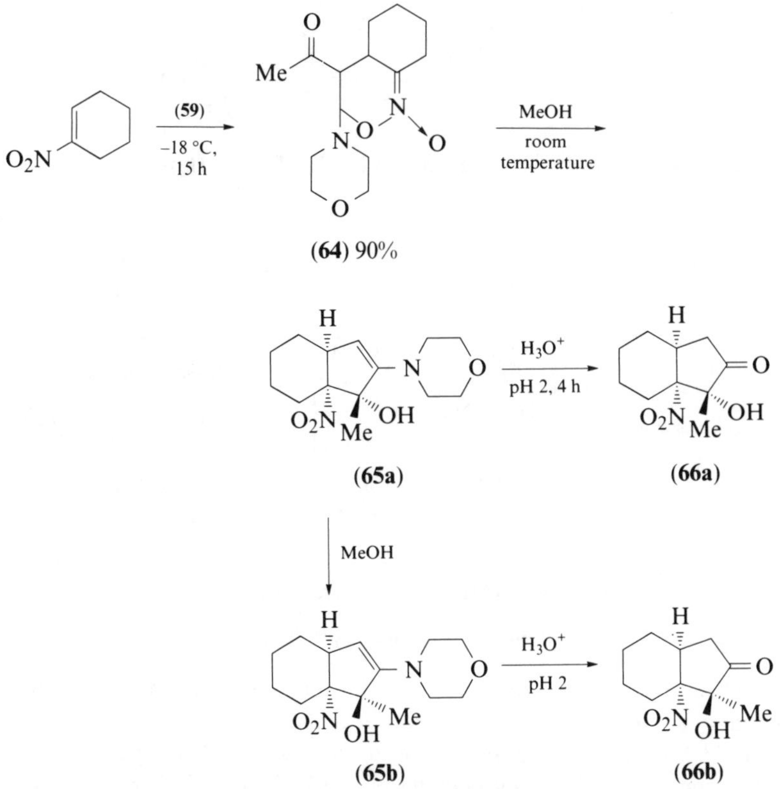

(18)

It is important that the continuous treatment of (65a) with MeOH effects the conversion to (65b). The structure of (66b) has been proven by X-ray analysis.

The same reaction of nitrocyclopentene takes place without the isolation of the 1,2-oxazine N-oxide and produces hexahydropentalene nitroalcohols – analogues of (66).

The reactions between some nitroalkenes and α-morpholinoacrylate methyl ester, which is the analogue of enamine (59), give 1,2-oxazine N-oxides. Their acidic hydrolysis leads to δ-nitro-α-keto esters and α,δ-diketo esters (equation 19) [366b].

Cyclic enaminones (67; equation 20) react with nitroalkenes at 20 °C to form a quantitative amount of the bicyclic oxazine N-oxide (68) [367]. The latter undergoes rearrangement upon heating in acetonitrile to give the hexahydro-1(2H)-pentalenones (69, 70) isolated as a mixture of stereoisomers. The structure

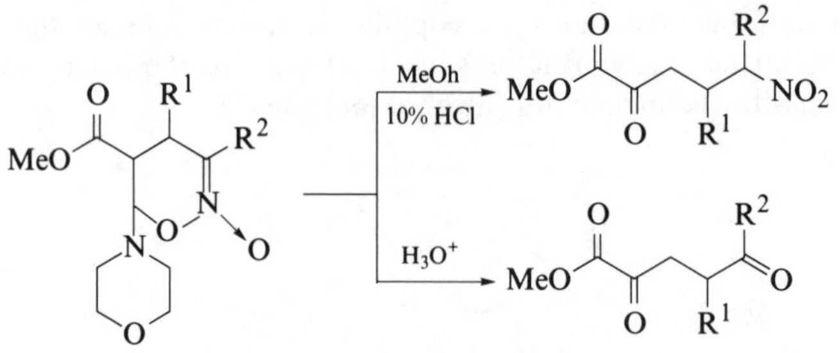

(19)

R^1 = Me, R^2 = H; R^1 = Me, R^2 = Ph; R^1, R^2 = $(CH_2)_4$

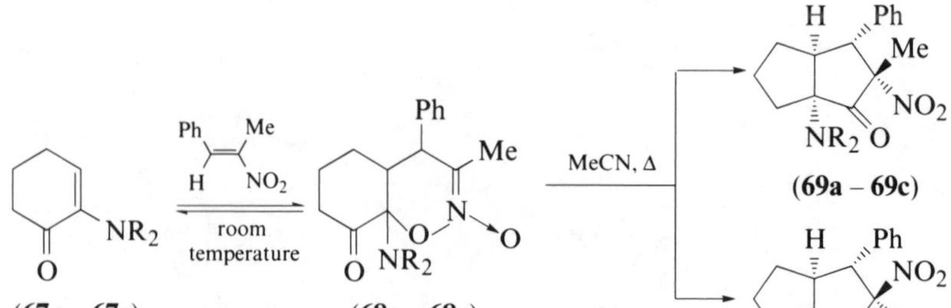

(67a – 67c) (68a – 68c) (69a – 69c) (70a – 70c)

(20)

(67a – 70a) **(67b – 70b)** **(67c – 70c)**

NR_2 = morpholion, piperidino, pyrrolidino

of **(69b)** has been proven by X-ray analysis. The rearrangement of **(68c)** takes place spontaneously in the solid state at room temperature.

Benzoxazine *N*-oxides formed in the reaction between 2-(1-piperidino)-2-cyclohexene-1-one and 1-nitrocyclopentene at −30 °C to 0 °C [368] undergo spontaneous rearrangement to pentalenone derivatives of type **(69)** and **(70)**, even at room temperature.

The rearrangement of the oxazine *N*-oxide obtained from 2-(1-pyrrolidino)-2-cyclohexen-1-one and 1-nitrocyclopentene is used in the synthesis of the pentalenone **(71)**. This compound is the basic structural block for the preparation of triquinane **(72)** [369].

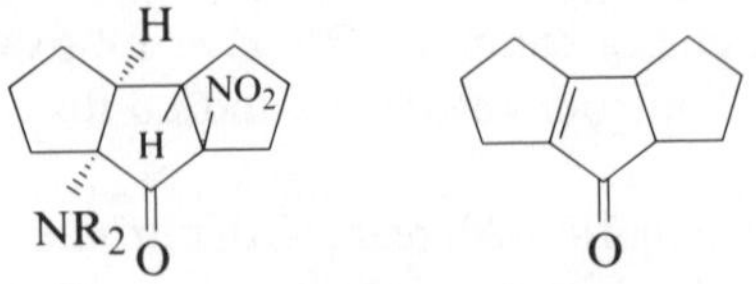

(71) **(72)**

NR_2 = pyrrolidino

The reaction of nitroalkenes with secondary α-ketoenamines [370] is a specific combination of several pathways (equation 21). Various conditions of the process as well as the nature of the radicals induced in the starting compounds direct the reaction towards the formation of three types of product.

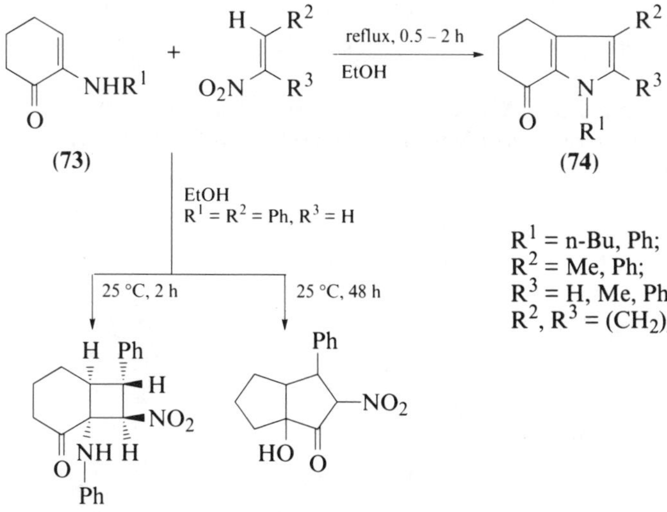

$$R^1 = \text{n-Bu, Ph;} \qquad (21)$$
$$R^2 = \text{Me, Ph;}$$
$$R^3 = \text{H, Me, Ph;}$$
$$R^2, R^3 = (CH_2)_n, n = 3, 4$$

(75) ~ 100%   (76) 10%

Nitroethenes react with butylaminocyclohexenes (73) to produce compounds of type (74). The interaction between nitrostyrene and anilinocyclohexenone leads to the product of [2 + 2] cycloaddition (75) along with the nitrocyclopentanone derivative (76). That a classical process involved is shown by the formation of Michael adducts (77) and 1,2-oxazine N-oxide systems (78).

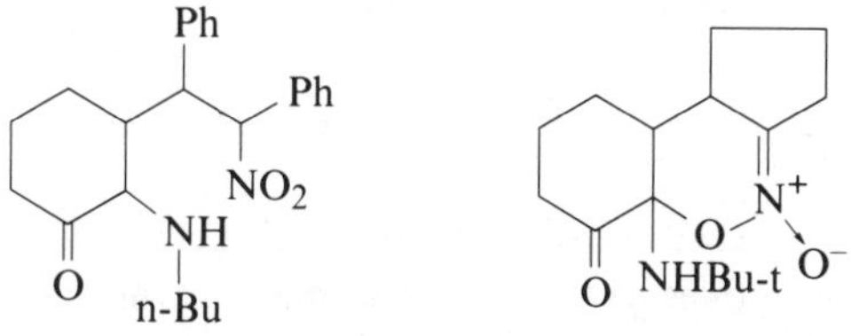

(77)   (78)

By varying the unsaturated nitro compounds and enamines along with the Michael adduct transformations, one can prepare a wide range of heterocyclic systems.

The products (79; equation 22) obtained from ethyl β-nitroacrylate and cyclohexanone morpholinoenamine in 81% yield [371] can be reduced into a 1:1 mixture of (80) and (81). Under harsher conditions the elimination of HNO_2 [372] takes place.

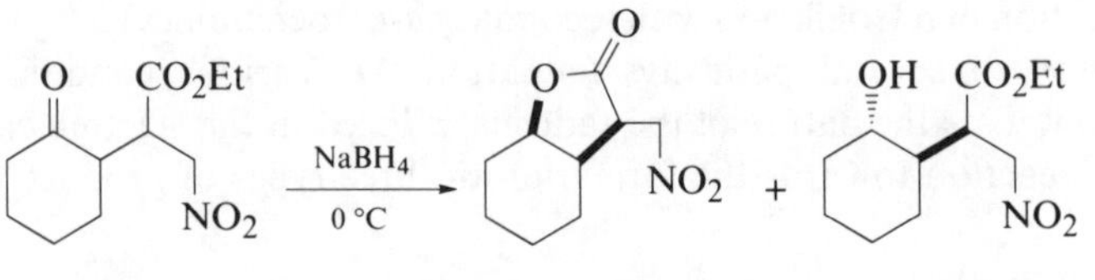

$$(22)$$

(79)   (80)   (81)

3-Aminocrotonic esters (82; equation 23) react with nitroalkenes containing the acetylsugar moiety to give Michael-type adducts (84, 85) isolated as one or two epimers [373].

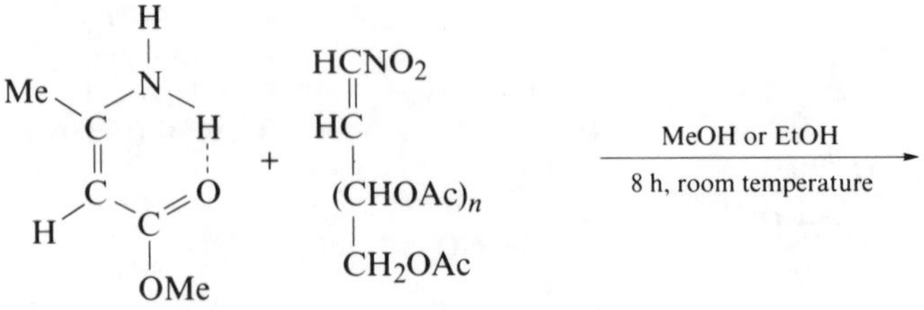

(82) $n = 4$, D-gluco-(83)

$$(23)$$

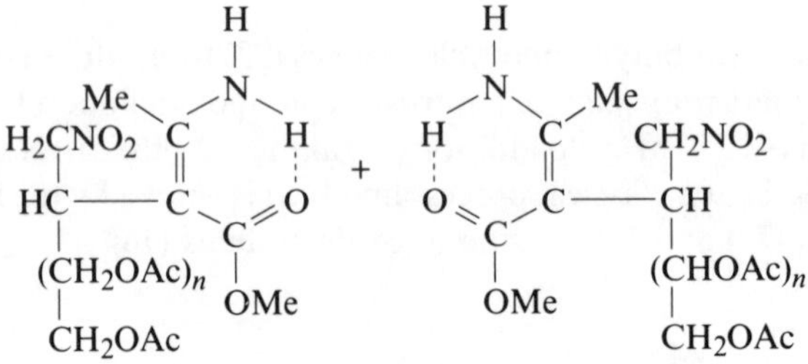

(84) 24%   (85) 55%

Under more severe conditions the primary adducts undergo heterocyclization to form pyrrole derivatives [374]. For example, several substituted pyrroles (88; equation 24) have been synthesized from 3-(alkylamino)crotonic esters (86) and 1-nitropropene or pentacetoxy-1-nitro-1-heptene [374, 375].

Meyer [376] has obtained a number of pyrroles, pyrrolequinolines and pyrrolethiazoles (89–92; equation 25) using different enamines, including those that are heterocyclic.

Some four-membered cyclic nitrones (93; equation 26) have been obtained by Wit et al. [377] and Pennings and Reinhoudt [378] in a smooth reaction between nitroalkenes and ynamines. The other product of the reaction is the 3-nitrocyclobutene (94).

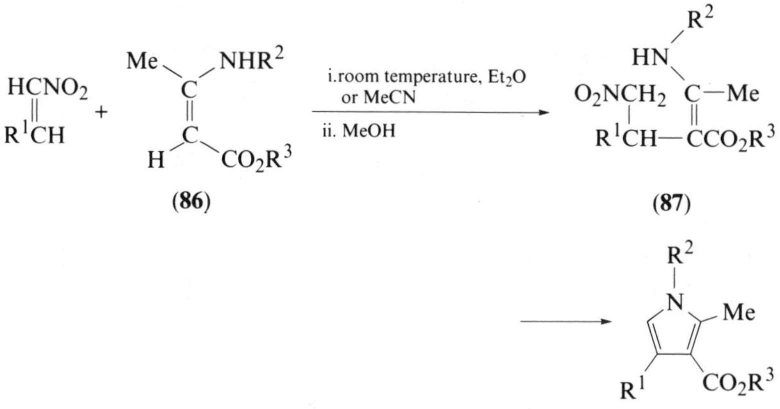

(86) (87) (88) (24)

R^1 = Me, CH(OAc) $(CHOAc)_3$ CH_2OAc; R^2 = H, Me, i-Pr, CH_2Ph

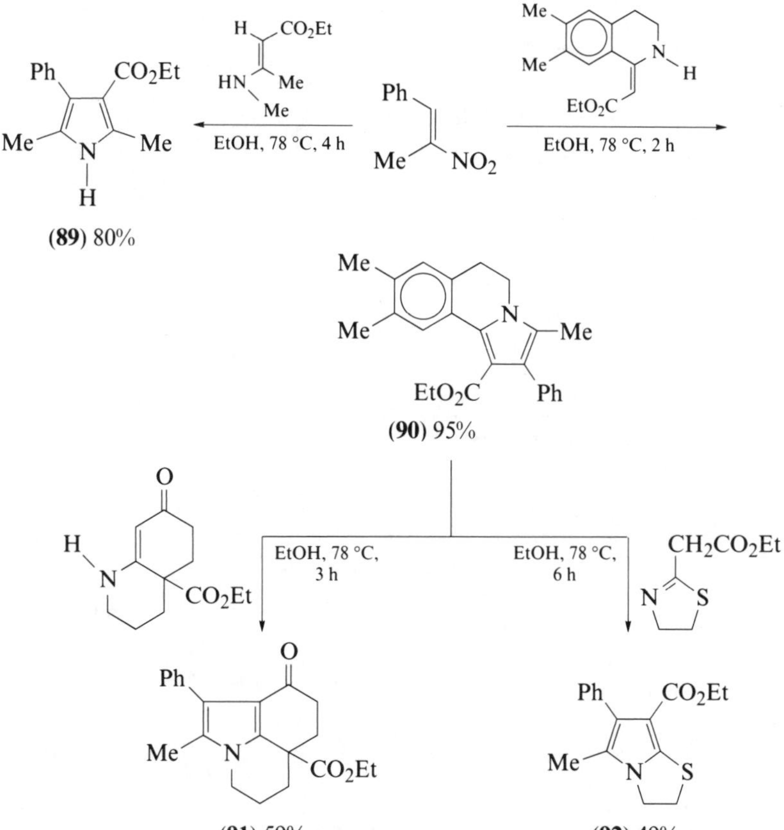

(89) 80%

(90) 95% (25)

(91) 59%

(92) 49%

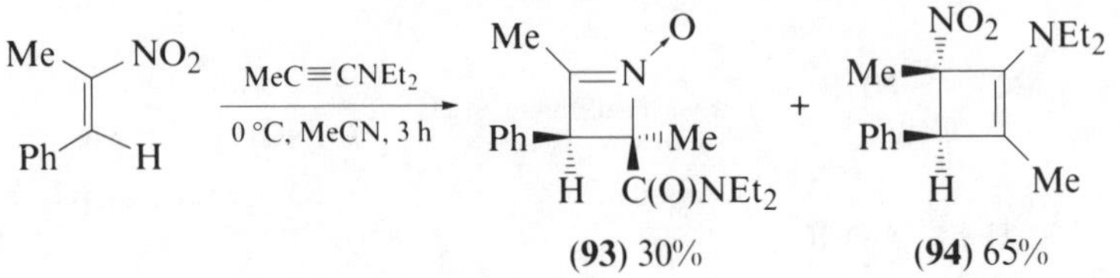

$$\text{(26)}$$

(93) 30%   **(94)** 65%

The reaction of 3-nitrobenzofuran with 1-diethylaminopropyne (equation 27) results in the formation of 3,3α-dihydrobenzofuro[3,2-c]isoxazole **(96)** [379]. The authors suggest that the reaction proceeds via the biradical **(95)**.

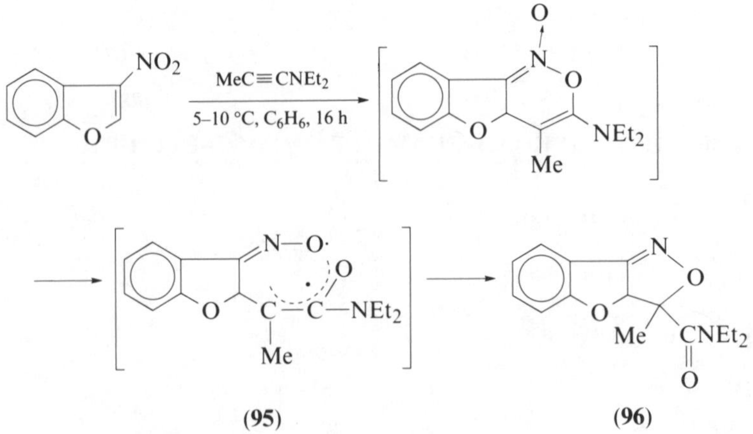

$$\text{(27)}$$

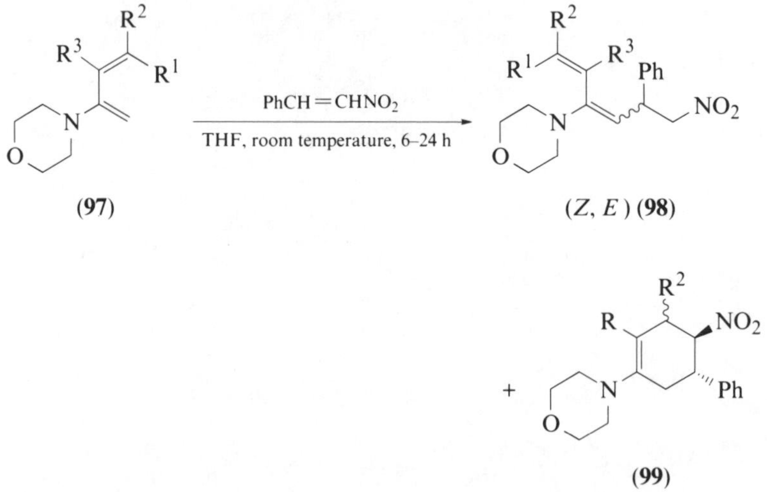

(95)   **(96)**

$$\text{(28)}$$

(97) (Z, E) **(98)**

+

(99)

$R^1 = H, Me, CH_2OMe;$ $R^2 = H, CH_2OMe;$ $R^2, R^3 = (CH_2)_4;$ $R^3 = Me, Et$

2-Morpholino-1,3-butadienes (97; equation 28) react with β-nitrostyrene with the formation of the products of either addition (98) or cyclization (99). The latter are isolated in the form of a diastereoisomeric mixture or the only isomer. The result depends on the nature and localization of the substituents [380].

The coupling of nitroalkenes with en(yn)amines depends not only on the structure of the unsaturated amine but also on the structure of the nitro compound. This has been illustrated by Seebach *et al.* [381] in the double Michael additions of 2-nitro-3-pyvaloyloxypropene (NPP) or its analogues (100) to ketones and keto esters [382]. The process starts in CH_2Cl_2 at $-78\,°C$ and completes at room temperature, or in some cases with heating, to give bicyclo[n.3.1]alkanones ($n = 2$–5) (101).

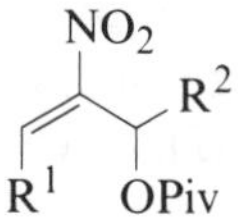

(100) 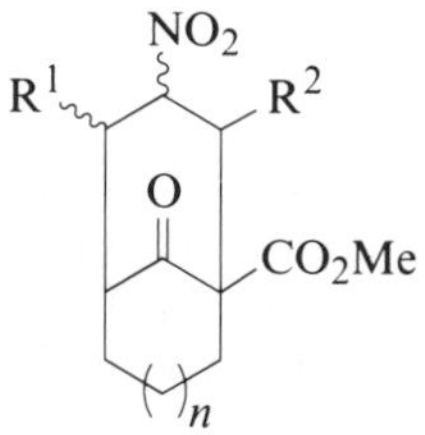   (101)

R^1, $R^2 = H$, $(CH_2)_3$ OPiv;   n = 0, 1, 2;   Piv = COBu-t

Lapirre and Gravel [382] have pointed out the high stereoselectivity of the reaction. Compounds of type (101) have five chiral centres and can be used in the synthesis of the alkaloid lycopodium.

2.7 THE DIELS–ALDER REACTION

The Diels–Alder reaction with nitroalkenes proceeds in the classical manner of diene synthesis. The structural frameworks of nitroalkenes applicable to the reaction either as dienes or dienophiles have been defined, along with the kinetics and stereochemistry of the process and the electronic and steric influences of the substituents in the starting materials. The latest trends in this reaction are in the synthesis of dienes with a reversal of electronic character and the application of nitroalkenes as heterodienes in catalytic syntheses.

2.7.1 NITROALKENES AS TYPICAL DIENOPHILES

The very first reaction of nitroalkenes with 1,3-dienes was carried out in 1938. Some works [130] describe the general procedure of the reaction between unsaturated nitro compounds, i.e. active dienophiles, and various dienes such as aliphatic, heterocyclic and cyclic.

2.7.1.1 Reaction conditions

Nitroethene and other simple nitroalkenes undergo this type of interaction under mild conditions. For example, 2-nitropropene reacts with the salt of (E,E)-4,6-heptadienecarbonic acid in water at 0 °C [383]; cyclopentadiene reacts in ether solution at room temperature [384]. Nitroethene reacts with 3-(p-tolylthio)-2-pyrone under similar conditions to give the bicyclo adduct (1) in 82% yield [385].

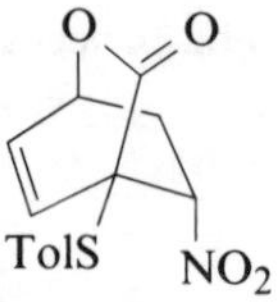

(1)

The high reactivity of nitroethene means that it can be combined under very mild conditions with extremely unstable dienes (equation 1) such as 5-substituted cyclopentadienes formed *in situ* [343, 383, 386, 387]. Such processes give 7-substituted, functionalized norbornenes (2).

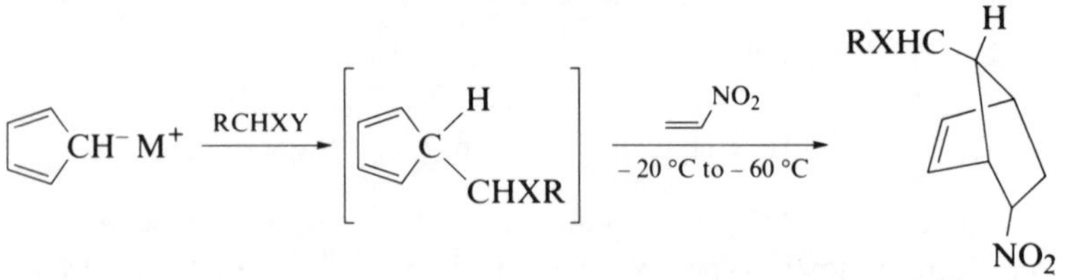

(1)

(2)

R = H, OEt;   X = OMe, OEt, OCH$_2$Ph, SiMe$_3$, —⟨S⟩ ;   Y = Cl, BF$_4$

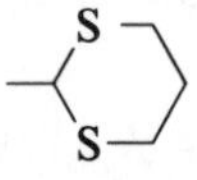

The reactions of dienes with nitroalkenes that contain activating, electron-withdrawing substituents in the β position to the nitro group have been widely investigated [31, 388–391]. Such nitroalkenes as β-nitroacrylic ester and β-cyanonitroethene react with furan and its homologues, which are normally inert in the reactions with unsubstituted nitroalkenes [390, 391]. Mixtures of *endo* and *exo* isomers (3, 4) are obtained by reaction of these dienophiles with

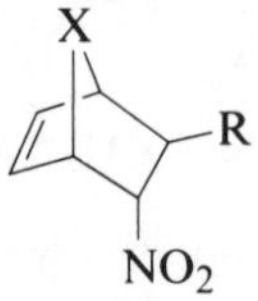

(3) 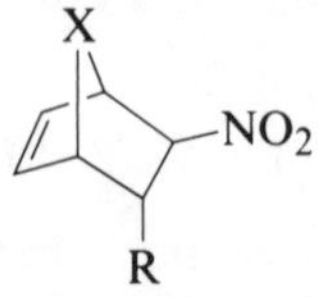   (4)

R = CO$_2$Me, X = CH$_2$, C=O, C=CPh$_2$, CHSiMe$_3$;   R = CN, X = (CH$_2$)$_2$

substituted and unsubstituted cyclopentadienes and cyclohexadienes under mild conditions [392–395].

The dienophilic activity is particularly high for α,β-dinitroethene and β-nitro-enones, which combine with cyclopentadiene to produce adducts at 20 °C within several minutes [142, 396, 397].

Representatives of highly reactive, cyclic β-nitroenones (3-nitrocycloalke-nones) are used in the reaction with some aliphatic dienes [397]. The resulting adducts (5; equation 2) are easily transformed into bicyclic, unsaturated ketones (6, 7) upon treatment with base, 1,5-diazabicyclo[4.3.0]non-5-ene (DBN).

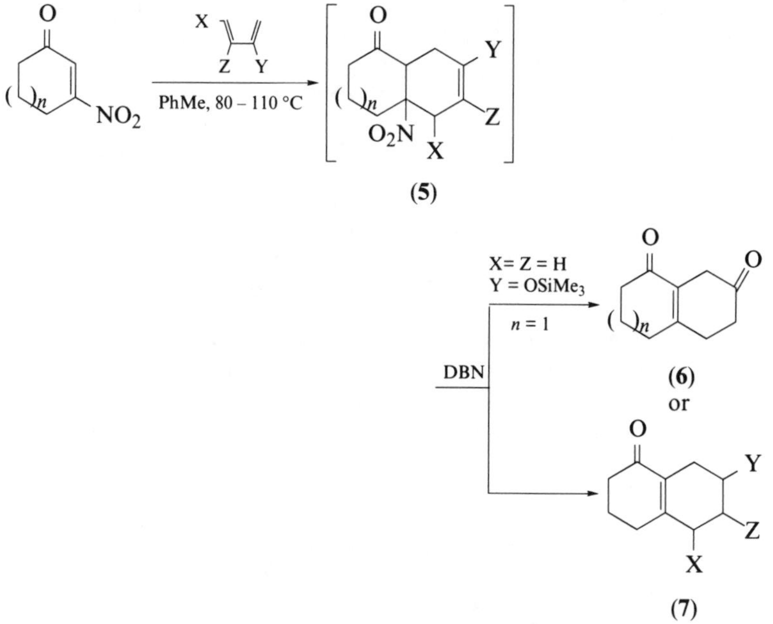

(2)

$n = 1;$ X = H, OMe;   Y = H, Me, OSiMe$_3$;   Z= H, Me

The high dienophilic activity of nitroalkenes allows them to react with both strongly nucleophilic and weakly nucleophilic dienes. The generation of dienes *in situ* from substituted Δ^3-thiolene 1,1-dioxides is a convenient method of their introduction into the diene synthesis. For example, butadiene cyclo-addition to arylnitroalkenes occurs upon heating the latter with Δ^3-thiolene 1,1-dioxide (equation 3) [398]. The adducts formed are key intermediates in the construction of the spiroheterocyclic skeleton of 3-benzazepine.

Nucleophilic dienes with oxygen-containing substituents as well as 1-amino-1,3-alkadienes and 2-amino-1,3-alkadienes are widely used in the Diels–Alder reaction. Adducts of nitroalkene reaction with 1-aminodienes are obtained (12; equation 4) [399].

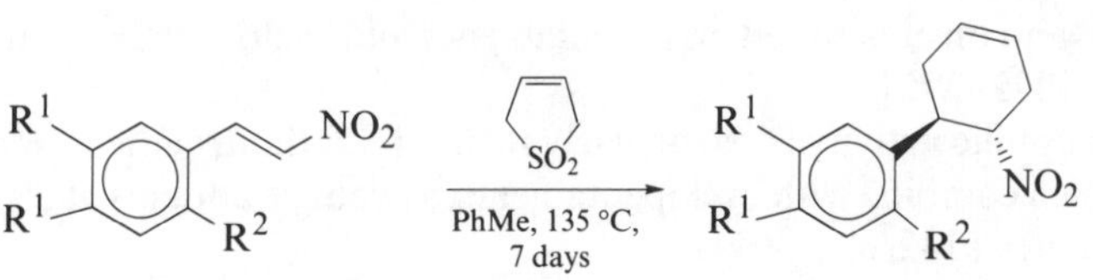

$$
\text{(8), (9)} \xrightarrow[\substack{\text{PhMe, 135 °C,} \\ \text{7 days}}]{\text{SO}_2} \text{(10) 75\%, (11) 80\%} \tag{3}
$$

(8), (10): R^1 = OMe, R^2 = CH$_2$CO$_2$Me
(9), (11): R^1, R^1 = OCH$_2$O, R^2 = NO$_2$

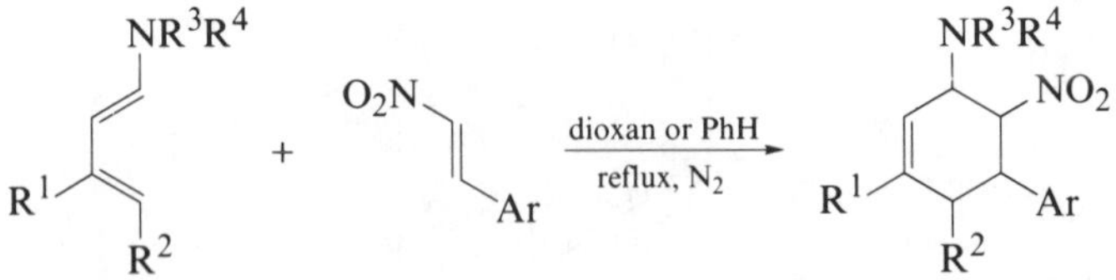

$$
\text{+} \quad \xrightarrow[\text{reflux, N}_2]{\text{dioxan or PhH}} \tag{4}
$$

(12)

R^1 = H, Me; R^2 = H, $\diagup$Me, Me
R^3 = R^4 = Et; R^3, R^4 = (CH$_2$)$_5$

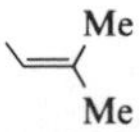

Depending upon the reaction conditions and diene structure, the reaction of dienes with 2-amino-1,3-alkadienes may proceed via one of two pathways, to give either products of bipolar cycloaddition (THF, 6 h; see Section 2.6) or classical adducts of diene synthesis (13; equation 5) (MeOH, 1 h) [380].

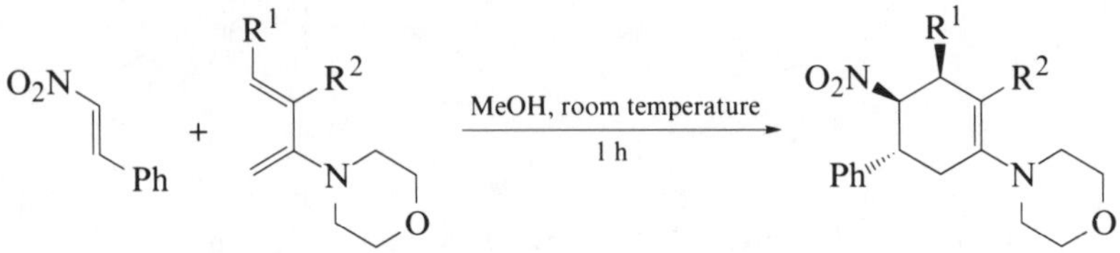

$$
\text{+} \quad \xrightarrow[\text{1 h}]{\text{MeOH, room temperature}} \tag{5}
$$

(13)

R^1 = Me; R^2 = H, CH$_2$OMe; R^1, R^2 = H, Alk

These results can be explained in terms of a stepwise cyclization mechanism.

The cyclic heterodiene 4-methyl-5-propoxyoxazole (14; equation 6) is an interesting starting material for the Diels–Alder reaction. The result of its reaction with nitroethene heterodiene is the unstable adduct (15). Subsequent aromatization gives 2-acyl-3-nitropyrrole (16). Such a method opens the way to nitropyrroles [400].

Another example of the [4 + 2] cycloaddition of heterodienes is the reaction of enaminothiones with nitroalkenes (equation 7) to give nitrothiopyran derivatives (17) [401].

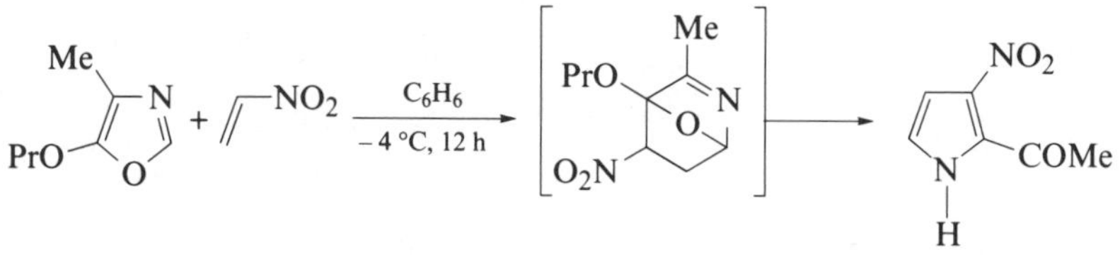

$$(6)$$

(14)   (15)   (16)

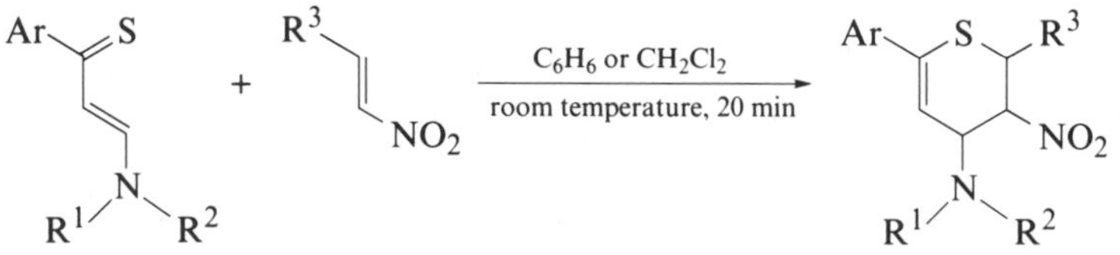

$$(7)$$

(17)

$$R^1 = R^2 = Me; \quad R^1 = H, R^2 = Ar; \quad R^3 = Ph, \textit{p}-MeOC_6H_4, C_4H_4O$$

2.7.1.2 Regioselectivity and stereoselectivity

The nitro group of a dienophile controls the regiochemistry in the formation of cyclohexene derivatives [397, 402, 403]. It is important that the orienting influence of the nitro group is prevalent when there are other electron-withdrawing groups in the β-position of the nitroalkene [397].

The highest regioselectivities are achieved in nitroalkenes reactions with highly nucleophilic dienes (alkoxy, acetoxy, trimethysilyloxy) (18; equation 8) [385, 397, 402, 403, 404a].

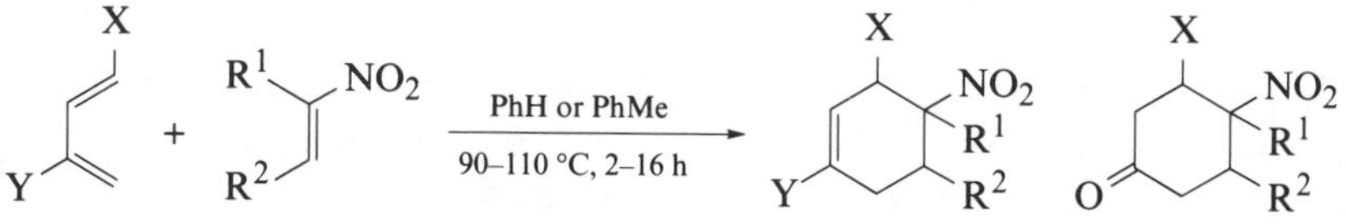

$$(8)$$

(18)   (19)   (20)

X = OAc, Y = H;   X = OMe, Y = H;   X = Ph, Y= OSiMe$_3$
X = OAc, Y = OSiMe$_3$;   X = OMe, Y = OSiMe$_3$;   R^1, R^2= H, Alk

The interactions between 1-dimethylamino-1,3-butadienes (21; equation 9) and β-(E)-nitrostyrene are regioselective and stereoselective and lead to compounds (22) that are characterized by the half-chair conformation [405].

The regioselectivity of the cycloaddition with the less nucleophilic dienes may depend upon the structure of the reagents and the process conditions. This can be illustrated by the reactions of nitroethene with (E)-1,3-pentadiene, isoprene and myrcene (equation 10), which under reflux in benzene produce mixtures of regioisomers (23, 24) consisting mainly of (23). At room temperature, however, isomer (23) is the only product of the process [403].

$$
\text{(21)} \quad \xrightarrow[\text{toluene, reflux, 8 h}]{\overset{\text{Ph}}{\underset{O_2N}{\diagup}}} \quad \text{(22) 26–70\%} \tag{9}
$$

R = Me, Et, Pr, i-Pr, Bu, C$_5$H$_{11}$

$$
\tag{10}
$$

(23)   (24)

R^1 = H, R^2 = Me; R^1 = Me, R^2 = H; R^1 = Me$_2$C=CHCH$_2$CH$_2$, R^2 = H

The stereochemical aspect plays an important part in diene synthesis with unsaturated nitro compounds. Such reactions are characterized by the formation of stereoisomers with *endo* nitro groups, particularly when the process occurs at relatively low temperature [384, 395, 406, 407]. The high stereoselectivity and regioselectivity of the process make it possible to synthesize natural products (alkaloids and antibiotics).

In adducts resulting from acyclic dienes, the configuration of the dienophile is preserved. For example, *trans*-arylnitroalkenes form *trans*-cyclohexenes [405, 408]. The stereospecific features of nitroalkene cycloadditions have been demonstrated in a study of the reactions of the (E) and (Z) isomers of β-nitrocrotonic ester with 1,3-butadienes.

The products of the reactions between (E)-β-nitroacrylic or nitrocrotonic esters (**25, 26**; equation 11) and butadienes, i.e. 1-acetoxybutadienes and 1,4-diacetoxy-butadienes, are stereoisomers with *trans*-oriented nitro and ethoxycarbonyl groups. Except for the interaction between (E)-(**26**) and 1-acetoxybutadiene, only one isomer is formed in all cases. The (Z) isomer of nitrocrotonic ester produces an additional stereoisomer with *cis*-oriented nitro and ethoxycarbonyl groups [409]. The formation of stereoisomers (**28a**) and (**28b**) from the (Z) isomer results from the partial isomerization of (Z)-(**26**) to (E)-(**26**) upon heating. The half-chair conformation is prevalent for all isomers.

Owing to the strongly fixed norbornene skeleton in the adducts of nitroalkene reactions with cyclopentadiene (nitrobicyclo[2.2.1]heptenes), it is possible to obtain the resulting compounds with definite regioorientation and stereo-orientation of the substituents [384, 410].

A stereochemical study of (Z)-3-nitro-2-alkenecarbonic and (E)-3-nitro-2-alkenecarbonic acids (crotonic and cinnamic) (**26, 29**; equation 12) has shown that the (Z) isomers form [2+4] bis-*endo* adducts (**30a, 31a**) with cyclopentadiene

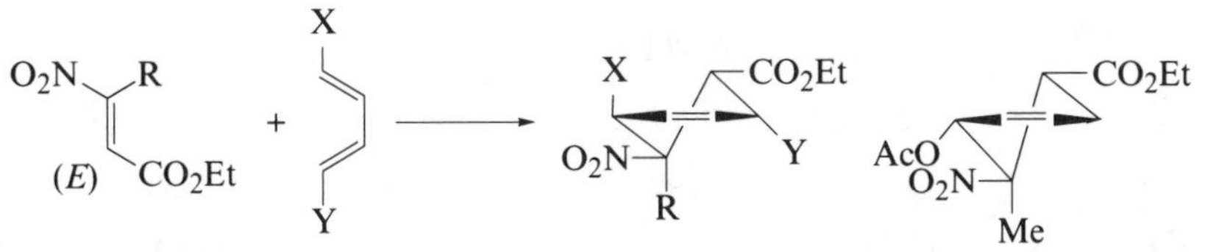

(25), **(26)** **(27a)**, **(28a)** **(28b)**

R = H (**25**, **27**), Me (**26**, **28**); X = Y = H; X = OAc, Y = H; X = Y = OAc (11)

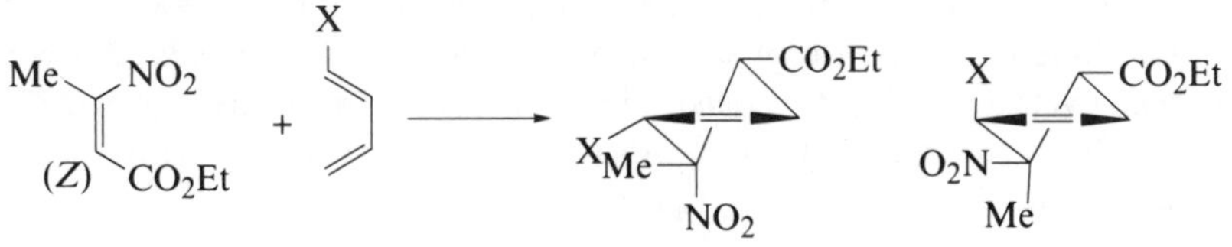

(**26**) X = H, OAc (**28c**, **28d**) (**28e**)

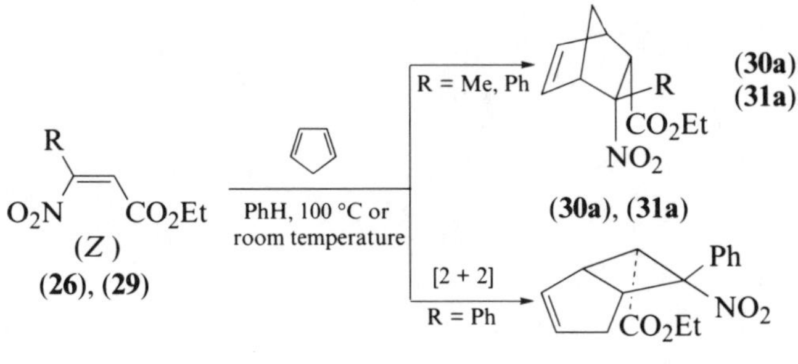

(**30a**): R = Me, 47%
(**31a**): R = Ph, 76%

(**26**), (**29**) (**30a**), (**31a**)

(**32**) 31% (12)

(**26**) (**30b**) 64% (**30c**) 15%

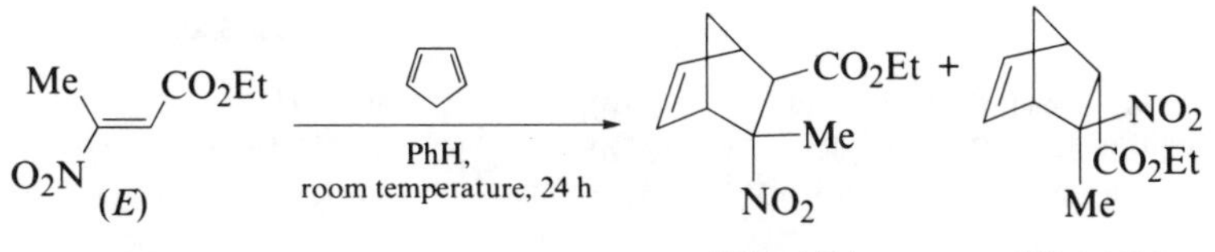

and the (*E*) isomers tend to produce a mixture of *endo* and *exo* isomers (**30b**, **30c**) [407]. In the case of (*Z*)-nitrocinnamic acid the product of [2+2] cycloaddition (**32**) was also isolated.

Endo isomerization and *exo* isomerization have been investigated for the adducts of cyclopentadiene combination with 3-nitroacrylonitrile [395]. A 9:1 mixture of 3-*endo*-nitro-2-*exo*-carbonitrilebicyclo[2.2.1]hept-5-ene (**33a**; equation 13) and 3-*exo*-nitro-2-*endo*-carbonitrilebicyclo[2.2.1]hept-5-ene (**33b**) under the action of an alkali followed by acidification transforms into a mixture of three isomers (**33a**–**33c**) in the ratio 46:22:32. The third isomer (**33c**) is 3-*endo*-nitro-2-*endo*-carbonitrilebicyclo[2.2.1]hept-5-ene [395].

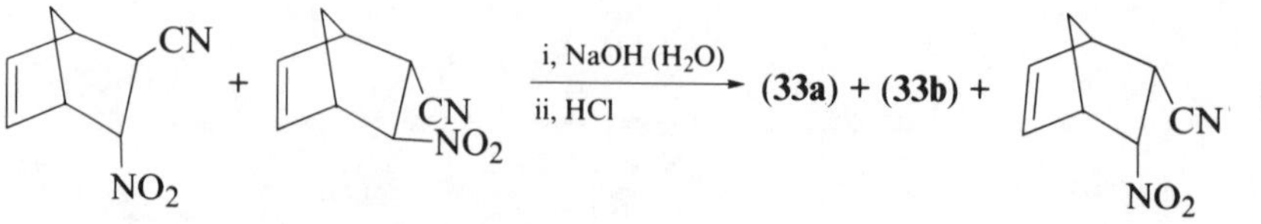

$$(33a) + (33b) + \quad (13)$$

(33a)   **(33b)**   **(33c)**

An original variation on the Diels–Alder reaction is the intramolecular cyclization of 1-nitrodeca-1,6,8-trienes [411] synthesized from isomeric non-adienols. The cyclization of the (E)-nitrotriene (**34**; equation 14) at 80 °C gives either (**35**) alone or a mixture of stereoisomeric bicyclic adducts (**35, 36**) with mostly *endo* cyclization (9:1 *trans/cis*). The same process with the isomeric (Z)-nitrotriene (**34**) at room temperature leads to a 1:1 mixture of the corresponding stereoisomers (**37, 38**) [411].

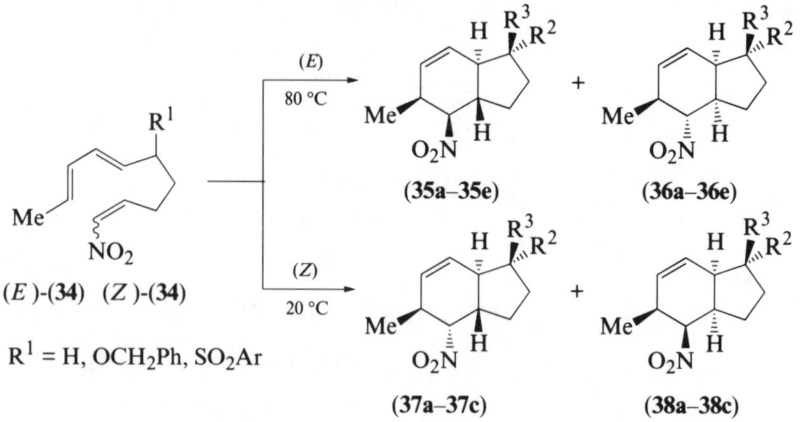

$$(14)$$

(35a), (36a): $R^1 = R^3 = H$ (37a), (38a): $R^2 = R^3 = H$
(35b, 35d), (36b, 36d): $R^1 = R^2, R^3 = H$ (37b), (38b): $R^2 = OCH_2Ph, R^3 = H$
(35c, 35e), (36c, 36e): $R^2 = H, R^1 = R^3$ (37c), (38c): $R^2 = H, R^3 = OCH_2Ph$

The cycloaddition reactions of chiral nitroalkenes are used in the synthesis of chiral cyclohexene derivatives. For example, such processes have been studied [404b, 409, 412] for nitroalkene sugars.

The Diels–Alder reactions of cyclopentadiene with *manno*-nitroalkenes lead to four stereoisomers of substituted nitrocyclohexenes with prevailing formation of the *endo* isomer. Similar reactions of *gluco*-nitroalkenes and *galacto*-nitro-alkenes are described in the literature [404b].

Adducts from the reactions of *manno*-sugar or *galacto*-sugar nitroalkenes (**39a, 39b**; equation 15) with 2,3-dimethyl-1,3-butadiene are the starting materials in the enantioselective syntheses of nitrocyclohexene aldehydes [412, 413]. These products are isolated as a diastereoisomeric mixture in good yield.

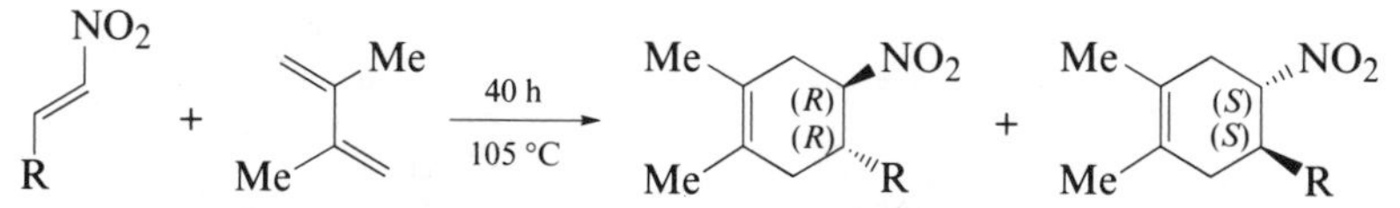

(15)

(39a, 39b) **(40a, 40b)** **(41a, 41b)**

R = D-*manno*-$(CHOAc)_4CH_2OAc$, **(39a)**–**(41a)**
R = D-*galacto*-$(CHOAc)_4CH_2OAc$, **(39b)**–**(41b)**

Hydrolysis of the adducts followed by degradative oxidation of the sugar side-chains leads to enantiomerically pure *trans*-cyclohexene or *cis*-cyclohexene nitroaldehydes (equation 16).

(40 a)

(42) **(43)**

R = D-*manno*-$(CHOH)_4CH_2OH$

(16)

(42)

(44)

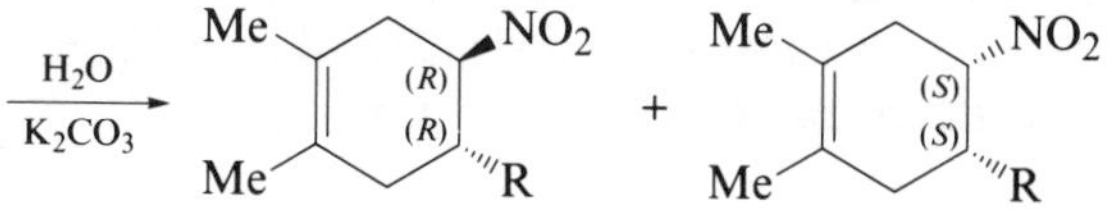

Optically pure cyclohexene derivatives are isolated from the cycloaddition reactions of D-*galacto*-nitroethene **(39b)** with 1-trimethylsilyloxy(acetoxy)-butadienes **(45)**, which proceed with complete regiofacial and diastereofacial selectivity [413, 414]. The reaction with 1-trimethylsilyloxy-1,3-butadiene leads to a 1:1 mixture of adducts **(46a, 47a)** in a total yield of 53% (equation 17). Similar coupling with 1-acetoxy-1,3-butadiene results in the formation of only one isomer **(47b)** in 75% yield.

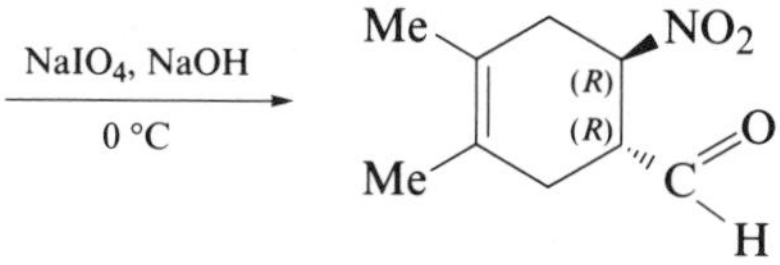

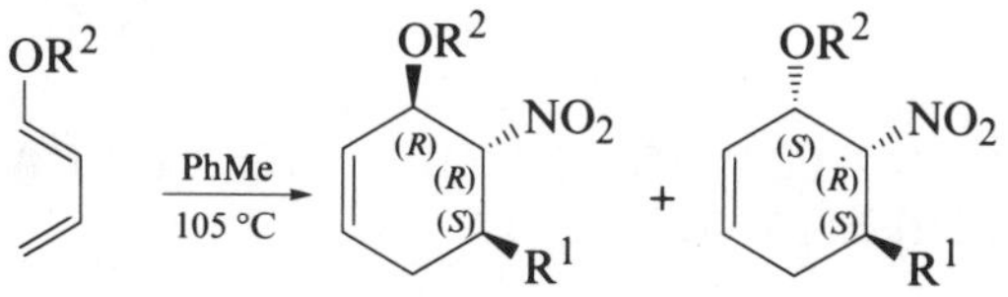

(39b) +

(17)

(45a, 45b) **(46a, 46b)** **(47a, 47b)**

R^1 = D-*galacto*-$(CHOH)_4CH_2OH$; $R^2 = SiMe_3$ **(46a, 47a)**, Ac **(46b, 47b)**

The cycloaddition of 3-nitrovinyl 1-δ-valerolactone **(48)** to substituted butadiene (**49**; equation 18) is the key step in the synthesis of the alkaloid (−)-physostiymine [415].

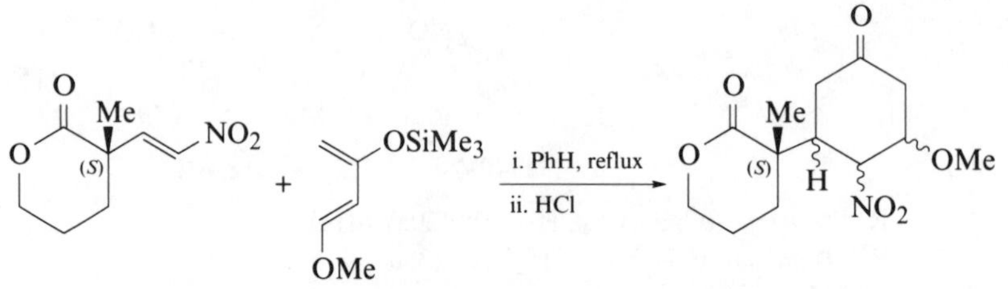

(48)   (49)   (50)

In contrast with this synthesis, the interactions between arylnitroethenes and such chiral reagents as the heteryl-substituted (S)-2-pyrrolidino-1,3-butadiene (51) are stereoselectively controlled and lead to adducts (52, 53; equation 19) with high diastereoisomeric and enantiomeric purity [416].

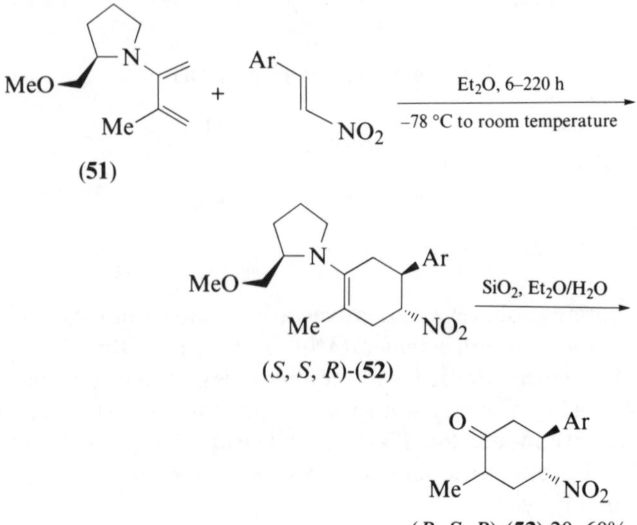

2.7.2 NITRODIENES AND NITROENYNES THAT ACT AS DIENOPHILES AND DIENES

Nitrodienes and nitroenynes are active dienophiles with low reactivity as dienes. 1-Nitro-1,3-butadiene forms the adduct with maleic anhydride in 4% yield after prolonged heating [417], while 3-nitro-1,3-pentadiene is passive in such a reaction [304].

1-Nitrodiene system of the 2-(β-nitrovinyl) furan (54; equation 20) appears to have sufficient reactivity in the reactions with alkyl acrylates and acrylonitrile. In the former case HNO_2 elimination takes place, and in the latter case dehydrogenation occurs. The process ends with aromatization and hence the formation of the corresponding monosubstituted and disubstituted benzofurans (**55, 56**) [418].

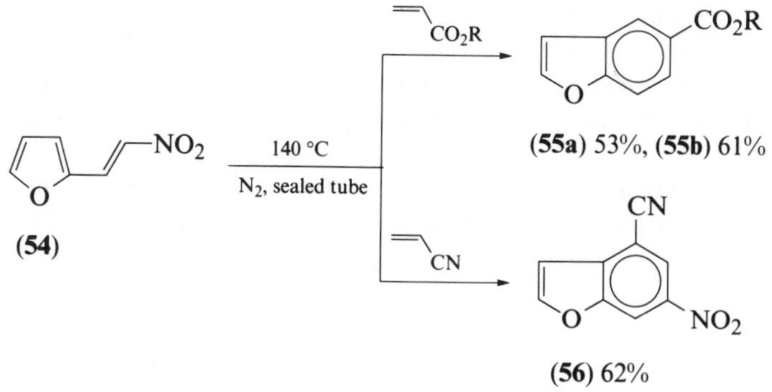

(20)

R = Me (**55a**), Et (**55b**)

The unsubstituted nitrovinyl fragment of 2-nitrodienes is an active dienophile and forms the corresponding products with cyclopentadiene, 1,3-butadiene and 2,3-dimethyl-1,3-butadiene. 2-Nitro-1,3-alkadienes also form the adducts (**57, 58**; equation 21) with maleic anhydride, nitroethene, acrolein and acrylonitrile [304]. Moreover, the combination of the properties of both diene and dienophile in 2-nitro-1,3-butadienes results in their spontaneous dimerization (**59**), even under normal conditions.

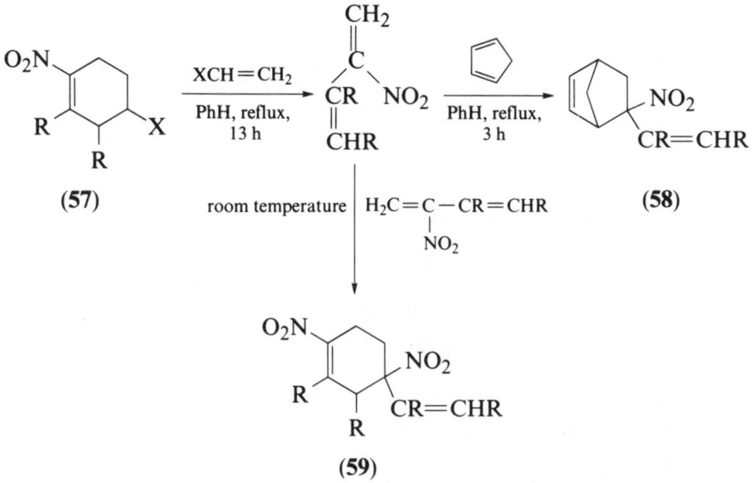

(21)

X = NO₂, CN, CHO;   R = H, Me

Owing to their ready dimerization, 2-nitro-1,3-alkadienes should be generated *in situ* in diene syntheses using such precursors as nitrothiolene 1,1-dioxides or β-nitro-β-vinyl ethyl ethers.

The thermal desulfonylation of 3-methyl-4-nitro-3-thiolene 1,1-dioxide in the presence of typical dienophiles, i.e. maleic anhydride and *p*-benzoquinone, results in the formation of 2-nitro-3-methyl-1,3-butadiene which then leads to the corresponding products of diene synthesis [319] (see Section 3.5). Use of the simplest 4-nitro-3-thiolene 1,1-dioxide in this reaction results in the well-known dimer (59) [419].

Convenient synthetic equivalents for 2-nitro-1,3-dienes are β-nitro-β-vinyl ethyl ethers, e.g. β-nitro-β-(1-cyclopentenyl)-α-benzoyloxyethane (60) and β-nitro-β-(2-propenyl)-α-benzoyloxyethane (61) [420].

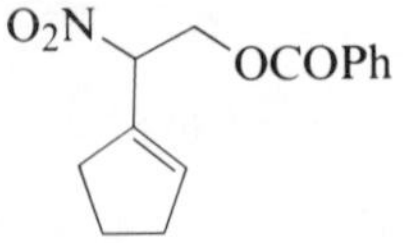

(60) 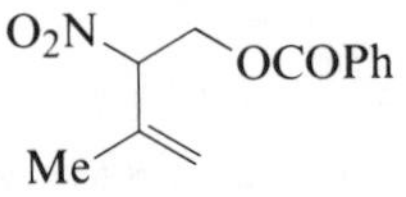   (61)

A 76:24 mixture of *endo* and *exo* adducts (63a, 63b) is formed upon heating the ether (60) and methyl acrylate (equation 22) in the presence of NaOAc [420]. Under these conditions 2-nitroisoprene, obtained from the corresponding precursor (61), gives only the *endo* isomer.

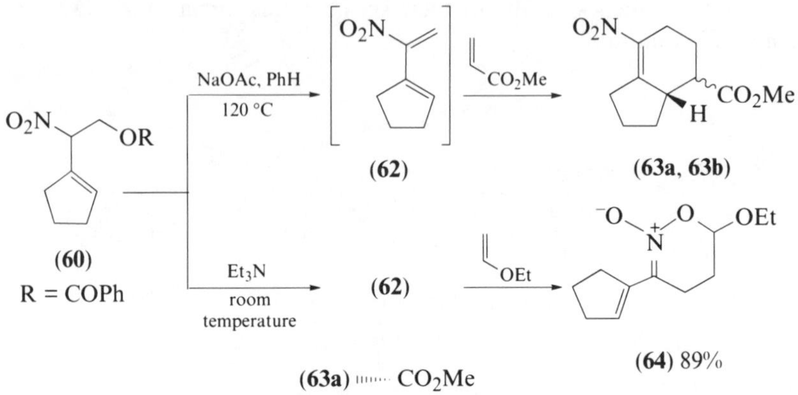

(22)

The interaction between (60) and ethyl vinyl ether, which is used also as a solvent, gives rise to another type of adduct, i.e. the cyclic nitrone ester (64), even under normal conditions.

These dienes interact with cyclopentadiene as active dienophiles. For example, the precursor of (60) reacts with cyclopentadiene (equation 23) in benzene at elevated temperature to give three products: the *endo* (65a) and *exo* (65b) adducts 40% and 12.5% yield respectively and the tricyclic adduct (66) 47.5% yield. The formation of (66) is the result of cyclopentadiene acting as a dienophile in the reaction. Upon heating in benzene, adducts (65a) and (65b) are transformed into (66) completely [420].

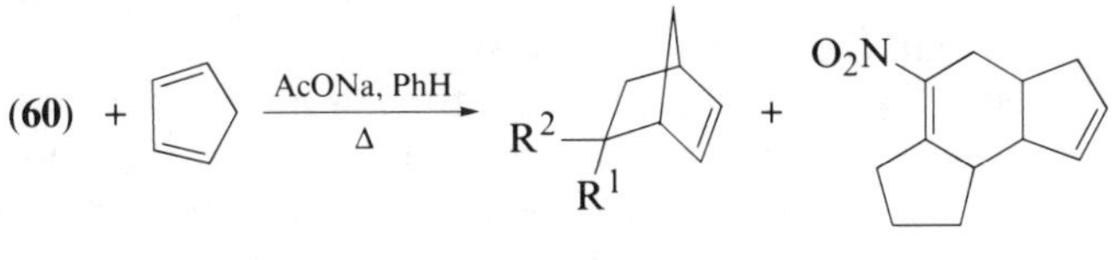

(65a): $R^1 = NO_2$, $R^2 = $

(65b): $R^1 = $, $R^2 = NO_2$

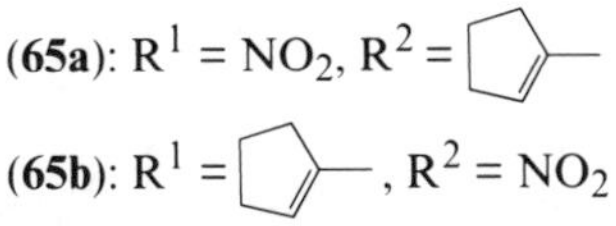

(65a) + (65b) $\xrightarrow{\text{PhH, reflux}}$ (66)

The reaction of the 2-nitroisoprene precursor (61) proceeds in a similar way. The ratio of the corresponding products is 44.8:19.2:36.

1-Nitro-1,3-enynes (67; equation 24) are active dienophiles as well. They add to 1,3-butadienes at the nitrovinyl fragment to form compounds of type (68) [211].

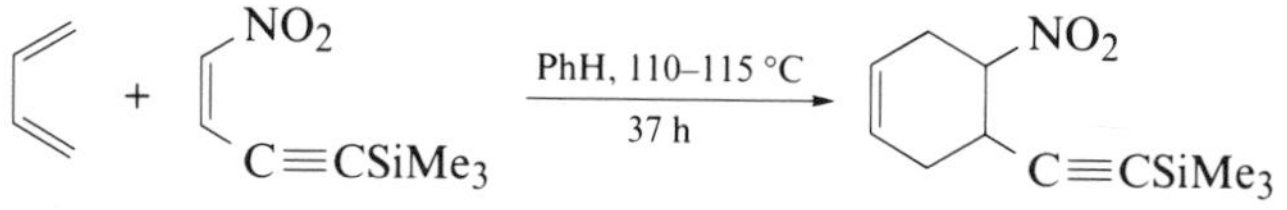

(67)   (68) 81%

1,4-Dinitrodienes (69; equation 25) do not enter the cycloaddition reaction with electrophiles as dienes; they are more reactive as dienophiles than mononi-troalkenes and produce mono-adducts (70) and bis-adducts (71) with dienes [311]. The smoothest process takes place with unsubstituted 1,4-dinitro-1,3-butadiene.

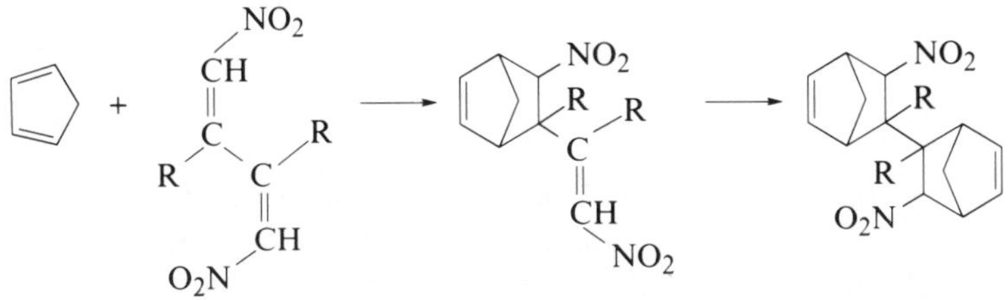

(69)   (70)   (71)

R = H, Me

Owing their high electrophilicity, 1,4-dinitro-1,3-dienes are active towards nucleophilic dienophiles. These compounds participate in the reaction of diene synthesis with a reversal of electronic character, in which the diene is the acceptor and the dienophile acts as the donor. This is true for reactions with styrene and its *para*-substituted derivatives (with substituents such as Hal, Me and OMe), indene, and acenaphthylene [421]. Such a diene synthesis requires more severe conditions than the ordinary diene synthesis.

The most reactive 1,4-dinitrodiene (1,4-dinitro-1,3-butadiene) forms (72) when heated in an excess of styrene or with styrene in a solvent (equation 26). The primary product (72) undergoes spontaneous denitration to give 1-nitrocyclo-hexadiene derivatives (73). Depending on the reaction conditions, further transformations may occur to form either (74), (75) or dimers.

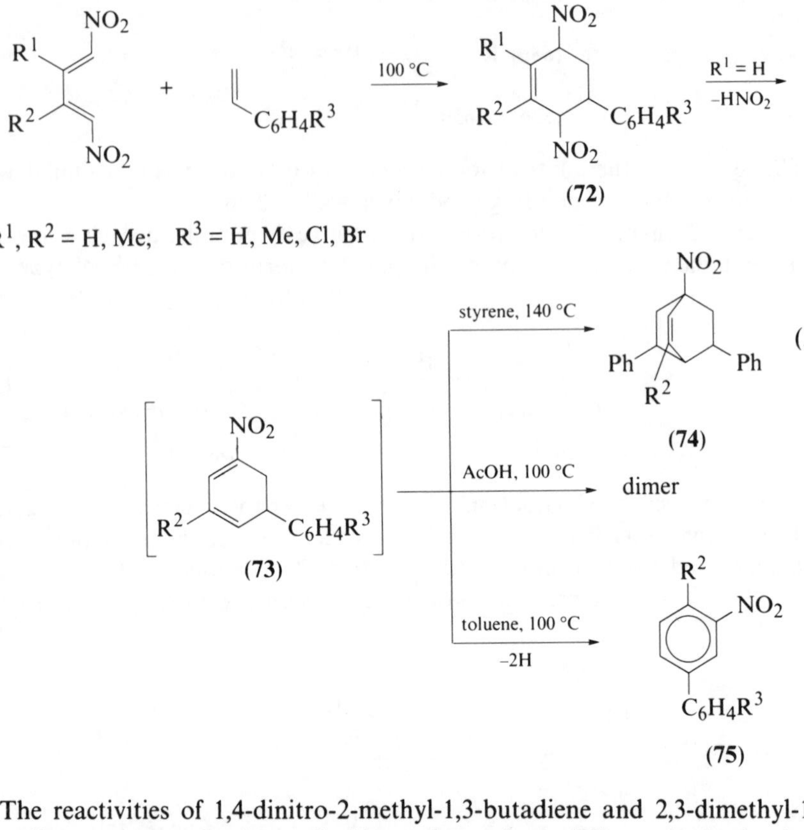

$$R^1, R^2 = H, Me; \quad R^3 = H, Me, Cl, Br$$

(26)

The reactivities of 1,4-dinitro-2-methyl-1,3-butadiene and 2,3-dimethyl-1,3-butadiene are somewhat lower. Stable products of type (72) are obtained only for the latter. 1,4-Dinitro-1,3-cyclohexadiene, which contains the *cis*-fixed structure, forms the primary products with styrene under more mild conditions and with higher yields in comparison with aliphatic dinitrodienes.

Kinetic investigations of the reactions of 1,4-dinitro-1,3-cyclohexadiene with substituted styrenes show quantitatively an inversion of properties for the reagents [421].

Similar activity in reactions with donor dienophiles is exhibited by 9,10-dinitroanthracene, which is extremely passive (as are other *meso*-substituted anthracenes) in the reaction with maleic anhydride [421].

Reactions of the type 'donor dienophile, acceptor diene' were carried out for a series of 1,4-dinitro-1,3-butadienes without product isolation. These dinitro-butadienes were prepared from 2,5-dinitro-3-thiolene 1,1-dioxides *in situ* and reacted with styrene, which was also used as the solvent. The reaction takes place under reflux and, depending on the starting materials and reaction time, results in either the products of addition, of denitration of adducts, or the products of further cycloaddition to the intermediate cyclohexadiene (see Section 3.5) [422].

2.7.3 NITROCYCLOHEXENE ADDUCTS USED AS SYNTHONES FOR THE SYNTHESIS OF BIOLOGICALLY ACTIVE COMPOUNDS

Mononitroalkenes as dienophiles and nitrodienes as dienophiles and dienes play an important part in the synthesis of various blocks for natural and biologically active compounds. For example, adducts (76) from the reaction between β-nitro-acrylic ester and furan are the starting materials for the preparation of 2,5-anhydroallose [390, 404c] and 2,5-anhydroglucose derivatives [423].

(76a): X = NO$_2$, Y = CO$_2$Me
(76b): X = CO$_2$Me, Y = NO$_2$

(76)

The resulting chiral cyclohexene derivatives are useful intermediates for the preparation of biological structures [412]. Nitroalkenyl-D-mannose (39a; equation 15) reacts with 2,3-dimethyl-1,3-butadiene to form a mixture of stereoisomers (40a, 41a) [409] with the configurations (4R,5R) for (40a) and (4S,5S) for (41a).

Reactions between nitroalkene sugars and dienes give adducts which can be used in the synthesis of C-nucleosides, racemic showdomycin and related compounds [423].

The styrene hydroxy derivative 3,4-methylenedioxy-β-nitrostyrene (77; equation 27) combines with substituted dienes to form compounds of type (78), which are used as starting materials in the synthesis of alkaloids [424–426].

Unusual polycyclic dienes, e.g. morphinan-6,8-dienes, are used in the produc-tion of diene adducts that are the synthetic blocks for drugs of the opium alkaloid series [427, 428].

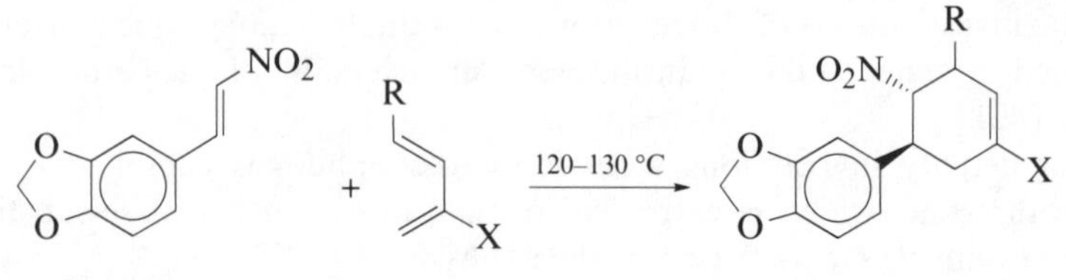

$$X = H, OEt; \quad R = H, OMe, Ac$$

The diene synthesis is the synthetic method for the introduction of the nitroalkyl fragment into alkaloid molecules of the morphine type. The reaction of 6-demethoxy-N-formyl-N-northebaine with nitroethene leads to a mixture of stereoisomers containing mostly the *endo*-product, i.e. 8α-nitro-6α,14α-etheno-isomorphinane (**79**), in 50% yield [428].

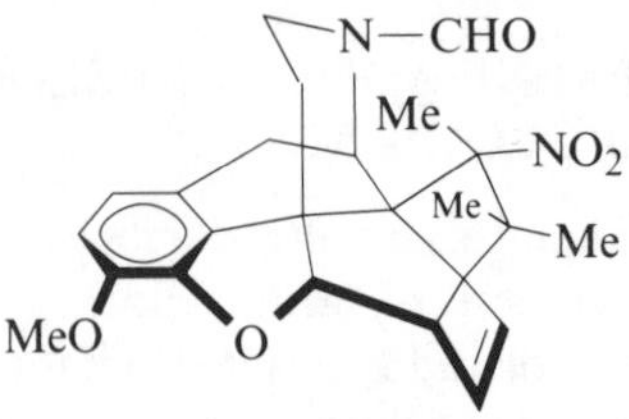

(**79**)

Nitrocyclohexane adducts prepared via the Diels–Alder reaction are able to undergo further chemical transformations such as reductive denitration, elimination of HNO$_2$ and reduction. The synthetic potential of the reaction is therefore extended and the way is opened to substituted, six-membered carbocycles with

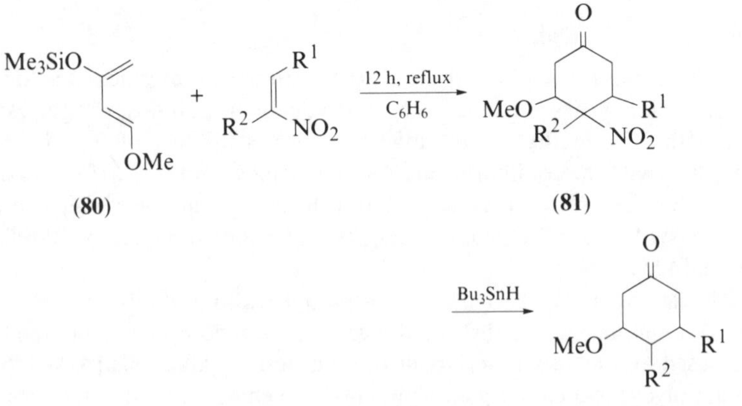

$$R^1 = Et, R^2 = Me; \quad R^1 = Me, R^2 = Et$$

fixed locations of substituents that are hardly available by other methods. The most widely applied reducing agent for such reactions is tributyltin hydride. The active diene system frequently used for introduction of the keto group is 1-methoxy-3-trimethylsilyloxy-1,3-butadiene (80; equation 28). The interaction of (80) with 3-nitro-2-pentene or 2-nitro-2-pentene proceeds regioselectively and leads to the adduct (81) which can be transformed further into 3-methyl-4-ethyl-5-methoxycyclohexanone or 4-methyl-3-ethyl-5-methoxycyclohexanone (82) [149, 402].

Equation (29) presents the synthetic approach, via intermediates (83a) and (84a), to the isomeric cyclohexenes (85) and (86). Two cyclohexanone isomers (87, 88) are obtained from compounds (83b) and (84b) containing the terimethyl-silyloxy group [403]. This type of reaction has been extended to other isomeric nitroalkenes [149, 403].

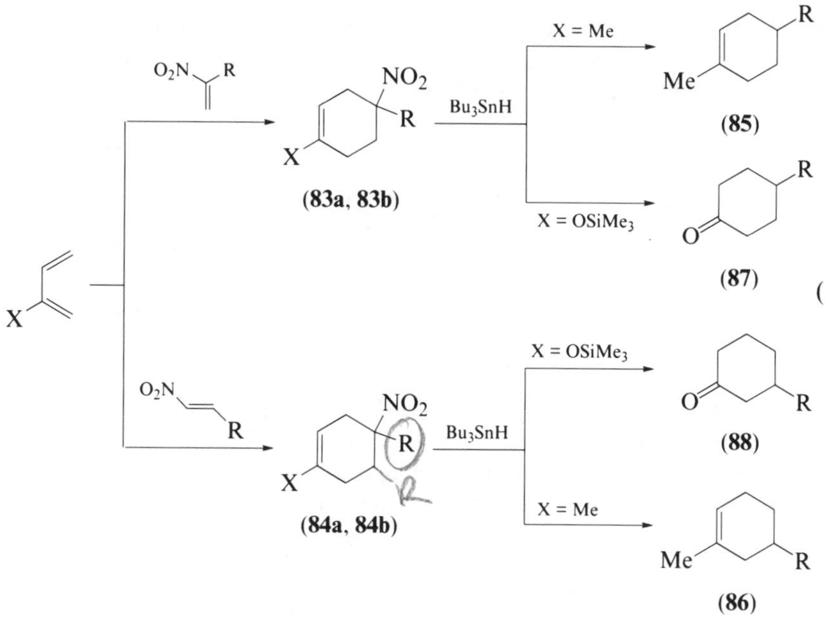

$$X = Me(83a, 84a),\ OSiMe_3\ (83b, 84b);\quad R = Et,\ CH_2Ph$$

The activating influence of the nitro group in a corresponding adduct can be used in the Michael or aldol condensation reaction. For example, the products of nitroethene addition to 2-alkyldienes (89; equation 30) may undergo nitro group substitution by alkyl groups in the process of the addition to vinyl sulfoxide followed by reductive denitration [429].

Under the action of base, diene adducts (substituted nitrocyclohexenes) eliminate HNO_2 to form substituted 1,4-cyclohexadienes [404a]. Such a pro-

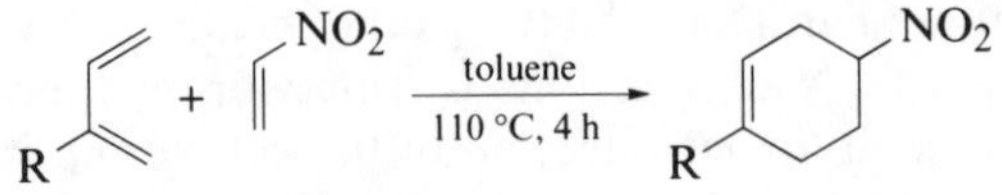

(89)

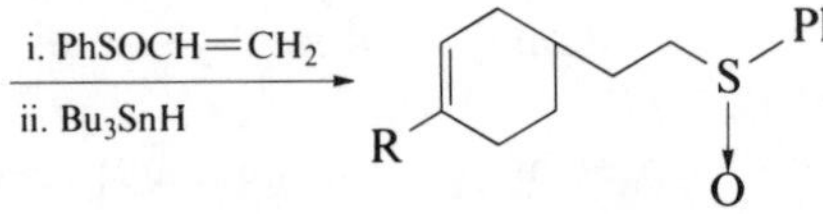

(30)

(90)

$R = Me, (CH_2)_2CH=CMe_2$

cedure was carried out for 1-t-butylcarbamate-1,3-butadiene and β-nitroacrylate in the synthesis of isogabaculine.

Bicyclic compounds (93; equation 31) are prepared in a two-step synthesis involving condensation of 1-vinylcyclohexenes (91) with β-nitroacrylic or β-nitrocrotonic esters (92) and denitration [430].

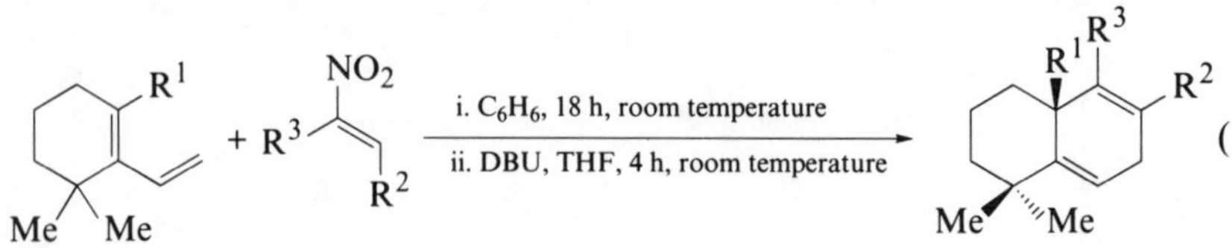

(91)   (92a)   (93)

$R^1, R^3 = H, Me; \quad R^2 = CO_2Et$

Condensation of β-nitroacrylic ester (92b) with 1,1-dimethoxy-3-trimethyl-silyloxy-1,3-butadiene (94) is regioselective (equation 32) and gives the nitro adduct (95). The adduct is aromatized in 99% yield to give 3-methoxy-5-hydroxybenzoic ester (96), which is difficult to prepare by other methods [404a].

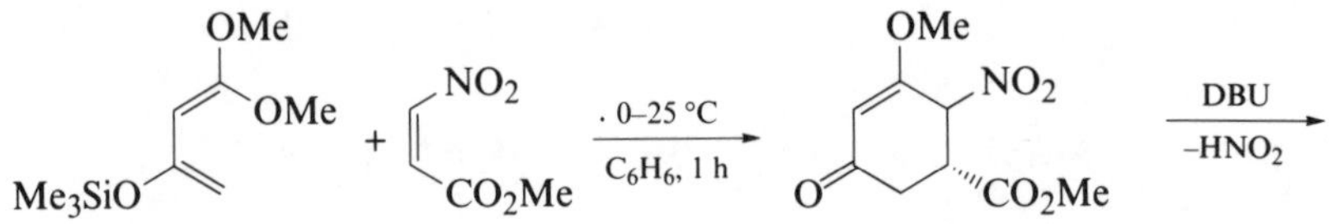

(94)   (92b)   (95)   (32)

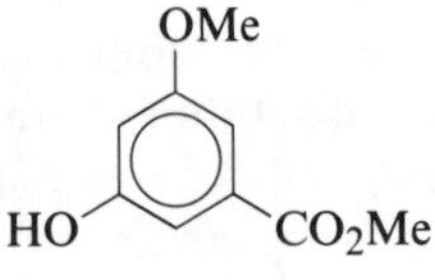

(96)

4-Nitro-3-phenyloxazole-5-carboxylate (**97**; equation 33) combines with 2,3-dimethyl-1,3-butadiene as a dienophile to form tetrahydro-1,2-benzoisoxazole (**98**). Subsequent denitration leads to 3-phenyl-4,5-benzoisoxazole (**99**) [431].

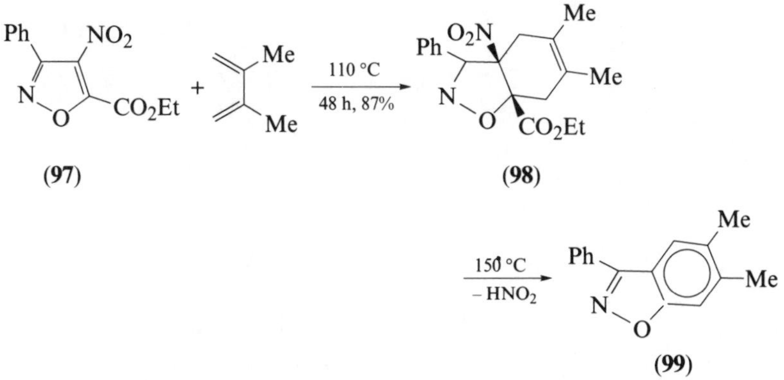

(33)

2.7.4 NITROALKENES USED AS HETERODIENES

The formation of cyclic nitronates (dihydrooxazine N-oxides) is a result of the coupling of nitroalkenes and nitrodienes with alkenes. This is a recent application of the Diels–Alder reaction, and it has been intensively developed over the last few decades. The process occurs by reaction of such highly reactive nitroalkenes as 2-nitro-1,3-dienes [420] or highly electron deficient α-nitrocinnamic esters [432] with nucleophilic dienophiles (vinyl ethers). The reaction occurs at room temperature to give the dihydrooxazine N-oxide (**100**; equation 34). Nitroalkenes react as heterodienes in this type of process.

Heterocycles similar to (**100**) are also prepared from the reaction of nitroalkenes with enamines or enolsilanes (see Section 2.6). However, the heterocyclic systems under discussion are obtained from alkoxyalkenes and alkenes and are more stable.

Mixutes of diastereoisomers are isolated in the reactions of cinnamic esters. A stereochemical study of the interaction between α,p-dinitrocinnamic acid and vinyl ether has demonstrated a stepwise cyclization mechanism. The coupling products from reactions of electron-deficient alkenes ($MeOCH{=}CH(NO_2)CO_2Me$, p-$O_2NC_6H_4CH{=}CHNO_2$) with vinyl ether cannot be isolated, probably because of their instability. Equation (35) shows that the reaction in the presence of acrylonitrile leads to the product of [4+2] and [3+2] double cycloaddition [432].

This type of reaction in the presence of a Lewis acid ($SnCl_4$) has been studied in detail [433, 434]. In this case the cycloaddition is faster and occurs at low temperature. Alkenes can be used instead of vinyl ethers and nitroalkenes without electron-withdrawing substituents.

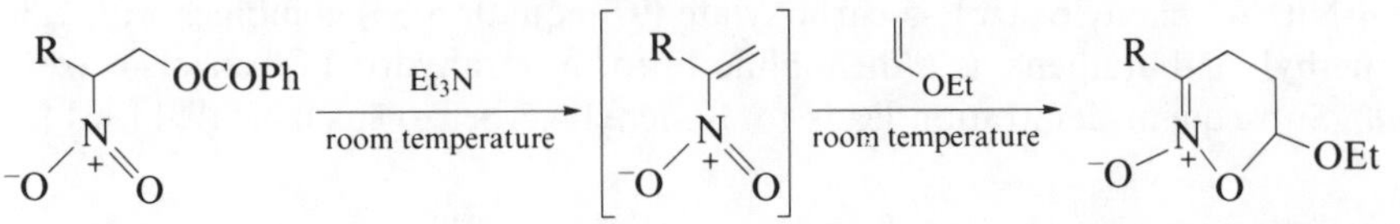

(100) 87%, (64) 89%

R =

(100)   (64)

(34)

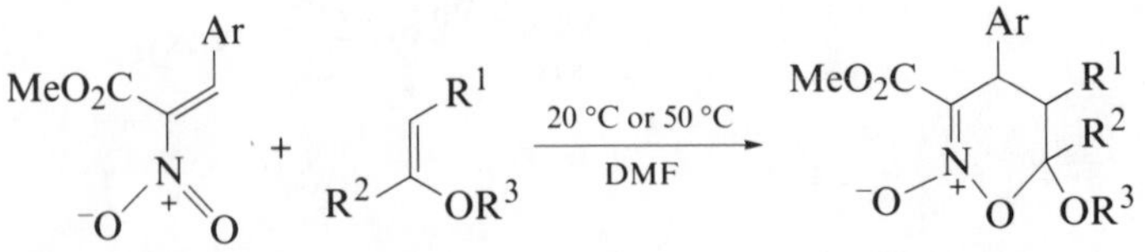

(101)   (102)   (103) 30–90%

Ar = p-NO$_2$C$_6$H$_4$

(102): CH$_2$=CHOEt, 2,3-dihydrofuran, 3,4-dihydro-2H-pyran,
1-methoxycyclopentene, 1-methoxycyclohexene

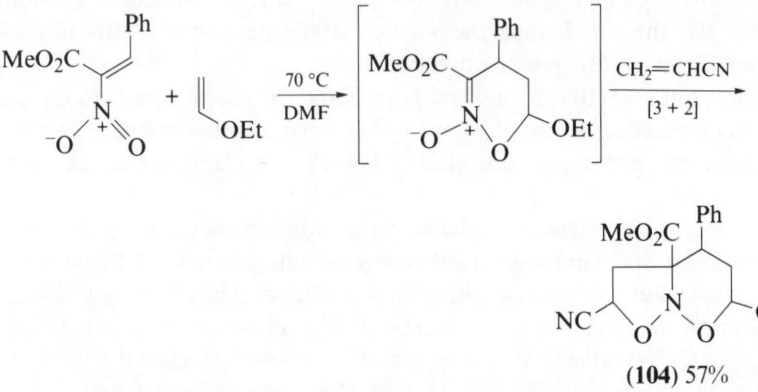

(35)

(104) 57%

Couplings between 1-nitrocyclohexene and cycloalkenes (equation 36) give mixtures of diastereoisomers (105, 106) with *exo/endo* selectivity. A small amount of nitronate (107) is also formed [433, 434].

Substituted cycloalkenes (1-methylcyclohexene) react faster to give two regioisomers and each isomer is a mixture of diastereoisomers. The application of cyclic nitroalkenes as starting materials makes the reaction proceed with high *anti* stereoselectivity (9:1). In the case of nitrostyrenes (equation 37), complete *anti* stereoselectivity is achieved [435].

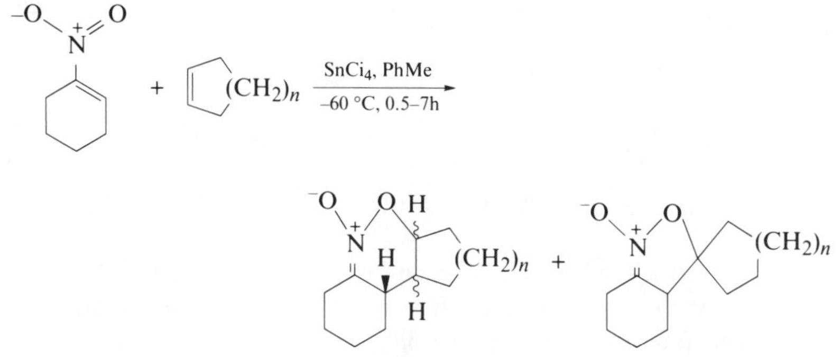

$$(36)$$

(105), (106)   **(107)**

$n = 1–3;$ **(105):** ⋯⋯ H (*exo*);   **(106):** ◄ H (*endo*)

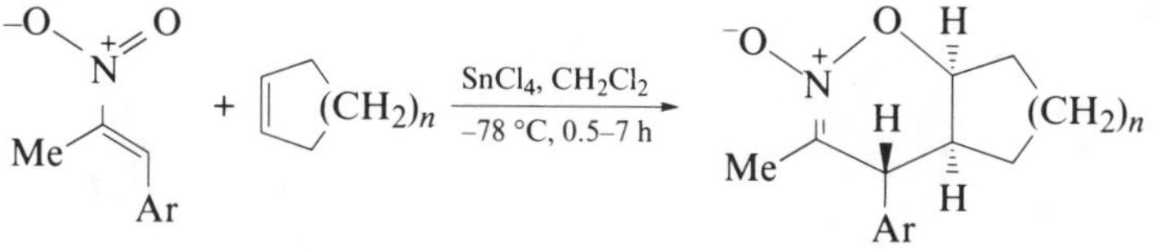

$$(37)$$

$n = 1–3$ **(108)** *anti*

$Ar = Ph, p\text{-}MeOC_6H_4, p\text{-}CF_3C_6H_4, 85–99\%;$ $Ar = p\text{-}N_2C_6H_4, 38\%$

The introduction of cycloalkadienes (dienophiles) into the reaction, in contrast with the above process, leads to diastereoisomeric mixtures (**109**; equation 38) consisting of mostly the *syn* adduct [435].

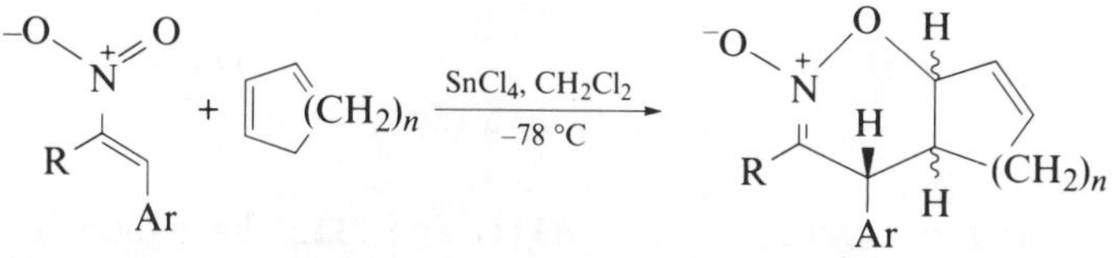

$$(38)$$

$n = 1, 2$ **(109)** 22–66%

$R = H, Me;$ $Ar = Ph, p\text{-}MeOC_6H_4$

It must be pointed out that under the conditions of this reaction the 'normal' nitro adduct, which can be obtained under thermic reaction conditions, was not formed.

A most detailed stereochemical discussion of this process has been presented for the intramolecular cycloaddition reactions (equation 39) of compounds of type (**110**) [436, 437].

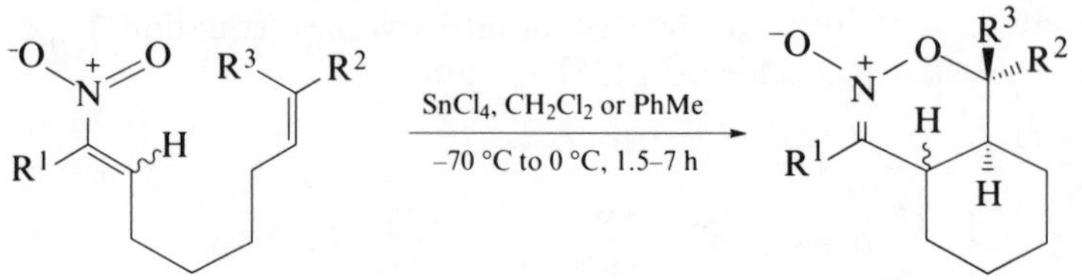

(110)   **(111)** 59–80%

$$R^1, R^2, R^3 = H, Me$$

When $R^1 = H$ (disubstituted (E)-nitroalkene fragment), a mixture of diastereo-isomers is formed with mostly *trans* ring fusion [435]. When $R^1 = Me$ (trisubsti-tuted nitroalkene fragment), the cycloaddition is highly stereoselective and practically only one isomer is formed. An (E)-nitroalkene gives the *trans* adduct, whereas a (Z)-nitroalkene gives the *cis* adduct.

When two fragments are connected via a benzene ring which is conjugated with the nitroalkene moiety (**112**; equation 40), the *cis* (*syn*) adduct is obtained (**113**) [435].

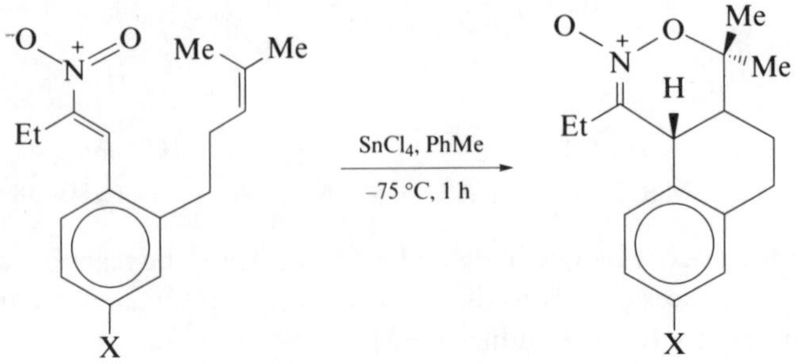

(112)   **(113)**

$$X = H (68\%), OMe (46\%)$$

The structures of adducts such as (**111**) and (**113**) have been elucidated through chemical transformations into known products and ^{1}H NMR.

The 'normal' adduct from the reaction of nitroalkenes with dienes is not formed in the presence of a Lewis acid (equation 38), and therefore the catalytic reaction must be dominant. To explain the behaviour of nitroalkenes as heterodienes in the presence of SnCl$_4$, Denmark *et al.* [435] have proposed the coordination complex shown in equation (41) (SnCl$_4$ at least loosely associated with a second oxygen atom in an antisymmetrical fashion). Equation (41) also shows that the interaction between nitrostyrenes and cycloalkenes proceeds with complete *anti* selectivity [435].

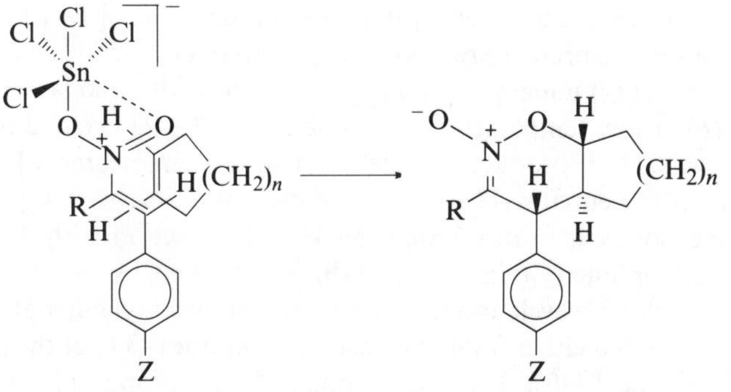

Cl
Sn
O
H
N
R
H(CH2)n
Z
O
H
(CH2)n
Z

exo-folding transition state

anti-nitronate

(41)

The preservation of dienophile geometry in the adduct signifies non-synchronized cycloaddition, but the periselective character is preserved. Kinetic data show an inverse electron demand in such diene syntheses. Thus, the formation of the complex between the nitroalkene and the Lewis acid lowers the energy of the lowest unoccupied molecular orbital (LUMO), which results in an acceleration of the reaction rate.

The catalytic reactions between nitroalkenes (**114**) which have double bonds separated by three or less carbon atoms and dienophile fragments with

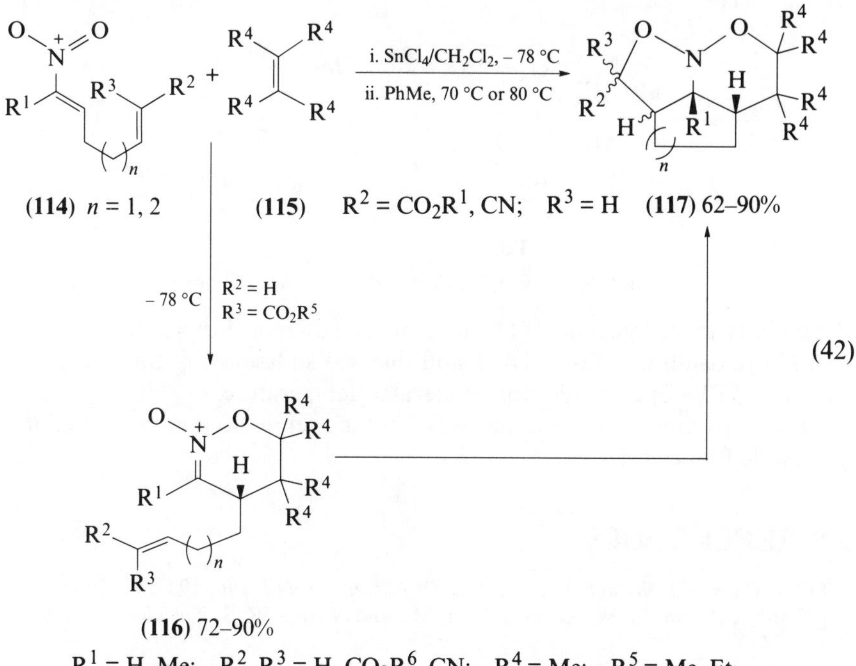

i. SnCl4/CH2Cl2, – 78 °C
ii. PhMe, 70 °C or 80 °C
(114) n = 1, 2
(115)
R2 = CO2R1, CN; R3 = H
(117) 62–90%
– 78 °C
R2 = H
R3 = CO2R5
(116) 72–90%

(42)

R^1 = H, Me; R^2, R^3 = H, CO_2R^6, CN; R^4 = Me; R^5 = Me, Et electron-withdrawing substituents proceed as tandem inter-$[4+2]$/intra-$[3+2]$ cycloadditions in the presence of 2,3-dimethyl-2-butene or vinyl ethers (equation 42). (Z)-enoates (114) undergo $[4+2]$ cycloaddition with alkenes (115) to form adducts (116) at low temperature. Upon heating, $[3+2]$ cycloaddition occurs to form tricyclic (117) products isolated as two diastereoisomers [438, 439].

In the case of (E)-enoates (118) the second cycloaddition occurs spontaneously and only one isomer is formed (equation 43). The reaction with the dienophile of higher nucleophilicity (vinyl ether 119) is catalysed by (i-PrO)$_2$TiCl$_2$. The products of the double cycloaddition were isolated for butyl ether and two chiral ethers that contained either fragments of (+)-camphor [439] or the (−)-*trans*-2-phenylcyclohexane [440]. These dienophiles form compounds (120) isolated as diastereoisomeric mixtures containing mostly the α anomer (R^3 = OX, R^4 = H).

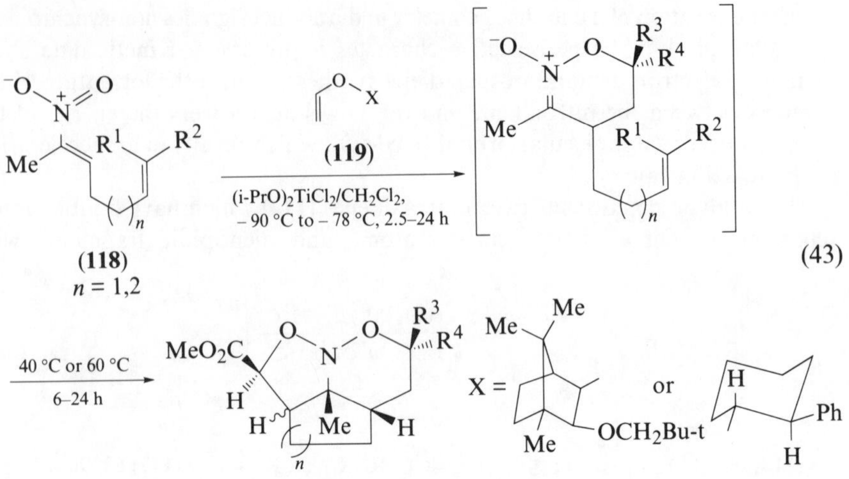

(43)

(120)
R^1, R^2 = H or CO$_2$Me;   R^3, R^4 = H, OBu or OX

Similarly to the reaction of alkenes, for the (Z)-enoate ($n=2$) the product of $[4+2]$ cycloaddition was isolated, and this was subsequently converted to the product of $[3+2]$ cycloaddition at elevated temperature.

These investigations open the way to the stereoselective construction of polycyclic frameworks.

2.8 REFERENCES

[1] Kabalka, G. W. and Varma, R. S. *Org. Prep. Proced. Int.*, 1987, **19**, 283.
[2] (a) Kabalka, G. W., Guindi, L. H. M., and Varma, R. S. *Tetrahedron*, 1990, **46**, 7443.
(b) Barrett, A. G. M. and Graboski, G. G. *Chem. Rev.*, 1986, **86**, 751.

[3] Shin, C., Masaki, M., and Ohta, M. *J. Org. Chem.*, 1967, **32**, 1860.
[4] Nightingale, D. and Janes, J. R. *J. Am. Chem. Soc.*, 1944, **66**, 352.
[5] Ahmad, M. S. and Rasa, S. K. *Indian J. Chem., Sect. B*, 1987, **26**, 579.
[6] Habib, R., Husain, M., Husain, M., and Khan, N. H. *Indian J. Chem. Sect. B*, 1984, **23**, 801.
[7] Hass, H. B., Susie, A. G., and Heider, R. L. *J. Org. Chem.*, 1950, **15**, 8; Hass, H. B. and Riley, E. F. *Chem. Rev.*, 1943, **32**, 373.
[8] Benigni, J. D. and Minnis, R. L. *J. Heterocycl. Chem.*, 1965, **2**, 387.
[9] Malesani, G., Galiano, F., Pietrogrande, A., and Rodighero, G. *Tetrahedron*, 1978, **34**, 2355.
[10] Sera, A., Yamouchi, H., Yamada, H., and Itoh, K. *Synlett*, 1990, 477.
[11] Dornow, A. and Müller, A. *Chem. Ber.*, 1960, **93**, 32.
[12] Smith, P. J., Sanghani, D. V., Bos, K. D., and Donaldson, J. D. *Chem. Ind.*, 1984, 167.
[13] Varma, R. S., Varma, M., and Kabalka, G. W. *Chem. Ind.*, 1985, 735.
[14] Kabalka, G. W. and Goudgaon, N. M. *Synth. Commun.*, 1988, **18**, 693.
[15] Varma, R. S. and Kabalka, G. W. *Chem. Lett.*, 1983, 243; Varma, R. S. and Kabalka, G. W. *Synth. Commun.*, 1985, **15**, 443.
[16] Singhal, G. M., Das, N. B., and Sharma, R. P. *J. Chem. Soc., Chem. Commun.*, 1990, 498.
[17] Varma, R. S., Varma, M., and Kabalka, G. W. *Heterocycles*, 1986, **24**, 2581.
[18] Kiprianov, A. I. and Verbovskaya, T. M. *Zh. Obshch. Khim.*, 1963, **33**, 479; 1962, **32**, 3703.
[19] Varma, R. S., Varma, M., and Kabalka, G. W. *Tetrahedron Lett.*, 1985, **26**, 6013.
[20] Varma, R. S., Varma, M., and Kabalka, G. W. *Synth. Commun.*, 1985, **15**, 1325.
[21] Varma, R. S., Varma, M., and Kabalka, G. W. *Tetrahedron Lett.*, 1985, **26**, 3777.
[22] Rao, T. S., Mathur, H. H., and Trivedi, G. K. *Tetrahedron Lett.*, 1984, **25**, 5561.
[23] Hanson, J. R. and Premuzic, E. *Tetrahedron Lett.*, 1966, 5441.
[24] Hanson, J. R. and Premuzic, E. *Tetrahedron*, 1967, **23**, 4105.
[25] Sera, A., Fukumoto, S., Tamura, M., Takabatake, K., Yamada, K., and Itoh, K. *Bull. Chem. Soc. Jpn*, 1991, **64**, 1787.
[26] Aizpurua, J. M., Oiarbide, M., and Palomo, C. *Tetrahedron Lett.*, 1987, **28**, 5365.
[27] Palomo, C., Aizpurua, J. M., Cossio, F. P., Garcia, J. M., Lopez, M. C., and Oiarbide, M. *J. Org. Chem.*, 1990, **55**, 2070.
[28] Cook, D. J., Pierce, O. R., and McBee, E. T. *J. Am. Chem. Soc.*, 1954, **76**, 83; McBee, E. T., Cook, D. J., and Pierce, O. R. *US Pat. 2.997 505*, 1961; *Chem. Abstr.*, 1962, **56**, 320e.
[29] Ramirez, F. A. and Burger, A. *J. Am. Chem. Soc.*, 1950, **72**, 2781.
[30] Terentiev, A. P. and Gracheva, R. A. *Zh. Obshch. Khim.*, 1962, **32**, 2231.
[31] Yuriev, Yu. K., Zefirov, N. S., and Ivanova, R. A. *Zh. Obshch. Khim.*, 1963, **33**, 3512.
[32] Shulgin, A. T. *J. Med. Chem.*, 1966, **9**, 445.
[33] Ho, B. T., McIsaac, W. M., An, R., Tansey, L. W., Walker, K. E., Englert, Jr, L. F., and Noel, M. B. *J. Med. Chem.*, 1970, **13**, 26; Ho, B. T., Tansey, L. W., Balster, R. L., An, R., McIsaac, W. M., and Harris, R. T. *J. Med. Chem.*, 1970, **13**, 134.
[34] Hamdan, A. and Wasley, J. W. F. *Synth. Commun.*, 1985, **15**, 71; Rizzacasa, M. A., Sargent, M. V., Skelton, B. W., and White, A. H. *Aust. J. Chem.*, 1990, 79.
[35] Kametani, T., Satoh, Y., and Fukumoto, K. *Chem. Pharm. Bull.*, 1977, **25**, 1129.
[36] Singh, C. and Kachru, C. N. *J. Indian Chem. Soc.*, 1978, **55**, 1314.
[37] Kupchan, S. M., Kubota, S., Fujita, E., Kobayashi, S., Block, J. H., and Telang, S. A. *J. Am. Chem. Soc.*, 1966, **88**, 4212.
[38] Vildavskaya, A. I. and Rall, K. B. *Zh. Org. Khim.*, 1968, **4**, 959.

[39] Yakhontov, L. N., Uritskaya, M. Ya., and Rubtsov, M. V. *Zh. Org. Khim.*, 1965, **1**, 2040.
[40] Fujiwara, A. N., Acton, E. M., and Goodman, L. *J. Heterocycl. Chem.*, 1969, **6**, 379.
[41] Kochetkov, N. K. and Dudykina, N. V. *Zh. Obshch. Khim.*, 1958, **28**, 2399.
[42] Meyers, A. I. and Sircar, J. C. *J. Org. Chem.*, 1967, **32**, 4134.
[43] Hassner, A., Kropp, J. E., and Kent, G. J. *J. Org. Chem.*, 1969, **34**, 2628.
[44] Goudgaon, N. M., Wadgaonkar, P. P., and Kabalka, G. W. *Synth. Commun.*, 1989, **19**, 805.
[45] Varma, R. S. and Kabalka, G. W. *Synth. Commun.*, 1984, **14**, 1093; 1985, **15**, 151.
[46] Hassner, A. and Heathcock, C. *J. Org. Chem.*, 1964, **29**, 1350.
[47] Bhattacharjya, A., Mukhopadhyay, R., and Pakrashi, S. C. *Synthesis*, 1985, 886.
[48] McDonald, E. and Martin, R. T. *Tetrahedron Lett.*, 1977, 1317.
[49] Sinhababu, A. K. and Borchardt, R. T. *Tetrahedron Lett.*, 1983, **24**, 227.
[50] Sakakibara, T., Tachimori, Y., and Sudoh, R. *Tetrahedron*, 1984, **40**, 1533.
[51] Baer, H. H. and Hanna, H. R. *Can. J. Chem.*, 1980, **58**, 1751.
[52] Barton, D. H. R., Motherwell, W. B., and Zard, S. Z. *Bull. Soc. Chim. Fr.*, 1983, II-61.
[53] Yamada, F., Makita, Y., Suzuki, T., and Somei, M. *Chem. Pharm. Bull.*, 1985, **33**, 2162.
[54] Varma, R. S., Kadkhodayon, M., and Kabalka, G. W. *Heterocycles*, 1986, **24**, 1647; Varma, R. S., Gai, Y. Z., and Kabalka, G. W. *J. Heterocycl. Chem.*, 1987, **24**, 767.
[55] Dauzonne, D. and Royer, R. *Synthesis*, 1988, 339.
[56] Osby, J. O. and Ganem, B. *Tetrahedron Lett.*, 1985, **26**, 6413.
[57] Feuer, H., Bartlett, R. S., Vincent, Jr, B. F., and Anderson, R. S. *J. Org. Chem.*, 1965, **30**, 2880.
[58] Mourad, M. S., Varma, R. S., and Kabalka, G. W. *J. Org. Chem.*, 1985, **50**, 133.
[59] Varma, R. S. and Kabalka, G. W. *Org. Prep. Proced. Int.*, 1985, **17**, 254; *Chem. Abstr.*, 1986, **104**, 88 196t.
[60] Varma, R. S. and Kabalka, G. W. *Synth. Commun.*, 1985, **15**, 843.
[61] Ranu, B. C. and Chakraborty, R. *Tetrahedron Lett.*, 1991, **32**, 3579.
[62] Ranu, B. C. and Chakraborty, R. *Tetrahedron*, 1992, **48**, 5317.
[63] Sowden, J. C. and Fischer, H. O. L. *J. Am. Chem. Soc.*, 1947, **69**, 1048.
[64] Seifert, W. K. *US Pat. 3 156 723*, 1964; *Chem. Abstr.*, 1965, **62**, 3954h.
[65] Freidlin, L. Kh., Litvin, E. F., and Chursina, V. M. *Catalytic Reduction and Hydrogenation in the Liquid Phase*, Ivanovo, 1970, p. 59; *Russ. Zh. Khim.*, 1970, **24**, B1029.
[66] Dyumaev, K. M. and Belostotskaya, I. S. *Zh. Obshch. Khim.*, 1962, **32**, 2661.
[67] Kazanskii, B. A. (Ed.) *The Synthesis of Organic Preparates*, Foreign Literature, Moscow, 1949, Coll. I, p. 357.
[68] Voronin, V. G., Kilikovskaya, G. D., and Magda, L. D. *Zh. Org. Khim.*, 1965, **1**, 719.
[69] Freidlin, L. Kh., Litvin, E. F., and Chursina, V. M. *Dokl. Akad. Nauk SSSR*, 1964, **155**, 1144.
[70] Konho, M., Sasao, S., and Muranashi, S.-I. *Bull. Chem. Soc. Jpn*, 1990, **63**, 1252; Murahashi, Sh. and Konho, M. *Jpn Kokai Tokkyo Koho, Jpn Pat. 0327, 345, 91, 27, 345* (Cl. CO7 C211/27), 1991; *Chem. Abstr.*, 1991, **114**, 246 938z.
[71] Battersby, A. R., Le Count, D. J., Garratt, S., and Thrift, R. I. *Tetrahedron*, 1961, **14**, 46.
[72] (a) Seifert, W. K. and Condit, P. C. *J. Org. Chem.*, 1963, **28**, 265.
(b) Coutts, R. T. and Malicky, J. L. *Can. J. Chem.*, 1974, **52**, 395.

[73] Kabalka, G. W., Pace, R. D., and Wadgaonkar, P. P. *Synth. Commun.*, 1990, **20**, 2453.
[74] Petrini, M., Ballini, R., and Rosini, G. *Synthesis*, 1987, 713.
[75] Noble, P., Borgardt, F. G., and Reed, W. L. *Chem. Rev.*, 1964, **64**, 19, 34.
[76] Bocharova, L. A., Perekalin, V. V., and Polyanskaya, A. S. *USSR Pat. 166 706*, 1964; *Chem. Abstr.*, 1965, **62**, 10 509a.
[77] Perekalin, V. V. and Temp, A. A. *USSR Pat. 209 465* (Cl. CO7), 1968; *Chem. Abstr.*, 1968, **69**, 77 732x.
[78] Babievskii, K. K., Belikov, V. M., and Tikhonova, N. A. *Izv. Akad. Nauk SSSR, Ser. Khim.*, 1965, 750.
[79] Belikov, V. M., Babievskii, K. K., and Tikhonova, N. A. *USSR Pat. 181 652* (Cl. CO7d), 1966; *Chem. Absrt.*, 1966, **65**, 18 683e.
[80] Babievskii, K. K., Belikov, V. M., and Tikhonova, N. A. *Izv. Akad. Nauk SSSR, Ser. Khim.*, 1965, 89.
[81] (a) Belikov, V. M., Babievskii, K. K., Filatova, I. M., Nalivaiko, E. V., and Nikitina, S. B. *Izv. Akad. Nauk SSSR, Ser. Khim.*, 1977, 1824.
(b) Dauzonne, D. and Royer, R. *Synthesis*, 1987, 399.
[82] Rodina, O. A. and Mamaev, V. P. *Zh. Obshch. Khim.*, 1964, **34**, 2146.
[83] Babievskii, K. K., Belikov, V. M., and Tikhonova, N. A. *Khim. Geterotsikl. Soedin. Sb. 1, Azotsoderzh. Heterocykli*, 1967, 46; *Chem. Abstr.*, 1969, **70**, 78 343d.
[84] Aboskalova, N. I., Polyanskaya, A. S., and Perekalin, V. V. *Dokl. Akad. Nauk SSSR*, 1967, **176**, 829.
[85] Hengartner, U., Valentine, D., Johnson, K. K., Larscheid, M. E., Pigott, F., Scheidl, F., Scott, J. W., Sun, R. C., Townsend, J. M., and Williams, T. H. *J. Org. Chem.*, 1979, **44**, 3741.
[86] Baxter, I. and Swan, G. A. *J. Chem. Soc. C*, 1967, 2446.
[87] Rajeswari, S., Drost, K. J., and Cava, M. P. *Heterocycles*, 1989, **29**, 415.
[88] Murphy, B. P. and Banks, H. D. *Synth. Commun.*, 1985, **15**, 321.
[89] Rogers, C. B., Blum, C. A., and Murphy, B. P. *J. Heterocycl. Chem.*, 1987, **24**, 941.
[90] Benington, F., Morin, R. D., and Clark, L. C. *J. Org. Chem.*, 1960, **25**, 1542.
[91] Hengartner, U., Batcho, A. D., Blount, J. F., Leimgruber, W., Larscheid, M. E., and Scott, J. W. *J. Org. Chem.*, 1979, **44**, 3748.
[92] Chikashita, H., Morita, Y., and Itoh, K. *Synth. Commun.*, 1985, **15**, 527; *Chem. Abstr.*, 1985, **103**, 141 543m.
[93] Chikashita, H., Nishida, S., Miyazaki, M., Morita, Y., and Itoh, K. *Bull. Chem. Soc. Jpn*, 1987, **60**, 737.
[94] Trefouel, T., Tintillier, P., Dupas, G., Bourguignon, J., and Quequiner, G. *Bull. Chem. Soc. Jpn*, 1987, **60**, 4492.
[95] Nakamura, K., Fuji, M., Oka, S., and Ohno, A. *Chem. Lett.*, 1985, 523.
[96] Fuji, M. *Bull. Chem. Soc. Jpn*, 1988, **61**, 4029.
[97] Ohta, H., Ozaki, K., and Tsuchihashi, G. *Chem. Lett.*, 1987, 191.
[98] Ohta, H., Kobayashi, N., and Ozaki, K. *J. Org. Chem.*, 1989, **54**, 1802.
[99] Mori, A., Ishiyama, I., Akita, H., Suzuki, K., Mitsuoka, T., and Oishi, T. *Chem. Pharm. Bull.*, 1990, **38**, 3449.
[100] Varma, R. S., Varma, V., and Kabalka, G. W. *Synth. Commun.*, 1986, **16**, 91.
[101] Singhal, G. M., Das, N. B., and Sharma, R. P. *J. Chem. Soc., Chem. Commun.*, 1989, 1470.
[102] Ruhrchemie, A.-G. *Br. Pat. 706 723*, 1954; *Chem. Abstr.*, 1955, **49**, 11 014d.
[103] Ban, V. *J. Pharm. Soc. Jpn*, 1954, **74**, 212; *Russ. Zh. Khim.*, 1956, **13**, 39 545.
[104] Convert, O., Bassinet, P., Pinson, J., and Armand, J. *C. R. Hebd. Seances Acad. Sci., Ser. C*, 1970, **271**, 1602.

[105] Torii, S., Tanaka, H., and Katoh, T. *Chem. Lett.*, 1983, 607.
[106] Shono, T., Hamaguchi, H., Mikami, H., Nogusa, H., and Kashimura, S. *J. Org. Chem.*, 1983, **48**, 2103.
[107] Convert, O., Bassinet, P., Moskowitz, H., and Armand, J. *C. R. Hebd. Seances Acad. Sci., Ser. C*, 1972, **275**, 1527.
[108] Armand, J. and Convert, O. *Collect. Czech. Chem. Commun.*, 1971, **36**, 351.
[109] Avrutskaya, I. A., Babievskii, K. K., Belikov, V. M., Zaporozhets, E. V., Filatova, I. M., and Fioshin, M. Ya. *Elektrokhimiya*, 1973, **9**, 1652.
[110] Zaporozhets, E. V., Avrutskaya, I. A., Fioshin, M. Ya., Babievskii, K. K., and Belikov, V. M. *Elektrokhimiya*, 1973, **9**, 270.
[111] Wessling, M. and Schäfer, H. J. *Chem. Ber.*, 1991, **124**, 2303.
[112] Zaporozhets, E. V., Avrutskaya, I. A., Babievskii, K. K., Belikov, V. M., and Fioshin, M. Ya. *Elektrokhimiya*, 1972, **8**, 1243.
[113] Avrutskaya, I. A., Fioshin, M. Ya., Babievskii, K. K., Belikov, V. M., and Zaporozhets, E. V. *Elektrokhimiya*, 1973, **9**, 1127.
[114] Zaporozhets, E. V., Avrutskaya, I. A., Fioshin, M. Ya., Babievskii, K. K., and Belikov, V. M. *Elektrokhimiya*, 1972, **8**, 1809.
[115] Avrutskaya, I. A., Babievskii, K. K., Belikov, V. M., Zaporozhets, E. V., Novikov, V. I., and Fioshin, M. Ya. *Elektrokhimiya*, 1975, **11**, 661.
[116] Zaporozhets, E. V., Avrutskaya, I. A., Fioshin, M. Ya., Babievskii, K. K., and Belikov, V. M. *Izv. Akad. Nauk SSSR, Ser. Khim.*, 1972, 1894.
[117] Avrutskaya, I. A., Babievskii, K. K., Belikov, V. M., Zaporozhets, E. V., and Fioshin, M. Ya. *Elektrokhimiya*, 1973, **9**, 1363.
[118] Zaporozhets, E. V., Avrutskaya, I. A., Fioshin, M. Ya., Babievskii, K. K., and Belikov, V. M. *USSR Pat. 432 132* (Cl. 7, O7c), 1974; *Chem. Abstr.*, 1974, **81**, 91 930q.
[119] Zaporozhets, E. V., Avrutskaya, I. A., Fioshin, M. Ya., Babievskii, K. K., and Belikov, V. M. *USSR Pat. 371 222* (Cl. CO7d), 1973; *Chem. Abstr.*, 1973, **79**, p. 31 867y.
[120] Zaporozhets, E. V., Avrutskaya, I. A., Fioshin, M. Ya., Babievskii, K. K., and Belikov, V. M. *Elektrokhimiya*, 1973, **9**, 72.
[121] Armand, J. and Convert, O. *C. R. Hebd. Seances Acad. Sci., Ser. C*, 1969, **268**, 842.
[122] Lipina, E. S., Stepanov, N. D., Bagal, I. L., Bodina, R. I., and Perekalin, V. V. *Zh. Org. Khim.*, 1980, **16**, 2404.
[123] Lipina, E. S., Kalb, G. L., and Kampar, V. E. *Sint. Issled. Nitrosoedin., Aminokislot., Leningrad Gos. Pedagog. Inst.*, Leningrad, 1983, 14; *Chem. Abstr.*, 1984, **100**, 5584m.
[124] Todres, Z. V. *Ion-radicals in Organic Synthesis*, Khimia, Moscow, 1986, p. 190.
[125] Todres, Z. V., Nekrasova, G. V., Lipina, E. S., and Chernyshova, T. M. *Izv. Akad. Nauk SSSR, Ser. Khim.*, 1979, 126.
[126] Novikov, S. S. *The Chemistry of Aliphatic and Alicyclic Nitrocompounds*, Khimia, Moscow, 1974, pp. 227–228.
[127] Lambert, A., Scaife, C. W., and Wilder-Smith, A. E. *J. Chem. Soc.*, 1947, 1474.
[128] Hopff, H. and Capaul, M. *Helv. Chim. Acta*, 1960, **43**, 1898.
[129] Shechter, H. and Cates, H. L. *J. Org. Chem.*, 1961, **26**, 51.
[130] Baer, G. and Urbas, L. in *The Chemistry of the Nitro and Nitroso Groups*, Part 2 (Ed. M. Feuer), Mir, Moscow, 1973, pp. 63–147.
[131] Burmistrov, V. I., Kozlov, L. M., and Solodov, A. V. *Izv. Vuzov Khimikotekhnol.* 1967, **10**, 895.
[132] Hwu, J. R. and Wang, N. *J. Chem. Soc., Chem. Commun.*, 1987, 427.
[133] Burmistrov, V. I., Fakhrutdinov, R. Z., and Matina, L. S. *Izv. Vuzov Khimikotekhnol.*, 1968, **11**, 1257; Burmistrov, V. I., Voronkova, V. A., Fakhrutdinov, R. Z., and Kozlov, L. M. *Zh. Org. Khim.*, 1972, **8**, 16.

[134] Rosenberg, M., Sturdic, E., Liptaj, T., Bella, J., Vegh, D., Povaznec, F., and Sitkey, V. *Collect. Czech. Chem. Commun.*, 1985, **50**, 470.

[135] (a) Hoz, S. and Speizman, D. *J. Org. Chem.*, 1983, **48**, 2904.
(b) Hoz, S., Gross, Z., and Cohem, D. *J. Org. Chem.*, 1985, **50**, 832.

[136] Kamimura, A. and Ono, N. *Tetrahedron Lett.*, 1989, **30**, 731.

[137] Kamimura, A., Sasatani, H., Hashimoto, T., Kawai, T., Hori, K., and Ono, N. *J. Org. Chem.*, 1990, **55**, 2437.

[138] Hori, K., Higuchi, S., and Kamimura, A. *J. Org. Chem.*, 1990, **55**, 5900.

[139] Colonge, J. and Lartigau, G. *Bull. Soc. Chim. Fr.*, 1964, 2436.

[140] Shin, C., Masaki, M., and Ohta, M. *Bull. Chem. Soc. Jpn*, 1970, **43**, 3219; Shin, C., Yonezawa, Y., Narukawa, H., Nanjo, K., and Yoshimura, J. *Bull. Chem. Soc. Jpn*, 1972, **45**, 3595.

[141] Burkett, H., Nelson, G., and Wright, W. *J. Am. Chem. Soc.*, 1958, **80**, 5812.

[142] Rybinskaya, M. I., Rybin, L. V., and Nesmeianov, A. N. *Izv. Akad. Nauk SSSR, Ser. Khim.*, 1963, 899.

[143] Yamamura, K., Watarai, S., and Kinugasa, T. *Bull. Chem. Soc. Jpn*, 1971, **44**, 2440.

[144] Russell, G. A. and Dedolph, D. *J. Org. Chem.*, 1985, **50**, 3878.

[145] Pavlova, Z. F. and Lipina, E. S. *Gertsenovskie Chteniya, Leningrad Gos. Pedagog. Inst., Leningrad*, 1973, **2**, p. 163, *Chem. Abstr.*, 1974, **81**, 151 395f.

[146] Burmistrov, V. I., Ryabova, L. E., Rogozhina, T. A., and Kozlov, L. M. *Tr. Khimikotekhnol., Khim. Nauk*, 1967, 515.

[147] Perekalin, V. V., Petryaeva, A. K., Zobacheva, M. M., and Metelkina, E. L. *Dokl. Akad. Nauk SSSR*, 1966, **166**, 1129.

[148] Ono, N., Miyake, H., Kamimura, A., Hamamoto, I., Tamura, R., and Kaji, A. *Tetrahedron*, 1985, **41**, 4013.

[149] Ono, N. and Kaji, A. *Synthesis*, 1986, 693.

[150] Dehaen, W. and Hassner, A. *J. Org. Chem.*, 1991, **56**, 897.

[151] Varma, R. S. and Kabalka, G. W. *Heterocycles*, 1985, **23**, 139.

[152] Sakakibara, T., Koezuka, M., and Sudon, R. *Bull Chem. Soc. Jpn*, 1978, **51**, 3095.

[153] Varma, R. S., Kadkhodayan, M., and Kabalka, G. W. *Synthesis*, 1986, 486.

[154] Neirallych, M. A., Koussini, R., and Guillaument, G. *Synth. Commun.*, 1990, **20**, 783.

[155] (a) Bowman, W. R., Brown, D. S., Shaffin J. D. E., Symons, M. C., Jackson, S. W., and Willcocks, N. A. *Tetrahedron Lett.*, 1991, **32**, 2285.
(b) Fisher, R. H. and Weitz, H. M. *Synthesis*, 1976, 53.

[156] Heath, R. L. and Lambert, A. *J. Chem. Soc.*, 1947, 1477.

[157] Cason, L. F. and Wanser, C. C. *J. Am. Chem. Soc.*, 1951, **73**, 142.

[158] Mustafa, A., Harnash, A. H. E., and Kamel, M. *J. Am. Chem. Soc.*, 1955, **77**, 3860.

[159] Ono, N., Kamimura, A., Sasatani, H., and Kaji, A. *J. Org. Chem.*, 1987, **52**, 4133.

[160] Kamimura, A., Sasatani, H., Hashimoto, T., and Ono, N. *J. Org. Chem.*, 1989, **54**, 4998.

[161] Ono, N., Kawai, S., Tanaka, K., and Kaji, A. *Tetrahedron Lett.*, 1979, 1733.

[162] Bernasconi, C. F. and Killion, R. B. *J. Am. Chem. Soc.*, 1988, **110**, 7506.

[163] Bernasconi, C. F. *Tetrahedron*, 1989, **45**, 4017.

[164] Bernasconi, C. F. and Schuck, D. F. *J. Org. Chem.*, 1992, **57**, 2365.

[165] Ono, N., Kamimura, A., and Kaji, A. *Tetrahedron Lett.*, 1984, **25**, 5319.

[166] Kamimura, A. and Ono, A. *J. Chem. Soc., Chem. Commun.*, 1988, 1278.

[167] Ono, N. in *Nitro Compounds* (Eds M. Feuer and A. T. Nielsen), VCH, New York, 1990, p. 115.

[168] Ono, N., Kamimura, A., Kawai, T., and Kaji, A. *J. Chem. Soc., Chem. Commun.*, 1987, 1550.

[169] Kobayashi, N. and Iwai, K. *Tetrahedron Lett.*, 1980, **21**, 2167.

[170] Kobayashi, N. and Iwai, K. *J. Org. Chem.*, 1981, **46**, 1823.
[171] Seebach, D. and Knochel, P. *Helv. Chim. Acta*, 1984, **67**, 261.
[172] Vildavskaya, A. I. and Rall, K. B. *Zh. Org. Khim.*, 1967, **3**, 434.
[173] Vildavskaya, A. I., Rall, K. B., and Petrov, A. A. *Zh. Obshch. Khim.*, 1971, **41**, 1279.
[174] Vildavskaya, A. I., Sulimov, I. G., and Rall, K. B. *Zh. Org. Khim.*, 1969, **5**, 1326.
[175] Bloom, A. J. and Mellor, J. M. *Tetrahedron Lett.*, 1986, **27**, 873.
[176] Russell, G. A., Ching-Fa, Y., Tashtoush, H. I., Russell, J. E., and Dedolph, D. F. *J. Org. Chem.*, 1991, **56**, 663.
[177] Park, K. P. and Ha, H.-J. *Bull. Chem. Soc. Jpn*, 1990, **63**, 3006.
[178] Pavlova, Z. F., Kazem, Ya. A., and Perekalin, V. V. *Sint. Issled. Nitrosoedin., Aminokislot., Leningrad Gos. Pedagog. Inst., Leningrad*, 1982, p. 74; *Chem. Abstr.*, 1983, **99**, 212 111f.
[179] Tamura, A. R., Kai, Y., Kakihana, M., Hayashi, K., Tsuji, M., Nakamura, T., and Oda, D. *J. Org. Chem.*, 1968, **51**, 4375.
[180] Tamura, A. R., Kamimura, A., and Ono, N. *Synthesis*, 1991, 423.
[181] Alekseev, D. I. *Zh. Org. Khim.*, 1975, **11**, 2621.
[182] Mukhina, E. S., Pavlova, Z. E., Berkova, G. A., Lipina, E. S., Mostyaeva, L. V., and Perekalin, V. V. *Zh. Org. Khim.*, 1991, **27**, 910.
[183] Aleksiev, D. I. *Zh. Org. Khim.*, 1975, **11**, 211.
[184] Health, R. L. and Piggott, H. A. *J. Chem. Soc.*, 1947, 1481.
[185] Gold, M. H., Druker, L. J., Yotter, R., Thor, C. J. B., and Lang, G. *J. Org. Chem.*, 1951, **16**, 1495.
[186] Samarenko, V. Ya., Kozmina, A. G., Passet, B. V., and Markova, N. R. *Zh. Org. Khim.*, 1977, **13**, 1707; Samarenko, V. Ya., Kozmina, A. G., and Passet, B. V. *Zh. Org. Khim.*, 1977, **13**, 1790.
[187] Shalygina, O. D., Kostyuchenko, N. P., Vinograd, L. N., and Suvorov, N. N. *Khim. Geterotsikl. Soedin.*, 1971, 205.
[188] Muratova, G. L. and Stepanov, V. M. *Zh. Obshch. Khim.*, 1964, **34**, 1687.
[189] Ranganathan, D., Ranganathan, S., and Bamezai, S. *Tetrahedron Lett.*, 1982, **23**, 2789.
[190] Barco, A., Benetti, S., Pollini, G. P., and Spalluto, G. *Synthesis*, 1991, 479.
[191] Dukhovskoi, V. P., Tiurin, V. D., and Shvehgeimer, G. A. *Khim. Geterotsikl. Soedin.*, 1969, 758.
[192] Reimlinger, H. and Oth, J. F. M. *Chem. Ber.*, 1964, **97**, 331.
[193] Tsin-iun, Ch., Gambarian, N. P., and Knuniants, I. L. *Dokl. Acad. Nauk SSSR*, 1960, **133**, 1113.
[194] Lough, C. E. and Currie, D. J. *Can. J. Chem.*, 1966, **44**, 1563.
[195] Southwick, P. L. and Anderson, J. E. *J. Am. Chem. Soc.*, 1957, **79**, 6222.
[196] Todres, Z. V., Diusengaliev, K. I., and Garbusova, I. A. *Zh. Org. Khim.*, 1986, **22**, 370.
[197] Bernasconi, C. F., Renfrow, R. A., and Tia, P. R. *J. Am. Chem. Soc.*, 1986, **108**, 4541; Bernasconi, C. F., Zitomer, J. L., and Schuck, D. F. *J. Org. Chem.*, 1992, **57**, 1132.
[198] Bernasconi, C. F. and Renfrow, R. A. *J. Org. Chem.*, 1987, **52**, 3035.
[199] Asaad, F. M., Grant, N., and Latif, N. *Liebigs Ann Chem.*, 1988, 183.
[200] Iton, K., Ishida, H., and Chikashita, H. *Chem. Lett.*, 1982, 1117.
[201] Simonov, A. M. and Dalgatov, D. D. *Zh. Obshch. Khim.*, 1964, **34**, 3052.
[202] Kada, R., Knoppova, V., Kovac, J., and Malenakova, I. *Collect. Czech. Chem. Commun.*, 1984, **49**, 2496.
[203] Rosenberg, M., Studrik, E., Liptaj, T., Bella, J., Vegh, D., Povszsnec, F., and Sitkey, V. *Collect. Czech. Chem. Commun.*, 1985, **50**, 470.

[204] Perepichka, I. F., Kostenko, L. I., Popov, A. F., and Chervinskii, A. Yu. *Zh. Org. Khim.*, 1988, **24**, 822; Popov, A. F., Perepichka, I. F., and Kostenko, L. I. *J. Chem. Soc., Perkin Trans. 2*, 1989, 395.
[205] Barco, A., Benetti, S., Casolari, A., Pollini, G. P., and Spalluto, G. *Tetrahedron Lett.*, 1990, **31**, 3039.
[206] Barco, A., Benetti, S., Pollini, G. P., Spalluto, G., and Zenirato, V. *J. Chem. Soc., Chem. Commun.*, 1991, 390.
[207] Barco, A., Benetti, S., Spalluto, G., Casolari, A., Pollini, G. P., and Zenirato, V. *J. Org. Chem.*, 1992, **57**, 6279.
[208] Baumberger, F., Beer, D., Christen, M., Prewo, R., and Vasella, A. *Helv. Chem. Acta*, 1986, **69**, 1191.
[209] Baumberger, F., Vasella, A., and Schauer, R. *Helv. Chim. Acta*, 1988, **71**, 429.
[210] Kitagawa, I., Cha, C. C., Nakae, T., Okaichi, Y., Takinami, Y., and Yoshikawa, M. *Chem. Pharm. Bull.*, 1989, **37**, 542.
[211] Vildavskaya, A. I., Rall, K. B., and Petrov, A. A. *Zh. Obshch. Khim.*, 1971, **41**, 1279.
[212] Bernasconi, C. F. and Carri, D. J. *J. Am. Chem. Soc.*, 1979, **101**, 2698.
[213] Allade, I., Dubois, P., Levillain, P., and Viel, C. *Bull. Soc. Chim. Fr.*, 1983, **11**, 339.
[214] Lipina, E. S., Pavlova, Z. F., Prikhodko, L. V., Paperno, T. Ya., and Perekalin, V. V. *Zh. Org. Khim.*, 1970, **6**, 1123.
[215] Rappoport, Z. and Topol, A. *J. Org. Chem.*, 1989, **54**, 5967.
[216] Mühlstadt, M. and Schulze, B. *J. Prakt. Chem.*, 1971, **313**, 205.
[217] Sakai, K., Oida, S., and Ohki, E. *Chem. Pharm. Bull.*, 1968, **16**, 1048; Hurd, C. D. and Patterson, J. *J. Am. Chem. Soc.*, 1953, **75**, 285.
[218] Akhtar, M. S., Sharma, V. L., Seth, M., and Bhaduri, A. P. *Indian J. Chem., Sect. B*, 1988, **27**, 448.
[219] Bulatov, N. N. and Suvorov, N. N. *Khim. Geterotsikl. Soedin.*, 1965, 813.
[220] (a) Gomez-Guillen, M., Maria, J., and Simon, L. *Carbohydr. Res.*, 1991, **210**, 175.
      (b) Gomez-Guillen, M. and Jimenez, J. L. C. *Carbohydr. Res.*, 1988, **180**, 1.
[221] Shibuya, M., Kuretani, M., and Kubotu, S. *Tetrahedron Lett.*, 1981, **22**, 4453.
[222] Hassner, A. and Chau, W. *Tetrahedron Lett.*, 1982, **23**, 1989.
[223] Tamura, R., Hayashi, K., Kai, Y., and Oda, D. *Tetrahedron Lett.*, 1984, **24**, 4437.
[224] Varma, R. S. and Kabalka, G. W. *Heterocycles*, 1986, **24**, 2645.
[225] Escribano, F. C., Alcantara, M. P. D., and Gomez-Sanchez, A. *Tetrahedron Lett.*, 1988, **29**, 6001.
[226] Klager, K. *J. Org. Chem.*, 1955, **20**, 650.
[227] Dornow, A. and Frese, A. *Liebigs Ann. Chem.*, 1953, **581**, 211.
[228] Ivanova, I. S., Konova, Yu. V., and Novikov, S. S. *Izv. Akad. Nauk SSSR, Ser Khim.*, 1962, 2078.
[229] Hass, H. B. and Riley, E. F. *Chem. Rev.*, 1943, **32**, 373.
[230] Feuer, H. and Miller, R. *J. Org. Chem.*, 1961, **26**, 1348.
[231] Feuer, H., Leston, G., Miller, R., and Nielsen, A. T. *J. Org. Chem.*, 1963, **28**, 339.
[232] Novikov, S. S., Babievskii, K. K., and Korsakova, I. S. *Dokl. Akad. Nauk SSSR*, 1959, **125**, 560.
[233] Solomonovici, A. and Blumberg, S. *Tetrahedron*, 1966, **22**, 2505.
[234] Oreshko, G. V. and Eremenko, L. T. *Izv. Akad. Nauk SSSR, Ser. Khim.*, 1989, 1107.
[235] Perekalin, V. V. and Polyanskaya, A. S. *Dokl. Akad. Nauk SSSR*, 1957, **112**, 441; *Zh. Obshch. Khim.*, 1957, **27**, 1933.
[236] McBee, E. T., Cook, D. J., and Pierce, O. R. US Pat. 2 997 505, 1961; *Chem. Abstr.*, 1962, **56**, 320e.
[237] Klager, K. *Monatsh. Chem.*, 1965, **96**, 1.
[238] Klager, K. and Frankel, M. B. *Monatsh. Chem.*, 1968, **99**, 1336.

162 CHEMICAL TRANSFORMATIONS OF UNSATURATED NITRO COMPOUNDS

[239] Melot, J. M., Texier-Boullet, F., and Foucaud, A. *Tetrahedron*, 1988, **44**, 2215.
[240] Nielsen, A. T. and Archibald, T. G. *Tetrahedron Lett.*, 1968, **30**, 3375.
[241] Perekalin, V. V. and Baier, K. *Zh. Obshch. Khim.*, 1960, **30**, 943.
[242] Shipchandler, M. T. *Synthesis*, 1979, 666.
[243] Umezawa, S, and Zen, S. *Bull. Chem. Soc. Jpn*, 1960, **33**, 1016; 1963, **36**, 1146, 1150.
[244] Mühlstadt, M. and Schulze, B. *J. Prakt. Chem.*, 1971, **313**, 745.
[245] Gross, Z. and Hoz, S. *J. Am. Chem. Soc.*, 1988, **110**, 7489.
[246] Corey, E. J. and Estreicher, H. *J. Am. Chem. Soc.*, 1978, **100**, 6294.
[247] Paulsen, H. and Greve, W. *Chem. Ber.*, 1974, **107**, 3013.
[248] Buckley, G. D., Hunt, F. G., and Lowe, A. *J. Chem. Soc.*, 1947, 1504.
[249] Elnagli, M. H., Sherif, S. M., and Mohareb, R. M. *Heterocycles*, 1987, **26**, 497.
[250] Nasakin, O. E., Lukin, P. M., Zilberg, S. P., Terentyev, P. B., Bulai, A. Kh., Dyachenko, O. A., Zolotoi, A. B., Konovalikhin, S. V., Gusev, A. I., Sheludyakov, V. D., and Atovmian, L. O. *Zh. Org. Khim.*, 1988, **24**, 1007.
[251] Bhaner, C. T. *US Pat. 2 426 158*, 1947; *Chem. Absrt.*, 1947, **41**, 7410.
[252] Perekalin, V. V. and Lerner, O. M. *Zh. Obshch. Khim.*, 1958, **28**, 1815.
[253] Shadrin, V. Yu. *Nitroacetonitrile in Reactions with Carbonyl Compounds*, dissertation abstract, Leningrad State Pedagogical Institute, Leningrad, 1987.
[254] Demireva, Z. I., Poiyanskaya, A. S., Mladenov, I., and Perekalin, V. V. *Zh. Org. Khim.*, 1976, **12**, 1192.
[255] Tsuge, O., Ueno, K., Kanemasa, S., and Yorozu, K. *Bull. Chem. Soc. Jpn*, 1987, **60**, 3347.
[256] Barton, D. H. R. and Zard, S. Z. *J. Chem. Soc., Chem. Commun.*, 1985, 1098.
[257] Ono, N., Kawamura, H., Bougauchi, M., and Maruyama, K. *J. Chem. Soc., Chem. Commun.*, 1989, 1580.
[258] Ono, N. and Maruyama, K. *Chem. Lett.*, 1988, 1511; Ono, N. and Maruyama, K. *Bull. Chem. Soc. Jpn*, 1988, **61**, 4470.
[259] Ono, N., Kawamura, H., Bougauchi, M., and Maruyama, K. *Tetrahedron*, 1990, **46**, 7483.
[260] Ono, N., Bougauchi, M., and Maruyama, K. *Tetrahedron Lett.*, 1992, **33**, 1629.
[261] Perekalin. V. V. and Sopova, A. S. *Zh. Obshch. Khim.*, 1958, **28**, 675.
[262] Polyanskaya, A. S., Bodina, R. I., Shadrin, V. Yu., and Aboskalova, N. I. *Zh. Org. Khim.*, 1984, **20**, 2481.
[263] Boberg, F., Garburg, K. H., Görlich, K.-J., Pipereit, E., Redelfs, E., and Ruhr, M. *J. Heterocycl. Chem.*, 1986, **23**, 1853.
[264] Yoshikoshi, A. and Miyashita, M. *Acc. Chem. Res.*, 1985, **18**, 284.
[265] Yanami, T., Ballatore, A., Miyashita, M., Kato, M., and Yoshikoshi, A. *Synthesis*, 1980, 407.
[266] Perekalin, V. V. and Parfenova, K. S. *Dokl. Akad. Nauk SSSR*, 1959, **124**, 592; *Zh. Obshch. Khim.*, 1960, **30**, 388; Danilova, E. M. and Perekalin, V. V. *Zh. Org. Khim.*, 1965, **1**, 1708.
[267] Stetter, H. and Hoehne, K. *Chem. Ber.*, 1958, **91**, 1344.
[268] Larson, H. O., Ooi, T.-C., Siu, A. K. Q., Hollenbeak, K. H., and Cue, F. L. *Tetrahedron*, 1969, **25**, 4005.
[269] Domimianni, S. J., Chaney, M. O., and Jones, N. D. *Tetrahedron Lett.*, 1970, 4735.
[270] Ansell, G. B., Moore, D. W., and Nielsen, A. T. *J. Chem. Soc. B*, 1971, 2376.
[271] Miyashita, M., Kumasawa, T., and Yoshikoshi, A. *J. Org. Chem.*, 1980, **45**, 2945.
[272] Miyashita, M., Kumasawa, T., and Yoshikoshi, A. *J. Chem. Soc. Chem. Commun.*, 1978, 362; Yanami, T., Kato, M., and Yoshikoshi, A. *J. Chem. Soc. Chem. Commun.*, 1975, 726.
[273] Yanami, T., Ballatore, A., Miyashita, M., Kato, M., and Yoshikoshi, A. *J. Chem. Soc., Perkin Trans. 1*, 1978, 1144.

[274] Konkova, V. A. and Perekalin, V. V. *Zh. Prikl. Khim.*, 1959, **32**, 1178; Kamlet, M. J. *J. Am. Chem. Soc.*, 1955, **77**, 4896.

[275] Kamlet, M. J. and Glover, D. J. *J. Am. Chem. Soc.*, 1956, **78**, 4556.

[276] Sapozhnikova, K. A., Sopova, A. S., and Moldaver, B. L. *Khim. Geterotsikl. Soedin.*, 1978, 963.

[277] Takeda, T., Hoshiko, T., and Mukaiyama, T. *Chem. Lett.*, 1981, 797.

[278] Gomez-Sanchez, A., Galan, J., Rico, M., and Bellanato, J. *J. Chem. Soc., Perkin Trans. 1*, 1985, 2695; Gomez-Sanchez, A., Mancera, M., and Rosado, F. *Carbohydr. Res.*, 1984, **134**, 75.

[279] Perekalin, V. V. and Sopova, A. S. *Zh. Obshch. Khim.*, 1954, **24**, 513.

[280] Sopova, A. S. and Temp, A. A. *Zh. Obshch. Khim.*, 1961, **31**, 1532.

[281] Grineva, V. S., Zobacheva, M. M., and Perekalin, V. V. *Zh. Org. Khim.*, 1967, **3**, 779.

[282] Grineva, V. S., Zobacheva, M. M., and Perekalin, V. V. *Zh. Org. Khim.*, 1972, **8**, 411.

[283] Gomez-Sanchez, A., Fernandez-Fernandez, R., Pascual, C., and Bellanato, J. *J. Chem. Res. (S)*, 1986, 318.

[284] Boberg, F., Marei, A., and Schultze, G. R. *Liebigs Ann. Chem.*, 1962, **655**, 102; Boberg, F., Marei, A., and Kirchkoff, K. *Liebigs Ann. Chem.*, 1967, **708**, 142; Boberg, F. and Schultze, G. R. *Chem. Ber.*, 1957, **90**, 1215.

[285] Berezovskii, V. M., Spiro, V. B., Denisova, L. Ya., and Sheiman, B. M. *Khim. Geterotsikl. Soedin.*, 1968, 628.

[286] Gomez-Sanchez, A., Stiefel, B. M., Fernandez-Fernandez, R., Pascual, C., and Bellanato, J. *J. Chem. Soc., Perkin Trans. 1*, 1982, 441.

[287] Boberg, F., Ruhr, M., and Garming, A. *Liebigs Ann. Chem.*, 1984, 223; Boberg, F., Garburg, K. H., Görlich, K.-J., Pipereit, E., and Ruhr, M. *Liebigs Ann. Chem.* 1984, 911; 1985, 239.

[288] Akthar, M. S., Sharma, V. L., and Bhaduri, A. P. *J. Heterocycl. Chem.*, 1987, **24**, 23.

[289] Bhaduri, A. P. *Synlett*, 1990, 557.

[290] Perekalin, V. V., Sopova, A. S., and Lipina, E. S. *Unsaturated Nitro Compounds*, Khimiya, Moscow, 1982.

[291] Sakakibara, T., Ikeda, Y., Miura, T., and Sudoh, R. *Chem. Lett.*, 1982, 99.

[292] Perekalin, V. V., Sopova, A. S., Zobacheva, M. M., Kruzite, M. P., Spunde, R. Ya., and Mikstais, U. Ya. *USSR Pat. 236 479*, 1968; *US Pat. 3 947 492*, 1976; *Br. Pat. 1 385 846*, 1975.

[293] Perekalin, V. V. and Sopova, A. S. *Dokl. Akad. Nauk SSSR*, 1954, **95**, 993; Perekalin, V. V. and Zobacheva, M. M. *Zh. Obshch. Khim.*, 1959, **29**, 2905.

[294] Perekalin, V. V. *Zh. Org. Khim.*, 1985, **21**, 1111.

[295] Vasilieva, O. S., Zobacheva, M. M., Smirnova, A. A., and Perekalin, V. V. *Zh. Org. Khim.*, 1978, **14**, 1420.

[296] Perekalin, V. V., Zobacheva, M. M., Smirnova, A. A., Vasilieva, O. S., Usik, N. V., Novikov, B. M., Andreeva, O. A., and Sinilina, I. P. in *The Chemistry of the Physiologically Active Compounds*, Chernogolovka, 1989, p. 194.

[297] Makosza, M. and Kwast, A. *Tetrahedron*, 1991, **47**, 5001.

[298] Williams, T. M., Hudcosky, R. J., Hunt, C. A., and Shepard, K. L. *J. Heterocycl. Chem.*, 1991, **28**, 13.

[299] Samoilovich, T. I., Polyanskaya, A. S., and Perekalin, V. V. *Zh. Org. Khim.*, 1969, **5**, 579.

[300] Sulimov, I. G., Samoilovich, T. I., Perekalin, V. V., Polyanskaya, A. S., and Usik, N. V. *Zh. Org. Khim.*, 1972, **8**, 1328.

[301] Samoilovich, T. I., Polyanskaya, A. S., and Perekalin, V. V. *Dokl Akad. Nauk SSSR*, 1974, **217**, 1355.

[302] Samoilovich, T. I., Polyanskaya, A. S., and Perekalin, V. V. *Zh. Org. Khim.*, 1967, **3**, 1532.

[303] Ono, N., Jun, T. Kh., and Kaji, A. *Synthesis*, 1987, 821.

[304] Aleksiev, D., Perekalin, V. V., and Lipina, E. S. *Dokl. Bolg. Akad. Nauk*, 1976, **29**, 1451; *Chem. Abstr.*, 1977, **86**, 155 224a.

[305] Berestovitskaya, V. M., Speranskii, E. M., Sulimov, I. G., and Trukhin, E. V. *Zh. Org. Khim.*, 1972, **8**, 1763.

[306] Rall, K. B. and Vildavskaya, A. I. *Zh. Org. Khim.*, 1975, **11**, 504.

[307] Pavlova, Z. F., Lipina, E. S., and Mostyaeva, L. V. *Metody Sint., Str. Khim. Prevraschen Nitrosoedin., Gertsenovskie Chteniya, Leningrad Gos. Pedagog. Inst., Leningrad*, 1980, 2; *Chem. Abstr.*, 1982, **96**, 19 742x.

[308] Severina, T. A., Ivanova, L. N., and Kucherov, V. F. *Izv. Adad. Nauk SSSR, Ser. Khim.*, 1967, 1111.

[309] Klager, K., Kispersky, J. P., and Hamel, E. *J. Org. Chem.*, 1961, **26**, 4368.

[310] Lipina, E. S., Perekalin, V. V., and Bobovich, Ya. S. *Zh. Obshch. Khim.*, 1964, **34**, 3635.

[311] Lipina, E. S. and Perekalin, V. V. *Zh. Obshch. Khim.*, 1964, **34**, 3644.

[312] Lipina, E. S. and Perekalin, V. V. *Acta Phys. Chem.*, 1973, **19**, 125.

[313] Seebach, D. and Ehrig, V. *Angew. Chem.*, 1974, **86**, 446.

[314] Ehrig, V. and Seebach, D. *Chem. Ber.*, 1975, **108**, 1961.

[315] Seebach, D. and Leitz, H. F. *Angew. Chem.*, 1969, **81**, 1047; 1971, **83**, 542.

[316] Richter, F. and Otto, H.-H. *Tetrahedron Lett.*, 1987, **28**, 2945.

[317] Calderari, G. and Seebach, D. *Helv. Chim. Acta*, 1985, **68**, 1592.

[318] Miyashita, M., Awen, B. Z. E, and Yoshikoshi, A. *J. Chem. Soc., Chem. Commun.*, 1989, 841.

[319] Miyashita, M., Awen, B. Z. E., and Yoshikoshi, A. *Tetrahedron*, 1990, **46**, 7569.

[320] Miyashita, M., Awen, B. Z. E., and Yoshikoshi, A. *Synthesis*, 1990, 563.

[321] Cory, R. M., Anderson, P. C., Bailey, M. D., McLaren, F. R., Renneboog, R. M., and Yamamoto, B. R. *Can. J. Chem.*, 1985, **63**, 2618.

[322] Cory, R. M., Anderson, P. C., McLaren, F. R., and Yamamoto, B. R. *J. Chem. Soc., Chem. Commun.*, 1981, 73.

[323] Zûger, M., Weller, T., and Seebach, D. *Helv. Chim. Acta*, 1980, **63**, 2005.

[324] Miyashita, M., Yamaguchi, R., and Yoshikoshi, A. *Chem. Lett.*, 1982, 1505.

[325] Posner, G. H. and Crouch, R. D. *Tetrahedron*, 1990, **46**, 7509.

[326] Miyashita, M., Yamaguchi, R., and Yoshikoshi, A. *J. Org. Chem.*, 1984, **49**, 2857.

[327] Seebach, D., Leitz, H. F., and Ehrig, V. *Chem. Ber.*, 1975, **108**, 1924.

[328] Ghera, E., Ben-Yaakov, E., Yechezkel, T., and Hassner, A. *Tetrahedron Lett.*, 1992, **33**, 2741.

[329] Schöllkopf, U., Kûhnle, W., Egert, E., and Dyrbusch, M. *Angew. Chem.*, 1987, **99**, 480; Busch, K., Groth, U. M., and Schöllkopf, U. *Tetrahedron*, 1992, **48**, 5607.

[330] Rowley, M., Leeson, P. D., Williams, B. J., Moore, K. W., and Baker, R. *Tetrahedron*, 1992, **48**, 3557.

[331] Tanaka, T., Hazato, A., Bannai, K., Okamura, N., Sugiura, S., Manabe, K., Toru, T., and Kurozumi, S. *Tetrahedron*, 1987, **43**, 813.

[332] Henning, R., Lehr, F., and Seebach, D. *Helv. Chim. Acta*, 1976, **59**, 2213.

[333] March, J. *Organic Chemistry*, Mir, Moscow, 1987, Vol. 3, p. 200, Vol. 2, p. 443.

[334] Miyashita, M., Yanami, T., Kumazawa, T., and Yoshikoshi, A. *J. Am. Chem. Soc.*, 1984, **106**, 2149.

[335] Miyashita, M., Yanami, T., and Yoshikoshi, A. *J. Am. Chem. Soc.*, 1976, **98**, 4679.

[336] Brook, M. A. and Seebach, D. *Can. J. Chem.*, 1987, **65**, 836.

[337] Seebach, D. and Brook, M. A. *Helv. Chim. Acta*, 1985, **68**, 319.

[338] Mateos, A. F. and Blanko, F. J. A. *J. Org. Chem.*, 1990, **55**, 1349.
[339] Stevens, R. W. and Mukaiyama, T. *Chem. Lett.*, 1985, 855.
[340] Risaliti, A., Forchiassin, M., and Valentin, E. *Tetrahedron*, 1968, **24**, 1889.
[341] Risaliti, A., Forchiassin, M., and Valentin, E. *Tetrahedron Lett.*, 1966, 6331.
[342] Risaliti, A., Marchetti, L., and Forchiassin, M. *Ann. Chim.*, 1966, **56**, 317.
[343] Ranganathan, D., Rao, C. B., Ranganathan, S., Mehrotra, A. K., and Iyengar, R. *J. Org. Chem.*, 1980, **45**, 1182.
[344] Valentin, E., Pitacco, G., and Colonna, F. P. *Tetrahedron Lett.*, 1972, 2837.
[345] Pitacco, G., Colonna, F. P., Valentin, E., and Risaliti, A. *J. Chem. Soc., Perkin Trans. 1*, 1974, 1625.
[346] Colonna, F. P., Valentin, E., Pitacco, G., and Risaliti, A. *Tetrahedron*, 1973, **29**, 3011.
[347] Valentin, E., Pitacco, G., Colonna, F. P., and Risaliti, A. *Tetrahedron*, 1974, **30**, 2741.
[348] Galliqaris, M., Manzini, G., Pitacco, G., and Valentin, E. *Tetrahedron*, 1975, **31**, 1501.
[349] Nielsen, A. T. and Archibald, T. G. *Tetrahedron*, 1970, **26**, 3475.
[350] Dancot, S., Pitacco, G., Risaliti, A., and Valentin, E. *Tetrahedron*, 1982, **38**, 1499.
[351] Nitti, P., Pitacco, G., and Rinaldi, V. *Croat. Chem. Acta*, 1986, **59**, 165.
[352] Huffman, J. W., Cooper, M. M., Miburo, B. B., and Pennington, W. T. *Tetrahedron*, 1992, **48**, 8213.
[353] Bradamante, P., Pitacco, G., Risaliti, A., and Valentin, E. *Tetrahedron Lett.*, 1982, **23**, 2683.
[354] Asaro, F., Pitacco, G., and Valentin, E. *Tetrahedron*, 1987, **43**, 3279.
[355] Brannock, K. C., Bell, A., Burpitt, R. D., and Kelly, C. A. *J. Org. Chem.*, 1964, **29**, 801.
[356] Kuehne, M. E. and Foley, L. *J. Org. Chem.*, 1965, **30**, 4280.
[357] Felluga, F., Nitti, P., Pitacco, G., and Valentin, E. *Tetrahedron*, 1989, **45**, 5667.
[358] Felluga, F., Nitti, P., Pitacco, G., and Valentin, E. *Tetrahedron*, 1989, **45**, 2099.
[359] Seebach, D. and Golinski, J. *Helv. Chim. Acta*, 1981, **64**, 1413.
[360] Blarer, S. J., Schweizer, W. B., and Seebach, D. *Helv. Chim. Acta*, 1982, **65**, 1637.
[361] Blarer, S. J. and Seebach, D. *Chem. Ber.*, 1983, **116**, 3086.
[362] Seebach, D., Beck, A. K., Golinski, J., Hay, J. N., and Laube, T. *Helv. Chim. Acta*, 1985, **68**, 162.
[363] Seebach, D. and Prelog, V. *Angew. Chem.*, 1982, **94**, 696.
[364] Pocar, D., Trimarko, P., Destro, R., Ortoleva, E., and Ballabio, M. *Tetrahedron*, 1984, **40**, 3579.
[365] Felluga, F., Nitti, P., Pitacco, G., and Valentin, E. *Tetrahedron Lett.*, 1988, **29**, 4165.
[366] (a) Felluga, F., Nardin, G., Nitti, P., Pitacco, G., and Valentin, E. *Tetrahedron*, 1988, **44**, 6921.
(b) Brussa, D., Felluga, F., Nitti, P., Pitacco, G., and Valentin, E. *Gazz. Chim. Ital.*, 1992, **122**, 85.
[367] Barbarella, G., Brückner, S., Pitacco, G., and Valentin, E. *Tetrahedron*, 1984, **40**, 2441.
[368] Nitti, P., Pitacco, G., and Valentin, E. *Vestn. Slov. Kem. Drus.*, 1986, **33**, 251.
[369] Cooper, M. M. and Huffman, J. W. *J. Chem. Soc., Chem. Commun.*, 1987, 348.
[370] Benedetti, F., Berti, F., Nitti, P., Pitacco, G., and Valentin, E. *Gazz. Chim. Ital.*, 1990, **120**, 25.
[371] Stevens, T. E. and Emmons, W. D. *J. Am. Chem. Soc.*, 1958, **80**, 338.
[372] Patterson, J. W. and McMurry, J. E. *J. Chem. Soc., Chem. Commun.*, 1971, 488.
[373] Gomez-Sanchez, A., Mancera, M., Rosado, F., and Bellanato, J. *Carbohydr. Res.*, 1984, **134**, 63.

[374] Gomez-Sanchez, A., Mancera, M., Rosado, F., and Bellanato, J. *J. Chem. Soc.*, *Perkin Trans. 1*, 1980, 1199.
[375] Grob, C. A. and Schad, H. P. *Helv. Chim. Acta*, 1955, **38**, 1121.
[376] Meyer, H. *Liebigs Ann. Chem.*, 1981, 1534.
[377] Wit, A. D., Pennings, M. L. M., Trompenaars, W. P., Reinhoudt, D. N., Harkema, S., and Nevestveit, O. *J. Chem. Soc.*, *Chem. Commun.*, 1979, 993.
[378] Pennings, M. L. M. and Reinhoudt, D. N. *J. Org. Chem.*, 1982, **47**, 1816.
[379] Wit, A. D., Trompenaars, W. P., Reinhoudt, D. N., Harkema, S., and Hummel, G. J. *Tetrahedron Lett.*, 1980, **21**, 1779.
[380] Barluenga, J., Aznar, F., Cabol, M.-P., and Valdes, C. *J. Chem. Soc.*, *Perkin Trans. 1*, 1990, 633.
[381] Seebach, D., Missbach, M., Calderari, G., and Eberle, M. *J. Am. Chem. Soc.*, 1990, **112**, 7625.
[382] Lapirre, J. M. and Gravel, D. *Tetrahedron Lett.*, 1991, **32**, 2319.
[383] Grieco, P. A., Yoshida, K., and Gurner, P. *J. Org. Chem.*, 1983, **48**, 3137.
[384] Michael, J. P., Maqutu, T. L., and Howard, A. S. *J. Chem. Soc.*, *Perkin Trans. 1*, 1989, 2389.
[385] Posner, G. H., Nelson, T. D., Kinter, Ch. M., and Johnson, N. *J. Org. Chem.*, 1992, **57**, 4083.
[386] Ranganathan, S., Ranganathan, D., and Mehrotra, A. K. *J. Am. Chem. Soc.*, 1974, **96**, 5241.
[387] Bartlett, P. A., Green, F. R., and Webb, T. R. *Tetrahedron Lett.*, 1977, 331.
[388] Danishefsky, S., Prisbylla, M. P., and Hiner, S. *J. Am. Chem. Soc.*, 1978, **100**, 2918.
[389] Danishefsky, S. and Hershenson, F. M. *J. Org. Chem.*, 1979, **44**, 1180.
[390] Just, G., Martel, A., Grozinger, K., and Ramjeesingh, M. *Can. J. Chem.*, 1975, **53**, 131.
[391] Grieco, P. A., Zelle, R. E., Lis, R., and Finn, J. *J. Am. Chem. Soc.*, 1983, **105**, 1403.
[392] Michael, J. P., Blom, N. F., and Glintenkamp, N. A. *J. Chem. Soc.*, *Perkin Trans. 1*, 1991, 1855.
[393] Michael, J. P., Blom, N. F., and Boeyens, J. C. A. *J. Chem. Soc.*, *Perkin Trans. 1*, 1984, 1739.
[394] Blom, N. F., Edwards, D. M. F., Field, J. S., and Michael, J. P. *J. Chem. Soc.*, *Chem. Commun.*, 1980, 1240.
[395] Michael, J. P. and Blom, N. F. *J. Chem. Soc.*, *Perkin Trans. 1*, 1989, 623.
[396] Lipina, E. S., Pavlova, Z. F., and Perekalin, V. V. *Zh. Org. Khim.*, 1969, **5**, 1312.
[397] Corey, E. J. and Estreicher, H. *Tetrahedron Lett.*, 1981, **22**, 603.
[398] Cardiner, J. M. and Brycl, M. R. *J. Org. Chem.*, 1990, **55**, 1261.
[399] Rao, T. V., Ravishankar, L., and Trivedi, G. K. *Indian J. Chem.*, *Sect. B*, 1990, **29**, 207.
[400] Stepanova, S. V., Lvova, S. D., Belikov, A. B., and Gunar, V. I. *Zh. Org. Khim.*, 1977, **13**, 889.
[401] Baruah, P. D., Makherjee, S., and Mahajan, M. P. *Tetrahedron*, 1990, **46**, 1951.
[402] Ono, N., Miyake, H., and Kaji, A. *J. Chem. Soc.*, *Chem. Commun.*, 1982, 33.
[403] Ono, N., Miyake, H., Kamimura, A., and Kaji, A. *J. Chem. Soc.*, *Perkin Trans. 1*, 1987, 1929.
[404] (a) Ono, N. in *Nitro Compounds* (Eds H. Feuer and A. Nielsen), VCH, New York, 1990, pp. 31, 98, 99.
(b) Ono, N. in *Nitro Compounds* (Eds H. Feuer and A. Nielsen), VCH, New York, 1990, p. 253.
(c) Ono, N. in *Nitro Compounds* (Eds H. Feuer and A. Nielsen), VCH, New York, 1990, p. 96.
[405] Fotthoff, B. and Breitmaier, E. *Chem. Ber.*, 1987, **120**, 255.

[406] Jain, P. C., Mikerjec, Y. N., and Anand, N. *J. Chem. Soc., Chem. Commun.*, 1971, 303.

[407] Shin, C., Kosuge, Y., Yamaura, M., and Yoshimura, J. *Bull. Chem. Soc. Jpn*, 1978, **51**, 1137.

[408] Binolra, J. S., Iain, P. G., and Anand, N. *Indian J. Chem.*, 1971, **9**, 388.

[409] Shin, C.-g., Yamaura, M., Inui, S., Ishida, Y., and Yoshimura, J. *Bull. Chem. Soc. Jpn*, 1978, **51**, 2618.

[410] Boeyens, J. C. A., Denner, L., and Michael, J. P. *J. Chem. Soc., Perkin Trans. 2*, 1984, 767, 1569.

[411] Kurth, M. J., O'Brien, M. J., Hope, H., and Yanuck, M. *J. Org. Chem.*, 1985, **50**, 2626.

[412] Moreno, M. C., Plume, J., Roman, E., Serrano, A. J., Rodriguer, M. L., and Puiz-Perez, C. *Tetrahedron Lett.*, 1989, **30**, 3179.

[413] Serrano, J. A., Moreno, M. C., Roman, E., Arjiona, O., Plumet, J., and Jimenez, J. *J. Chem. Soc., Perkin Trans. 1*, 1991, 3207.

[414] Serrano, J. A., Caceres, L. E., and Roman, E. *J. Chem. Soc., Perkin Trans. 1*, 1992, 941.

[415] Node, M., Hao, X., and Fuji, K. *Chem. Lett.*, 1991, 57.

[416] Enders, D., Meyer, O., and Raabe, G. *Synthesis*, 1992, 1242.

[417] Kataev, E. *Soobshch. Vses. Khim. Obshch. D.I. Mendeleeva*, 1955, 49.

[418] Kusurkar, R. S. and Bhosale, D. K. *Synth. Commun.*, 1990, **20**, 101.

[419] Speranskii, E. M., Berestovitskaya, V. M., Sulimov, I. G., and Trukhin, E. V. *Zh. Org. Khim.*, 1972, **8**, 1763; Berestovitskaya, V. M., Speranskii, E. M., and Perekalin, V. V. *Zh. Org. Khim.*, 1976, **12**, 2256.

[420] Barko, A., Benetti, S., Pollini, G. P., Spaiuto, G., and Zanirato, V. *Tetrahedron Lett.*, 1991, **32**, 2517.

[421] Podgornova, N. N., Lipina, E. S., and Perekalin, V. V. *Zh. Org. Khim.*, 1974, **10**, 409; 1976, **12**, 25.

[422] Berestovitskaya, V. M., Titova, M. V., and Perekalin, V. V. *Zh. Org. Khim.*, 1980, **16**, 891.

[423] Just, G., Liak, G. J. T., Lim, M.-I., Potvin, P., and Tsantrizos, Y. S. *Can. J. Chem.*, 1980, **58**, 2024.

[424] Hendrickson, J. B., Alder, R. W., Dalton, D. R., and Heg, D. G. *J. Org. Chem.*, 1969, **34**, 2667.

[425] Landeryou, V. A., Grabowski, E. J. J., and Autrey, R. L. *Tetrahedron*, 1969, **25**, 4307.

[426] Bryce, M. R., Gardiner, J. M., and Hursthouse, M. B. *Tetrahedron Lett.*, 1987, **28**, 577.

[427] Prazere, M. A., Peters, J. A., Linders, J. T. M., and Maat, L. *Recl Trav. Chim. Pays-Bas*, 1986, **105**, 554.

[428] Linders, J. T. M., Briel, P., Fog, E., Lie, T. S., and Maat, L. *Recl Trav. Chim. Pays-Bas*, 1989, **108**, 268.

[429] Ono, N., Miyake, H., Kamimura, A., Tsukui, N., and Kaji, A. *Tetrahedron Lett.*, 1982, **23**, 2957.

[430] Tanis, S. P. and Abdollah, Y. M. *Synth. Commun.*, 1986, **16**, 251.

[431] Nesi, R., Giomi, D., Papaleo, S., and Quartara, L. *J. Chem. Soc., Chem. Commun.*, 1986, 1536.

[432] Tohda, Y., Yamawaki, N., Matsui, H., Kawashima, T., Ariga, M., and Mori, Y. *Bull. Chem. Soc. Jpn*, 1988, **61**, 461.

[433] Denmark, S. E., Cramer, C. J., and Sternberg, J. A. *Tetrahedron Lett.*, 1986, **27**, 3693.

[434] Denmark, S. E., Cramer, C. J., and Sternberg, J. A. *Helv. Chim. Acta*, 1986, **69**, 1971.

[435] Denmark, S. E., Kesler, B. S., and Moon, Y.-Ch. *J. Org. Chem.*, 1992, **57**, 4912.

[436] Denmark, S. E., Dappen, M. S., and Cramer, C. J. *J. Am. Chem. Soc.*, 1986, **108**, 1306.

[437] Denmark, S. E., Moon, Y.-Ch., Cramer, C. J., Dappen, M. S., and Senanayake, C. B. W. *Tetrahedron*, 1990, **46**, 7373.

[438] Denmark, S. E., Moon, Y.-Ch., and Senanayake, C. B. W. *J. Am. Chem. Soc.*, 1990, **112**, 311.

[439] Denmark, S. E., Senanayake, C. B. W., and Ho, G. D. *Tetrahedron*, 1990, **46**, 4857.

[440] Denmark, S. E. and Schnute, M. E. *J. Org. Chem.*, 1991, **56**, 6738.

3 FUNCTIONALIZED DERIVATIVES OF UNSATURATED NITRO COMPOUNDS

3.1 HALONITROALKENES

3.1.1 METHODS OF SYNTHESIS

The geminal and vicinal monohalonitroalkenes and polyhalonitroalkenes are very useful in the syntheses of various classes (including some that are extraordinary) of organic compounds. Some halonitroethenes are used as fungicides and biologically active substances [1, 2].

Vicinal halonitroalkenes are prepared through nitration–chlorination of acetylene hydrocarbons (1; equation 1). The method is of considerable importance because it provides an approach to different 2-chloro-1-nitroethenes (2) in a one-step synthesis [3, 4].

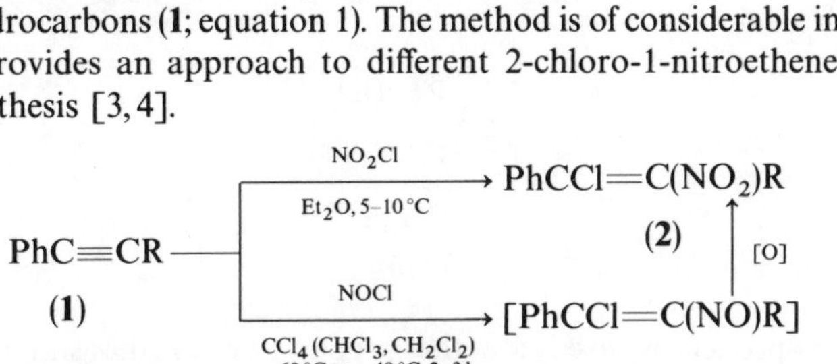

$$PhC\equiv CR \begin{array}{c} \xrightarrow[Et_2O,\,5-10\,°C]{NO_2Cl} PhCCl\!=\!C(NO_2)R \quad (2) \; \Big\uparrow {}_{[O]} \\ \\ \xrightarrow[\substack{CCl_4\,(CHCl_3,\,CH_2Cl_2) \\ -60\,°C\,to\,-40\,°C,\,2-3\,h}]{NOCl} [PhCCl\!=\!C(NO)R] \end{array} \qquad (1)$$

$$(1)$$

$$R = H,\ Me,\ Ph$$

Nitration–chlorination is also used as a synthetic method for vicinal alkyl(aryl)-halonitroethenes [5–7].

Freeman and Emmons [8] isolated (3) and (4) in a free radical reaction (equation 2)

$$(1) \xrightarrow[Et_2O,\,12\,h]{NO_2Cl} PhCCl\!=\!CHNO_2 + PhC(O)CHCl_2 \qquad (2)$$

$$(3)\ 34\% \qquad\qquad (4)\ 9\%$$

Simultaneous action of N_2O_4 and I_2 upon acetylenes (5) [9] results in the formation of vicinal iodonitroethenes (6; equation 3) [10a].

$$R^1C\equiv CR^2 \xrightarrow{N_2O_4+I_2} R^1CI{=}C(NO_2)R^2 \qquad (3)$$

$$\text{(5)} \qquad\qquad\qquad \text{(6)}$$

$$R^1 = n\text{-Bu, }t\text{-Bu, Ph;} \quad R^2 = H, Ph$$

Dehydrohalogenation (equation 4) of adducts of type (8) produces compounds of type (9) [11–14].

$$RCCl{=}CH_2 \longrightarrow
\begin{cases}
\xrightarrow{NOCl} [RCCl_2CH_2] \xrightarrow{[O]} \\
\qquad\qquad\quad\mid \\
\qquad\qquad\; NO \\
\xrightarrow[NO_2Cl,\,CHCl_3,\,0-5\,°C]{}
\end{cases}
RCCl_2CH_2NO_2 \longrightarrow RCCl{=}CHNO_2$$

$$\text{(7)} \qquad\qquad\qquad\qquad\qquad\qquad\qquad \text{(8)} \qquad\qquad \text{(9)}$$

$$\qquad\qquad\qquad\qquad\qquad\qquad\qquad\qquad\qquad\qquad\qquad\qquad\qquad (4)$$

$$R = Me, Et, i\text{-Pr}$$

The synthesis of α-bromo-β-nitro-β-alkyl(aryl)acrylates involves the dehydrobromination (TEA) of α,β-dibromo-β-nitrocarboxylates in C_6H_6 [15]. An alternative approach to such products is through nitration of α-bromoacrylates with HNO_3 [16].

The synthetic methods for geminal halonitroethenes consist mainly in the dehydration of halonitroalcohols (10), deacylation of the corresponding acylated derivatives of (10), dehydrohalogenation of dihalonitroethanes, or nitration of haloethenes.

$$
\begin{array}{c}
R^1 \quad Hal \\
\mid \quad\; \mid \\
R^2{-}C{-}C{-}NO_2 \\
\mid \quad\; \mid \\
OH \;\; H
\end{array}
$$

$$\text{(10)}$$

The first approach is used for the preparation of the simplest aliphatic halonitroethenes. The halonitroethanols are the starting materials in the synthesis of 1-bromo-1-nitroethenes, 1-chloro-1-nitroethenes and 1-fluoro-1-nitroethenes [17–19]. Denitration of 2-chloro-2-nitroethyl nitrate was the first synthetic method used to prepare 1-chloro-1-nitroethene [19]. Halonitroethenes with C_3–C_8 chains (equation 5) are formed in the deacylation of compounds such as (11) under the action of dry $KHCO_3$ [20] or Na_2CO_3 [21].

$$MeC(O)OC(R)HCH(Hal)NO_2 \xrightarrow[Et_2O]{dry\,KHCO_3\,or\,Na_2CO_3} RCH{=}C(Hal)NO_2 \qquad (5)$$

$$\text{(11)} \qquad\qquad\qquad\qquad\qquad\qquad\qquad\qquad \text{(12)}$$

$$R = Alk; \quad Hal = Cl, Br$$

Dehydrohalogenation of (14) results in the alkene (15; equation 6) in high yield under the action of sodium or potassium acetate, TEA, pyridine or other base [22–26].

$$\text{ArCH}\!=\!\text{CHNO}_2 \xrightarrow{\text{Br}_2,\,\text{CHCl}_3} \text{ArCHBrCHBrNO}_2 \xrightarrow{\text{AcOK, EtOH}} \text{ArCH}\!=\!\text{CBrNO}_2 \qquad (6)$$

$$\quad\;\;(13) \qquad\qquad\qquad\quad (14) \qquad\qquad\qquad\quad (15)$$

Geminal chloronitroethenes are prepared by a similar method [27] from dichloronitroalkanes.

This method has been used successfully for the synthesis of geminal bromonitroenones [28]. 1,3-Bis(β-bromo-β-nitrovinyl)benzene and 1,4-bis(β-bromo-β-nitrovinyl)benzene are easily obtainable by dehydrohalogenation of the corresponding tetrabromo derivatives [24, 29].

Kono et al. [30] suggested the original preparative method for 2-ferrocenyl-1-iodo-1-nitroethene (17; equation 7).

$$\text{FcCH}\!=\!\text{CHNO}_2 \xrightarrow{\text{MeONa}} \text{FcCH(OMe)CH}\!=\!\text{NO}_2^-\,\text{Na}^+ \xrightarrow{\text{I}_2,\,\text{Na}_2\text{S}_2\text{O}_3}$$

$$(16) \qquad\qquad\qquad\qquad\qquad\qquad\qquad\qquad \text{FcCH}\!=\!\text{CINO}_2 \qquad (7)$$

$$\qquad\qquad\qquad\qquad\qquad\qquad\qquad\qquad\qquad\qquad (17)$$

1-Chloro-1-nitro-2-furylethenes are formed in a one-step synthesis by coupling of furfurol or its derivatives with chloronitromethane [31, 32].

The condensations of salicyclic aldehyde derivatives (18; equation 8) with halonitroethanes lead to substituted 1-halo-1-nitroethenes (19) [33].

$$(18) \qquad\qquad\qquad\qquad\qquad\qquad\qquad (19)$$

$$R^1 = \text{H, H, H, H, H, H, H, Br;}$$
$$R^2 = \text{H, Cl, Br, H, H, NO}_2\text{, Br, H;}$$
$$R^3 = \text{H, H, H, OMe, H, H, H, H;}$$
$$R^4 = \text{H, H, H, H, OMe, H, OMe, OMe}$$

The reaction of dinitrogen tetroxide with 1-bromo-2,2-diphenylethene in ether at 10 °C over six hours gave 1-bromo-1-nitro-2,2-diphenylethene in 50% yield [34].

Dihalonitroethenes and polyhalonitroethenes are obtained by a very simple direct nitration of unsaturated hydrocarbons. 1,2-Dichloro-1-nitroethene (equation 9) can be synthesized from vicinal dichloroethene or acetylene and N_2O_4 [35].

$$\text{ClCH}\!=\!\text{CHCl} \xrightarrow{\text{N}_2\text{O}_4} \text{HCCl}\!=\!\text{CClNO}_2 \xleftarrow{\text{N}_2\text{O}_4} \text{ClC}\!\equiv\!\text{CCl} \qquad (9)$$

$$\quad\;\;(20) \qquad\qquad\qquad (22) \qquad\qquad\qquad (21)$$

It is possible to prepare 2,2-dichloro-1-nitroethenes (25; equation 10) in the reactions of geminal dichloroethenes with fuming nitric acid, nitrile chloride or nitrosyl chloride followed by dehydrogenation [12, 13, 36].

$$Cl_2C{=}CHR \xrightarrow[\text{iii. NO}_2\text{Cl}]{\substack{\text{i. Concentrated HNO}_3 \\ \text{ii. NOCl}}} Cl_3CCHRNO_2 \xrightarrow[\text{ii. NEt}]{\text{i. AcONa}} Cl_2C{=}CRNO_2 \quad (10)$$
$$\quad\;\; \textbf{(23)} \qquad\qquad\qquad\qquad\quad \textbf{(24)} \qquad\qquad\quad \textbf{(25)}$$

The action of a nitrating mixture on chlorovinylidene (26; equation 11) leads to (27) along with a small amount (10%) of 2,2-dichloro-1-nitroethene (28) [37].

$$CCl_2{=}CH_2 \xrightarrow{\text{HNO}_3 + \text{H}_2\text{SO}_4} O_2NCH_2C(O)Cl + CCl_2{=}CHNO_2 \quad (11)$$
$$\quad\;\; \textbf{(26)} \qquad\qquad\qquad\qquad \textbf{(27)} \qquad\qquad \textbf{(28)}$$

Recently, 2,2-dibromo-1-nitroethene was synthesized in 50% yield by interaction between nitric acid (57–68%) and tribromoethene [38]. Trichloroethene (29; equation 12) transforms into (30) under the action of 70% HNO_3 [39].

$$Cl_2C{=}CHCl \xrightarrow[85\text{–}90\,°C,\,2\text{–}3\,h]{70\%\,\text{HNO}_3} Cl_2C{=}CClNO_2 \quad (12)$$
$$\quad\;\; \textbf{(29)} \qquad\qquad\qquad\qquad \textbf{(30)}$$

A systematic investigation of the relative activities of halogens at the double bond in the reactions of symmetrical dichlorodiiodoethenes, dibromodiiodoethenes and dichlorodibromoethenes with N_2O_4 was performed by Buevich et al. [40]. The study exposed a tendency for preferential iodine substitution by the nitro group (equation 13) in compounds of type (31).

$$IC(Hal){=}C(Hal)I \xrightarrow[\text{Hal = Cl, Br}]{N_2O_4} IC(Hal){=}C(Hal)NO_2 \quad (13)$$
$$\quad\;\; \textbf{(31)} \qquad\qquad\qquad\qquad \textbf{(32)}$$

The nitration of tetraiodoethene by HNO_3 leads to the formation of 2,2-diiodo-1,1-dinitroethene [41].

3.1.1.1 Halonitrodienes

The first synthesis of 1,4-dibromo-1,4-dinitro-1,3-butadiene, reported by Lipina and Perekalin [42], included dehydrobromination of 1,2,3,4-tetrabromo-1,4-dinitrobutane (34; equation 14), which proceeded smoothly with heating for a short period in ethanol. Subsequently, the method was developed by using potassium acetate [43] or aqueous MeOH (20 h) [44].

Compound (35) was also isolated in low yield after bromination [45] from the reaction mixture of 1,4-dinitro-2-butene and butyllithium. Dehydrochlorination of 1,2,3,4-tetrachloro-1,4-dinitrobutane with potassium acetate gives 1,4-dichloro-1,4-dinitro-1,3-butadiene in quantitative yield [43]. The same compound can be synthesized by nitration of 1,4-dichloro-1,3-butadiene [46].

$$O_2NCH=CHCH=CHNO_2 \xrightarrow{2Br_2} \underset{\underset{Br}{|}}{\overset{\overset{Br\ Br}{|\ \ |}}{O_2NCHCHCHCHNO_2}} \underset{-2HBr}{\longrightarrow}$$

(33)

(34)

$$O_2NC\underset{\overset{|}{Br}}{=}CHCH\underset{\overset{|}{Br}}{=}CNO_2 \qquad (14)$$

(35)

An original synthetic approach to 1-halo-1,3-dinitro-2,4-diphenyl-1,3-buta-dienes by means of the thermal isomerization of 3-halo-1,3-dinitro-2,4-diphenyl-1-cyclobutenes (equation 15) has been reported by Miller *et al.* [47].

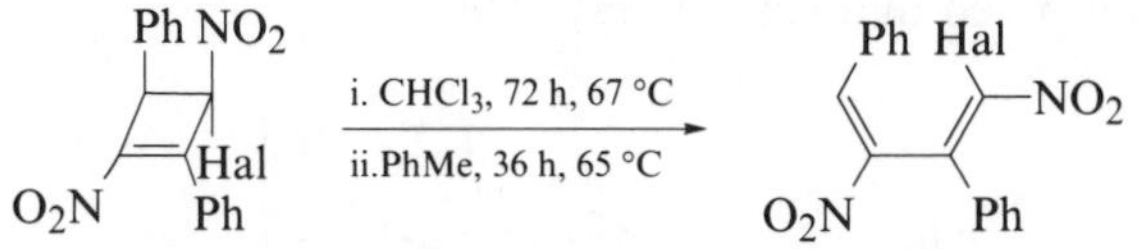

$$\underset{\underset{(36)}{O_2N\quad Ph}}{\overset{Ph\ NO_2}{\diagdown\diagup}}\underset{Hal}{\diagup} \xrightarrow[\text{ii.PhMe, 36 h, 65 °C}]{\text{i. CHCl}_3,\ 72\ h,\ 67\ °C} \underset{\underset{(37)}{O_2N\quad Ph}}{\overset{Ph\ Hal}{\diagup\diagdown}}NO_2 \qquad (15)$$

Hal = Br, Cl

The substitutional nitration of 2-*H*-pentachloro-1,3-butadiene (38; equation 16) with 57–68% HNO_3 at 80–95 °C leads to (39) [48].

$$Cl_2C=CClCH=CCl_2 \xrightarrow[\text{80–95 °C}]{\text{HNO}_3\ (57–68\%)} Cl_2C=CClC(NO_2)=CCl_2 \qquad (16)$$

(38) (39)

A similar synthetic pathway has been proposed for 1-nitropentachloro-1,3-butadiene [49].

3.1.2 REACTIONS OF HALONITROETHENES

The nucleophilic substitution of a halogen atom is characteristic for vicinal halonitroalkenes. The process is illustrated by the interaction between 2-halo-1-nitroethenes (40) and amines (equation 17), which results in substitution of the halogen by the primary or secondary amine [3–5, 8, 10a, 13].

$$R^1C(Hal)=CR^2NO_2 \xrightarrow{R^3R^4NH} R^3R^4NCR^1=CR^2NO_2 \qquad (17)$$

(40) (41)

$$R^1, R^2, R^3, R^4 = H,\ Alk,\ Ar;\quad Hal = Cl,\ I$$

The reactions with other nucleophiles such as phenolates, thiolates, thio-isocyanates, salts of sulfinic acids, etc. [6, 10a, 50a] proceed in a similar way (equation 18).

$$R^1C(Hal){=}CR^2NO_2 + Nu^- \longrightarrow \begin{matrix} R^1 \\ Nu \end{matrix}{>}C{=}C{<}\begin{matrix} NO_2 \\ R^2 \end{matrix} \qquad (18a)$$

$$Hal = Cl, I$$

The kinetics and stereochemistry of the reactions of 2-halo-1-1nitrostilbenes with various nucleophiles have been studied in detail [10a, 10b]. The reactions are not generally stereospecific, and the only case of specific formation of the (E) isomer proves that the reactions proceed via the formation of an ionic inter-mediate which has a tendency towards intramolecular rotation.

The reactions of vicinal halonitroethenes with difunctionalized nucleophiles are more complex. For example, reactions between 3-nitro-4-chlorocoumarins and 1,2-ethanediol in the presence of TEA are accompanied by intramolecular nucleophilic attack and give spirocyclic compounds that are isolated in the form of the corresponding salts. The reactions of pyrocatechol derivatives (equation 18b) proceed in a similar way [50b].

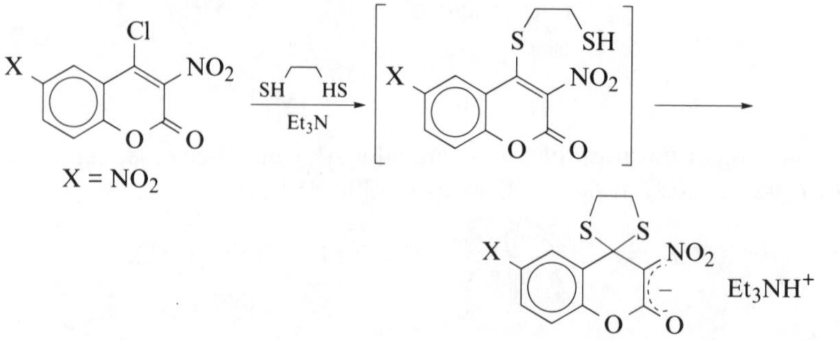

$$(18b)$$

$$X = NO_2$$

Additional activating groups such as acyl or alkoxy in the vicinal position in halonitroethenes initiate halogen atom and nitro group substitution under the action of a nucleophile [16].

Acetylides (43; equation 19) substitute the halogen atom and the nitro group in compounds such as (42) [51].

$$Ph(O_2N)C{=}ClPh + 2NaC{\equiv}CR \xrightarrow[70\,°C,\,2\,h]{DMF} RC{\equiv}CCPh{=}CPhC{\equiv}CR \qquad (19)$$

$$(42) \qquad\qquad (43) \qquad\qquad\qquad (44)$$

$$R = H, Ph$$

The reactions of halonitroethenes (45; equation 20) with MeONa give compounds such as (46) (R = H [13], Ph [52]).

$$HalCR{=}CHNO_2 \xrightarrow{MeO^-} (MeO)_2CRCH_2NO_2 \qquad (20)$$

$$(45) \qquad\qquad\qquad (46)$$

$$Hal = Cl, I; \quad R = H, Ph$$

The reaction of 2-iodo-1,2-diphenyl-1-nitroethene with NaN_3 results in the formation of furoxan [9, 10a].

Halonitroethenes (47; equation 21) combine with non-enolizing CH acids in accordance with an addition–elimination mechanism via the intermediate carbanion and give the substituted nitroallyls (48) [52].

$$PhC(Hal){=}CHNO_2 \xrightarrow[-NaHal]{RCH_2CO_2Me,\ MeONa} MeO_2CCR{=}CPhCH_2NO_2 \quad (21)$$
$$\text{(47)} \hspace{6.5cm} \text{(48)}$$

$$Hal = Cl, I; \quad R = CO_2Me, CN$$

2-Chloro-1-nitroethene reacts with malonic ester to produce the dimeric product (49), the structure of which has been elucidated by $^1H\,NMR$ spectroscopy [53].

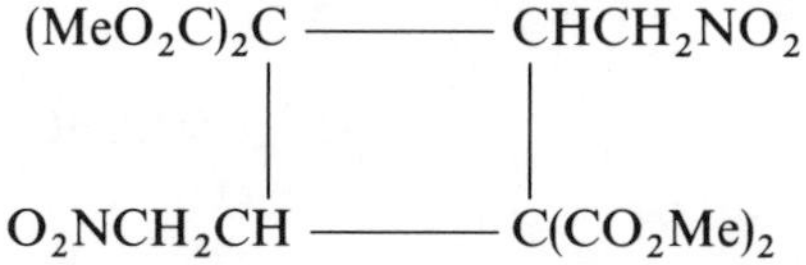

$$(MeO_2C)_2C \text{———} CHCH_2NO_2$$
$$\mid \hspace{3.5cm} \mid$$
$$O_2NCH_2CH \text{———} C(CO_2Me)_2$$
$$\text{(49)}$$

On the other hand, Volynskii et al. [52] have found that enolizing CH acids (acetoacetic esters, acetylacetone, benzoylacetone, dimedone) in combination with nitrohalo alkenes (50; equation 22) give substituted nitrovinyls (51).

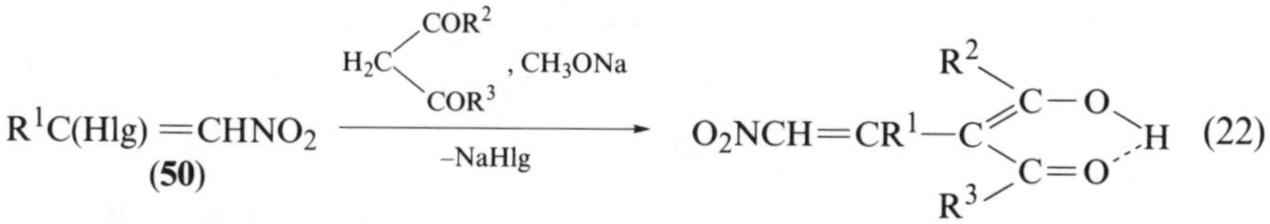

$$R^1C(Hlg){=}CHNO_2 \xrightarrow[-NaHlg]{H_2C\langle^{COR^2}_{COR^3},\ CH_3ONa} O_2NCH{=}CR^1{-}C\langle^{R^2{-}C-O}_{R^3{-}C=O}\rangle H \quad (22)$$
$$\text{(50)} \hspace{8.5cm} \text{(51)}$$

$$Hlg = Cl, I; \quad R^1 = H, Ph; \quad R^2 = R^3 = Me; \quad R^2 = Me, R^3 = Ph;$$
$$R^2 = Me, R^3 = OEt$$

Armand and Convert [54] have described some of the chemical reactions of geminal halonitroethenes, such as oxidation, hydrolysis, reduction and pyrolysis, many of which are accompanied by degradation. Geminal halonitroethenes participate in electrophilic reactions with difficulty. For example, their halogenation either takes place only under severe conditions or not at all [21].

Halogen atoms in the geminal position in halonitroethenes (52; equation 23) redirect the reaction pathway with nucleophilic reagents in comparison with the vicinal analogues. For example, primary and secondary amines (ammonia, aniline, piperidine and others) react with geminal bromonitroethenes to give the corresponding hydrochlorides (53) [21].

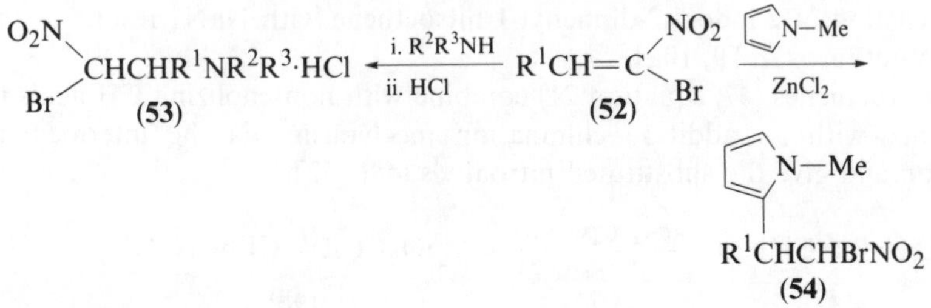

$$O_2N \diagdown \atop Br \diagup CHCHR^1NR^2R^3 \cdot HCl \xleftarrow[\text{ii. HCl}]{\text{i. } R^2R^3NH} R^1CH = C \diagup NO_2 \diagdown \atop Br \quad \xrightarrow{ZnCl_2} \quad (23)$$

(53)   (52)

R^{1}CHCHBrNO$_2$
(54)

Bromonitrostyrene reacts with N-methylpyrrole in the presence of anhydrous ZnCl$_2$ to form the adduct (54) [55].

The reactions of sulfinic acids with geminal halonitrostyrenes (55; equation 24) depend upon the substituents in the halonitrostyrene aromatic ring and result in the formation of either the adducts (56) or the products of their dehydrohalogenation (57) [56].

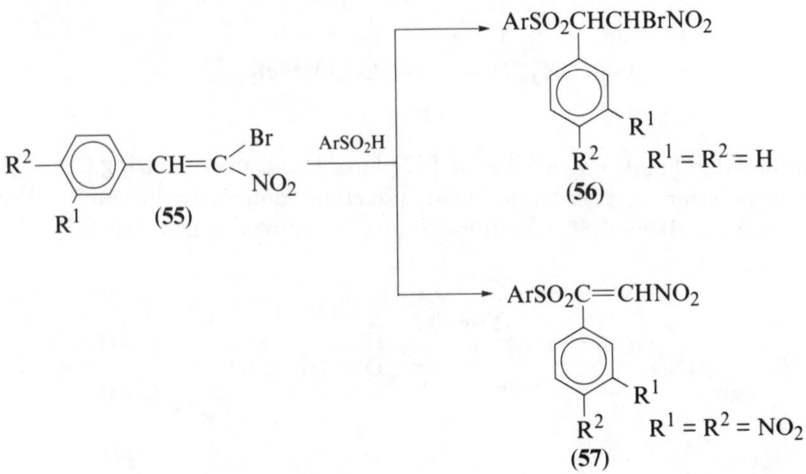

ArSO$_2$CHCHBrNO$_2$

$$R^2 \diagdown \bigcirc \diagup CH = C \diagup Br \diagdown NO_2 \xrightarrow{ArSO_2H}$$

R^1 R^1
(55)   R^2 R^1 = R^2 = H   (24)
                                (56)

ArSO$_2$C=CHNO$_2$

R^1
R^2 R^1 = R^2 = NO$_2$
(57)

As it was shown by Sopova et $al.$ compounds such as (52) react with various CH acids (equation 25), including β-dicarbonyl compounds (58) (acetylacetone, benzoylacetone, dibenzoylmethane, acetoacetic esters), to give the Michael products (59), nitrocyclopropanes (60) and nitrodyhydrofurans (61) [57–63].

The reactions of 1-bromo-1-nitro-2-alkyl(aryl)ethenes with cyanoacetic ester or malonitrile initiated by an equimolar quantity of MeONa do not end in nucleophilic addition but progress further with intramolecular C-alkylation to give the corresponding substituted nitrocyclopropanes (62) [64]. Geminal bromonitroethenes (52) combine with 1,3-indandione and 1-phenyl-3-methyl-5-pyrazolones to give spironitrocyclopropanes (63).

The reaction of dimedone and dihydroresorcinol with (52) leads to the formation of (64) [65, 66].

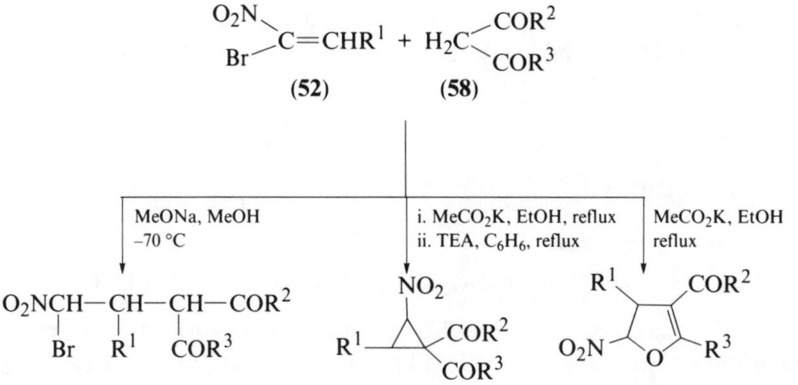

$$
\begin{array}{cc}
\underset{Br}{\overset{O_2N}{>}}C\!\!=\!\!CHR^1 + H_2C\!\!\underset{COR^3}{\overset{COR^2}{<}} \\
(52) \qquad\qquad (58)
\end{array}
$$

| MeONa, MeOH −70 °C | i. MeCO$_2$K, EtOH, reflux ii. TEA, C$_6$H$_6$, reflux | MeCO$_2$K, EtOH reflux |

$$
\begin{array}{ccc}
\underset{\underset{Br}{|}\ \underset{R^1}{|}\ \underset{COR^3}{|}}{O_2NCH\!-\!CH\!-\!CH\!-\!COR^2} &
R^1\!\!-\!\!\triangle\!\!\overset{NO_2}{\underset{COR^3}{<}COR^2} &
\underset{O_2N}{\overset{R^1}{>}}\!\!\diagdown\!\!\overset{COR^2}{\underset{R^3}{<}} \qquad (25)\\
(59) & (60) & (61)
\end{array}
$$

(59), (60): R^1 = Alk, Ar; R^2 = R^3 = OMe, OEt

(60): R^1 = Alk, Ar; R^2 = R^3 = Me; R^2 = OEt, R^3 = Me; R^2 = Me, R^3 = Ph

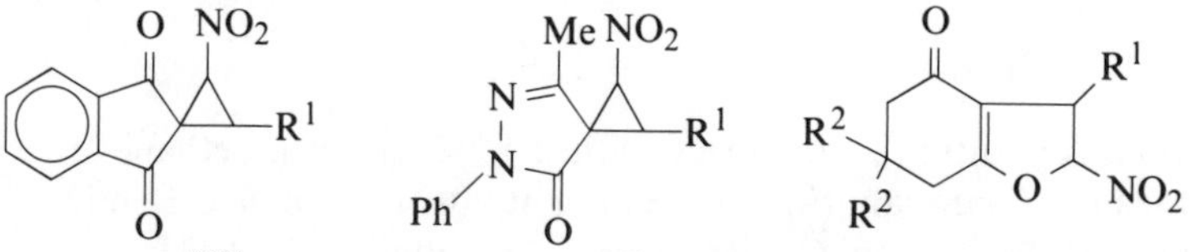

$$
\begin{array}{ccc}
(62) & (63) & (64)
\end{array}
$$

R^1 = Alk, Ar; R^2 = H, Me

Intramolecular cyclization is also observed in the reactions of 1-chloro-1-nitro-2-arylethenes with 4-hydroxy-2-benzopyranone [67] and substituted salicylic aldehydes [68a], leading to the corresponding compounds (65) and (66).

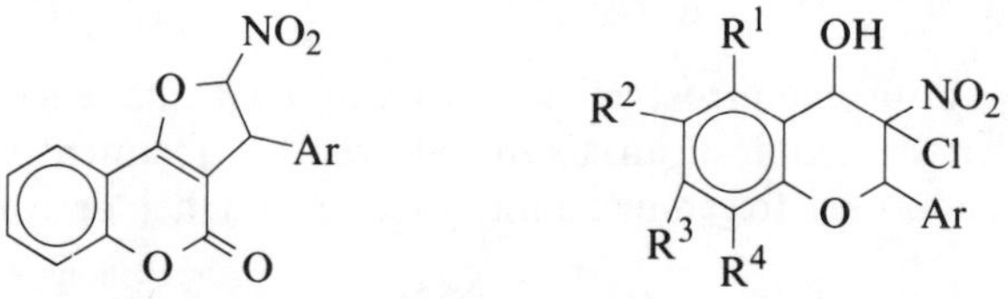

$$
\begin{array}{cc}
(65) & (66)
\end{array}
$$

A similar reaction takes place with 2,4-dihydroxypyrimidine [68b]. The reactions between nitroalkanes and 1-bromo-1-nitro-2-alkyl(aryl)ethenes result in the formation of the products of Michael condensation (68; equation 26) [69].

$$
\underset{(67)}{(52) + R^2CH(NO_2)R^3} \xrightarrow{\text{base}} \underset{(68)}{O_2NCHBrCHR^1CR^2R^3NO_2} \qquad (26)
$$

R^1 = Me, Et, Ph; R^2 = H, Me; R^3 = H, Me, Et

The characteristic feature of the reactions between geminal bromonitroethenes and nitroacetic esters or nitroacetonitrile (equation 27) is the formation of isoxazoline N-oxides [70].

$$\underset{\substack{\text{Br} \\ (52)}}{\overset{\text{O}_2\text{N}}{\diagdown}}\text{C}=\text{CHR}^1 + \text{H}_2\text{C}\underset{(69)}{\overset{\text{NO}_2}{\diagup_{\text{R}^2}}} \xrightarrow[\substack{0-25\,^\circ\text{C}}]{\text{MeONa, MeOH}} \underset{(70)}{\text{isoxazoline}} \tag{27}$$

$R^1 = \text{Et, Ph, } p\text{-MeOC}_6\text{H}_4, m\text{-NO}_2\text{C}_6\text{H}_4, p\text{-NO}_2\text{C}_6\text{H}_4; \; R^2 = \text{CO}_2\text{Et, CN}$

Substituted isoxazoline N-oxides (72; equation 28) are formed in a similar way under the action of ketoselenonium ketoylides (71) on bromonitrostyrenes (15) [71].

$$\underset{\substack{\text{Ar}^1 = \text{Ar}^2 = \text{Ph} \\ (15)}}{\text{Ar}^1\text{CH}=\text{C}\underset{\text{Br}}{\overset{\text{NO}_2}{\diagup}}} \xrightarrow[\text{NaOH}]{\overset{+}{\text{Me}_2\text{SeCH}_2\text{COAr}^2\text{Br}^-}} \underset{(72)}{\text{structure}} \tag{28}$$

Geminal halonitroalkenes react with 1,1,2,2-tetracyanoethane under mild conditions in a 2-propanol/water medium at 40–50 °C to give 2-amino-3-nitro-1,5,5-tricyano-1-cyclopentene derivatives (75; equation 29) [72].

$$\underset{(73)}{\text{RCH}=\text{C}\underset{\text{NO}_2}{\overset{\text{Cl}}{\diagup}}} \xrightarrow[\substack{40\,^\circ\text{C, 4 h} \\ \text{R = Me, Et}}]{\overset{(\text{CN})_2\text{CHCH}(\text{CN})_2}{(74)}} \underset{(75)}{\text{cyclopentene}} \tag{29}$$

The reaction of bromonitrostyrene with tetranitromethane (equation 30) in methanol under the action of an alkali followed by treatment with acid is a novel synthetic approach to geminal dinitroalkanes and their derivatives [73].

$$\underset{(15)}{\text{PhCH}=\text{C}\underset{\text{NO}_2}{\overset{\text{Br}}{\diagup}}} \xrightarrow[\text{NaOH/MeOH}]{\overset{\text{C(NO}_2)_4}{(76)}} \underset{\text{OMe}}{\text{PhCHC}=\text{NO}_2^-\text{Na}^+} \xrightarrow{\text{HCl}} \underset{\substack{\text{OMe} \\ (77)}}{\text{PhCHCH(NO}_2)_2} \tag{30}$$

Neuman and Seebach [74] have described the 'self-condensation' reaction of (78) initiated by t-BuLi (equation 31). In this case one molecule of the reagent acts as an electrophile and the other as a nucleophile.

$$\underset{(78)}{\text{MeCH}=\text{CBrNO}_2} \quad \xrightarrow[\text{ii. AcOH}]{\text{i. t-BuLi, }-130\,^{\circ}\text{C to }-120\,^{\circ}\text{C}} \quad \left[\text{MeCH}=\text{C}\underset{\text{Li}}{\overset{\text{NO}_2}{\diagup}} \right] \qquad (31)$$

$$\Bigg\downarrow \begin{array}{l}\text{i. t-BuLi, }-115\,^{\circ}\text{C to }-110\,^{\circ}\text{C} \\ \text{ii. AcOH}\end{array} \qquad\qquad\qquad \Bigg\downarrow \;\; \text{MeCH}=\text{CBrNO}_2$$

$$\underset{(79)}{\text{Me(CH}_2)_3\text{CHMeCHBrNO}_2} \qquad\qquad \underset{(80)}{\text{MeCH}=\text{C(NO}_2)\text{CHMeCHBrNO}_2}$$

Reactions between 1-bromo-1-nitro-2-arylethenes and sodium azide lead to the formation of 4-aryl-5-nitro-1,2,3-triazoles [75]. Diazomethane reacts with 1-bromo-1-nitro-2-arylethenes in ether at room temperature to form 3-bromo-3-nitro-4-arylpyrazolines [76].

3.1.2.1 Reactions of polyhalonitroalkenes

The specific structure of a polyhalonitroalkene with vicinal and geminal halogen atoms defines the reactivity of such a compound with nucleophilic reagents. Interaction between 2,2-dichloro-1-nitroethene and MeONa or PhONa results in the formation of nitroacetic acid orthoesters [13]. The same reaction with trichloronitroethene gives chloronitroacetic acid orthoester [77].

Investigation of the reactions of polyhalonitroethenes with aliphatic and aromatic monoamines, o-phenylenediamine and o-aminophenol has opened the way to original synthetic approaches to substituted nitroindoles, benzimidazoles, benzoxazoles and their halo derivatives.

Series of nitroenamines [13] and nitroacetamidines [78] can be prepared in the reactions of amines with 1,2-dichloro-1-nitroethene or 2,2-dichloro-1-nitroethene (2,2-dibromo)-1-nitroethene [13, 38, 39, 51, 78] and 1,2,2-trichloro-1-nitroethene, (equation 32) [78]. It is probable that the last of these transforms in to the halonitroethene (30) which then reacts further with the amine.

$$\underset{\substack{(81)}}{\text{R}^1\text{R}^2\text{NCH}=\underset{\underset{\text{NO}_2}{|}}{\text{CNR}^1\text{R}^2}} \xleftarrow{\underset{(22)}{\text{ClCH}=\text{CClNO}_2}} \;\; \text{R}^1\text{R}^2\text{NH} \;\; \xrightarrow{\underset{(28)}{\text{Cl}_2\text{C}=\text{CHNO}_2}} \underset{(82)}{(\text{R}^1\text{R}^2\text{N})_2\text{C}=\text{CHNO}_2}$$

$$\Bigg\downarrow \begin{array}{c}\text{Cl}_2\text{C}=\text{CClNO}_2 \\ (30)\end{array} \qquad\qquad\qquad\qquad (32)$$

$$\underset{(83)}{(\text{R}^1\text{R}^2\text{N})_2\text{C}=\text{CClNO}_2}$$

$$\text{R}^1 = \text{H, R}^2 = \text{Ph}, o\text{-MeC}_6\text{H}_4, o\text{-MeOC}_6\text{H}_4, p\text{-MeOC}_6\text{H}_4, p\text{-BrC}_6\text{H}_4;$$
$$\text{R}^1 = \text{H, Ph, R}^2 = \text{Et, t-Bu}$$

On the basis of literature analogues, the synthesis of the nitroindoles (84; equation 33) (79) was developed using the reaction of aniline with (30).

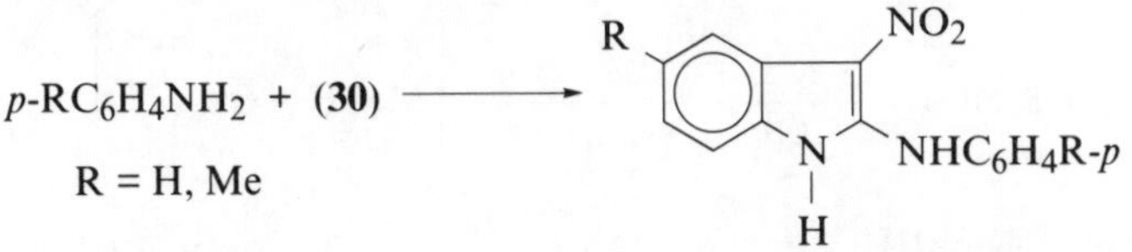

$$p\text{-}RC_6H_4NH_2 + (30) \longrightarrow \quad \text{(84)} \qquad (33)$$

R = H, Me

(84)

Reactions of **(27)** or **(30)** wirh *o*-phenylenediamines lead to a series of substituted nitrobenzimidazoles **(86;** equation 34) stabilized by intramolecular hydrogen bonding [80].

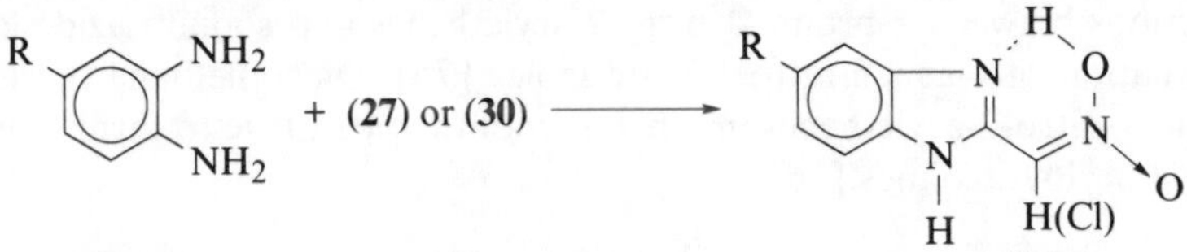

$$\text{(85)} + (27) \text{ or } (30) \longrightarrow \text{(86)} \qquad (34)$$

(85)   **(86)**

R = H, Me, NO$_2$

Pyrimidine derivatives are formed through the reactions of 1,8-naphthylene-diamines with **(27)** or **(30)** [81]. The same alkenes **(27, 30)** combine with bifunctional *o*-aminophenols (equation 35) to give products such as **(88)** [82].

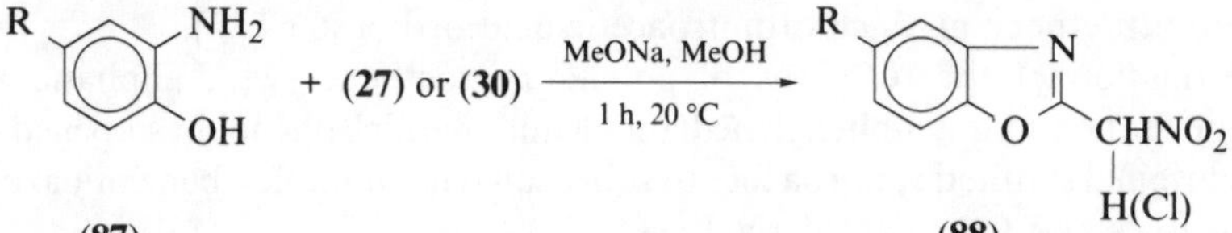

$$\text{(87)} + (27) \text{ or } (30) \xrightarrow[\text{1 h, 20 °C}]{\text{MeONa, MeOH}} \text{(88)} \qquad (35)$$

(87)   **(88)**

R = H, Me

Nucleophiles react with 2,2-diiodo-1,1-dinitroethene to give products which vary with the nature of the nucleophile. The reactions result in the formation of either the products of disubstitution **(90;** equation 36) or disubstitution–addition **(91, 92)** [41, 83].

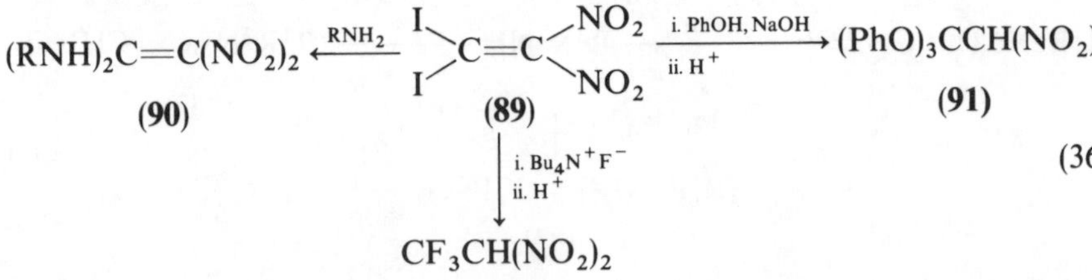

$$(RNH)_2C{=}C(NO_2)_2 \xleftarrow{\text{RNH}_2} \text{(89)} \xrightarrow[\text{ii. H}^+]{\text{i. PhOH, NaOH}} (PhO)_3CCH(NO_2)_2$$

(90)   **(89)**   **(91)**

$$\Big\downarrow \begin{array}{l} \text{i. Bu}_4\text{N}^+\text{F}^- \\ \text{ii. H}^+ \end{array} \qquad (36)$$

$$CF_3CH(NO_2)_2$$

(92)

Reactions of polyhalonitroethenes with enolizing CH acids (equation 37) give substituted nitrofurans probably in accordance with a nucleophilic addition–elimination mechanism [84].

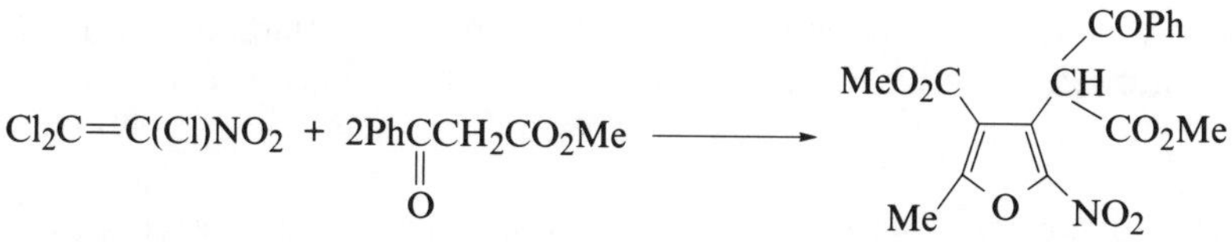

$$Cl_2C = C(Cl)NO_2 + 2PhCCH_2CO_2Me \longrightarrow$$

$$\underset{O}{\overset{\|}{}}$$

(30)   (93)

(94)   (37)

The data of Buevich *et al.* [85] show that a similar mechanism is responsible for the products of the reaction of trichloronitroethene with malonic ester.

3.1.2.2 Reactions of halonitrodienes

From its structure, perchloro-2-nitro-1,3-butadiene (95) will have a tendency towards nucleophilic substitution of the halogen atom in the nitrovinyl fragment. The reaction of (95) with KNO_2 under the action of triethylbenzylammonium chloride (TEBAC) gives a mixture of (E) and (Z) isomers (96; equation 38). Reaction with KSCN leads to (97) [86].

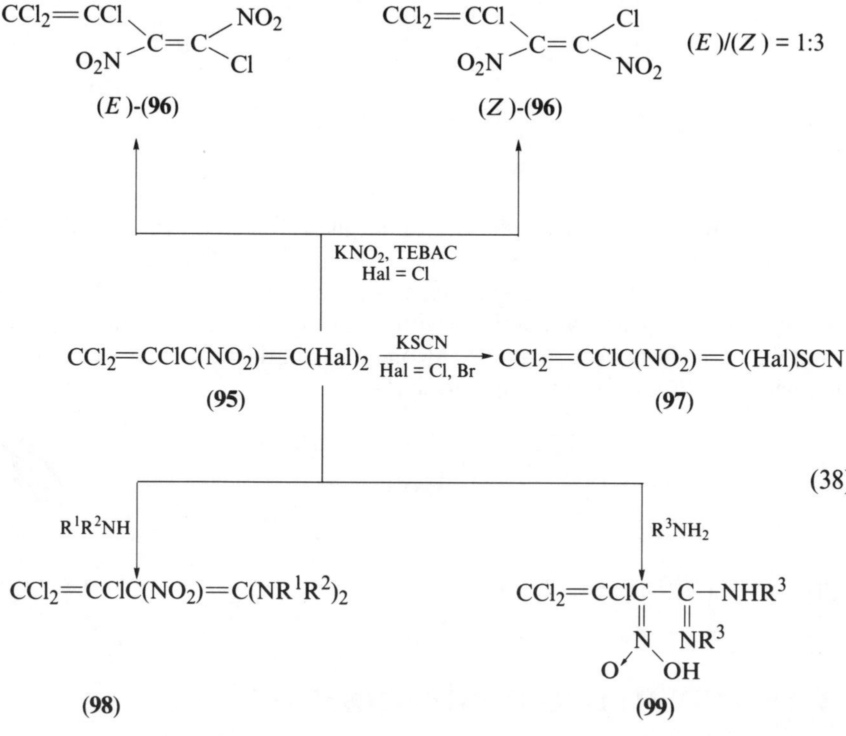

(E)-(96) (Z)-(96) $(E)/(Z) = 1:3$

KNO$_2$, TEBAC
Hal = Cl

$$CCl_2 = CClC(NO_2) = C(Hal)_2 \xrightarrow[\text{Hal = Cl, Br}]{\text{KSCN}} CCl_2 = CClC(NO_2) = C(Hal)SCN$$

(95)   (97)

(38)

R^1R^2NH R^3NH_2

$$CCl_2 = CClC(NO_2) = C(NR^1R^2)_2$$

(98)

$$CCl_2 = CClC - C - NHR^3$$

(99)

$R^1 = R^2 = Et, n\text{-}Bu;$ $R^1, R^2 = CH_2(CH_2)_3CH_2, (CH_2)_2O(CH_2)_2;$
$R^3 = Ph, o\text{-}ClC_6H_4$

Perchloro-2-nitro-1,3-butadiene reacts with aliphatic amines to form 1,1-diamino-2-nitro-3,4,4-trichloro-1,3-butadienes (98) (90% yield), whereas coupling with aromatic amines gives rise to trichlorovinylnitroacetamidines (99) (97% yield) [87].

Similarly to polychloronitroalkenes, polyhalonitrobutadienes react with o-phenylenediamine and o-aminophenol to produce imidazoles and oxazoles [88].

Interaction between (95) and a primary or secondary alcohol leads to the α-nitro-β,γ,γ-trichlorocrotonic ester or 1,1,2-trichloro-3-nitro-1-propene [89], respectively. Under the action of AlBr$_3$ (equation 39) in ethyl bromide, (95) gives compounds such as (100) and (101). The latter product forms (102) in the reaction with RSH [90].

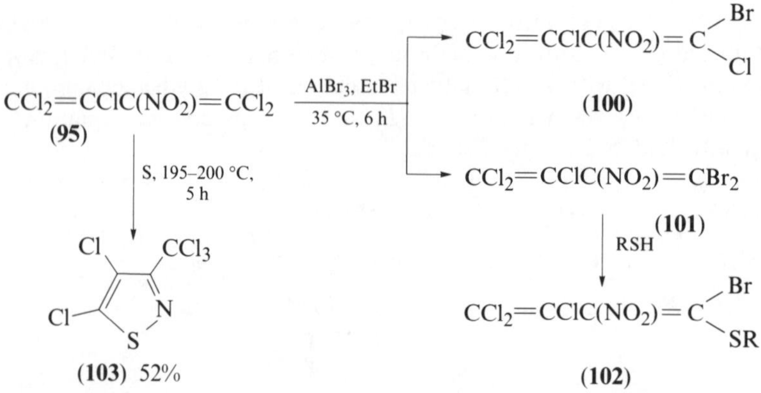

$$CCl_2=CClC(NO_2)=C\begin{smallmatrix}Br\\Cl\end{smallmatrix}$$
(100)

$$CCl_2=CClC(NO_2)=CCl_2$$
(95)

$$\xrightarrow[35\,°C,\,6\,h]{AlBr_3,\,EtBr}$$

$$CCl_2=CClC(NO_2)=CBr_2 \qquad (39)$$
(101)

S, 195–200 °C, 5 h

RSH

$$CCl_2=CClC(NO_2)=C\begin{smallmatrix}Br\\SR\end{smallmatrix}$$
(102)

(103) 52%

Under heating, a mixture of (95) with sulfur gives (103), the structure of which has been elucidated by mass spectrometry, IR and ^{13}C NMR [91].

In contrast with (95), its structural isomer pentachloro-1-nitro-1,3-butadiene (104; equation 40) reacts with amines with substitution of the chlorine atom in position 2 of the butadiene fragment. Alcoholates, on the other hand, react with (104) with substitution of the nitro group [49].

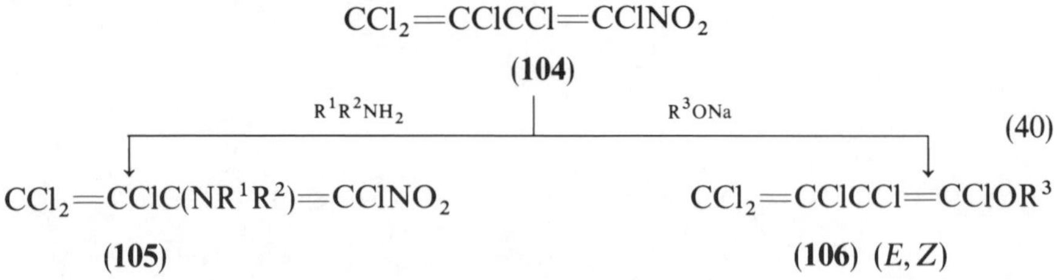

$$CCl_2=CClCCl=CClNO_2$$
(104)

R^{1}R^{2}NH$_2$ R^{3}ONa   (40)

$$CCl_2=CClC(NR^1R^2)=CClNO_2$$
(105)

$$CCl_2=CClCCl=CClOR^3$$
(106) (E, Z)

3.2 ALKOXY(HYDROXY)NITROETHENES

There are two synthetic approaches to 2-alkoxy-1-nitroalkenes: (i) through the reactions of 1,2-dinitroalkenes [92, 10a] or 2-halo-1-nitroalkenes [93, 10a] with alcoholates (phenolates) (equation 1) and (ii) through the pyrolysis of nitroacetals (equation 2). The nitroacetals are obtained by condensation of an orthoformic ester with a nitroalkane or by reaction of a geminal bromonitrostyrene with an alcoholate [94].

$$O_2NCR^1=CR^1NO_2 \xrightarrow[-50\,°C\,to\,0\,°C]{R^3ONa,\,MeOH} O_2NCR^1=CR^2OR^3 \qquad (1)$$

$\quad$ **(1)** $\hspace{5cm}$ **(2)** 75–87%

$\quad$ (Z) or (E)

$$R^1 = H, Me, Ph; \quad R^2 = Me, Ph; \quad R^3 = Alk$$

$$O_2NCPh=CPh(Hal) \xrightarrow{RONa} O_2NCPh=CPh(OR) \qquad (2)$$

$\quad$ **(3)** $\hspace{5cm}$ **(4)**

$\quad$ (E)

$\quad$ R = Me ((E) and (Z)), Hal = I;
$\quad$ R = p-MeC$_6$H$_4$, p-ClC$_6$H$_4$(E), Hal = Cl

The unsubstituted 2-nitrovinyl ether (**6**) was prepared for the first time in 10% yield by condensation of nitromethane with diethoxymethyltriethylammonium tetrafluoroborate followed by alcohol elimination [95]. The method developed later (equation 3) [94] made it possible to increase the yield of (**6**) to 40%.

$$RCH_2NO_2 \xrightarrow[ZnCl_2\,or\,Zn,\,\Delta]{(EtO)_3CH} (EtO)_2CHCHRNO_2 \xrightarrow[-EtOH]{p\text{-}TSA,\,xylene,\,\Delta} EtOCH=CRNO_2 \quad (3)$$

$\hspace{4cm}$ **(5)** $\hspace{5cm}$ **(6)**

$\quad$ R = H, Me, Et $\quad$ p-TSA = p-toluenesulfonic acid

A novel and convenient method of synthesis for (**6**) is based on the nitration of vinyl ether by acetyl nitrate, formed *in situ* (equation 4), with subsequent elimination of AcOH [96].

$$CH_2=CHOEt \xrightarrow{HNO_3,\,Ac_2O,\,-33\,°C} O_2NCH_2CH(OAc)OEt \xrightarrow[temperature]{Et_3N,\,CH_2Cl_2,\,room} (6)\ 80\%$$

$\quad$ **(7)** $\hspace{5cm}$ **(8)** $\hspace{4cm}$ (4)

The synthesis of the aryl-substituted ethoxynitroalkene (**10**) was carried out in accordance with equation (5). Sodium ethoxide was used for the alcohol elimination [97].

$$PhCH_2NO_2 \xrightarrow{(EtO)_3CH}_{ZnCl_2} (EtO)_2CHCH(Ph)NO_2 \xrightarrow[ii.\,HCl]{i.\,EtONa} EtOCH=C(Ph)NO_2 \quad (5)$$

$\quad$ **(9)** $\hspace{4cm}$ **(5)** $\hspace{4cm}$ (Z + E) **(10)** 67%

Isomeric alkoxynitrostyrenes (**12**; equation 6) can be obtained by pyrolysis of nitroacetals in quantitative yield [98].

$$PhCH{=}CBrNO_2 \xrightarrow{RONa} PhC(OR)_2CH_2NO_2 \xrightarrow{\Delta, N_2} PhC(OR){=}CHNO_2 \quad (6)$$

$$\textbf{(11)} \qquad\qquad\qquad\qquad \textbf{(12)}$$

$$(E)/(Z) = 3{:}1$$

$$R = Me, n\text{-}Pr, Bu$$

Reflux of a nitroacetic ester with an orthoformic ester in acetic anhydride results in the formation of the α-nitro-β-ethoxyacrylic ester (66% yield) [99]. Nitroacetic ester condensation with ketene diethyl acetal gives rise to the α-nitro-β-ethoxycrotonoate in 68% yield [100].

Studies of the chemical transformations of 2-alkoxy-1-nitroalkenes have revealed that alkoxy group substitution is the dominant feature of the chemistry of these compounds. Amines and their derivatives substitute the alkoxy group in alkoxynitroalkenes to give nitroenamines [101–105] in the (Z) form for primary amines and the (E) form for secondary amines [97].

Similarly, some β-amino derivatives (13) have been prepared from β-alkoxy-α-nitroacrylic esters (equation 7) [102–104]. Some products derived from hydrazine and urea were isolated in the form of the (E) isomers.

$$R^1OCR^2{=}CR^3NO_2 \xrightarrow{R^4R^5NH} R^4R^5NCR^2{=}CR^3NO_2 \qquad (7)$$

$$\textbf{(2)} \qquad\qquad\qquad\qquad \textbf{(13)}$$

$$R^1 = Me, Et; \quad R^2 = H, Ph; \quad R^3 = H, Ph, CO_2Me;$$
$$R^4, R^5 = H, Alk, Ph, CONHAlk, NHCOMe, NHCOPh$$

Bernasconi et al. [106, 107] have reported kinetic investigations of the interaction between β-methoxy-α-nitrostilbene and N-butylamine or cyclic amines (morpholine, piperidine, pyrrolidine) (equation 8).

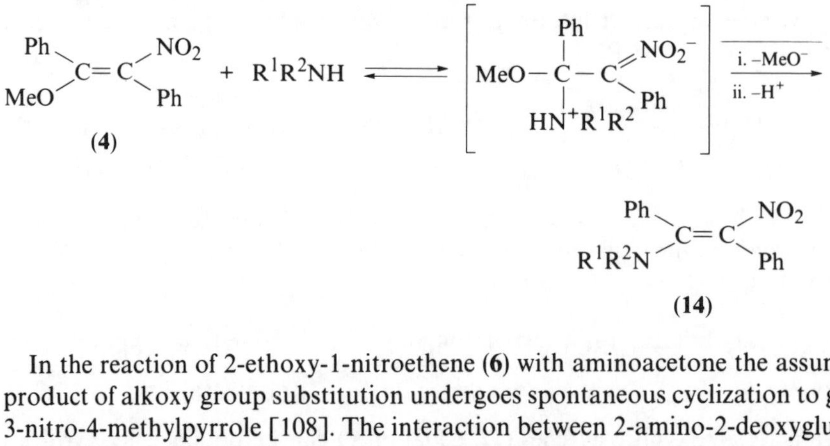

$$(8)$$

In the reaction of 2-ethoxy-1-nitroethene (6) with aminoacetone the assumed product of alkoxy group substitution undergoes spontaneous cyclization to give 3-nitro-4-methylpyrrole [108]. The interaction between 2-amino-2-deoxygluco-pyranose 15α, 15β (or its N-butyl derivative) and (6) depends on the experimental conditions and results in the formation of the products of substitution (α-**16**, β-**16**; equation 9) or cyclization (**17**).

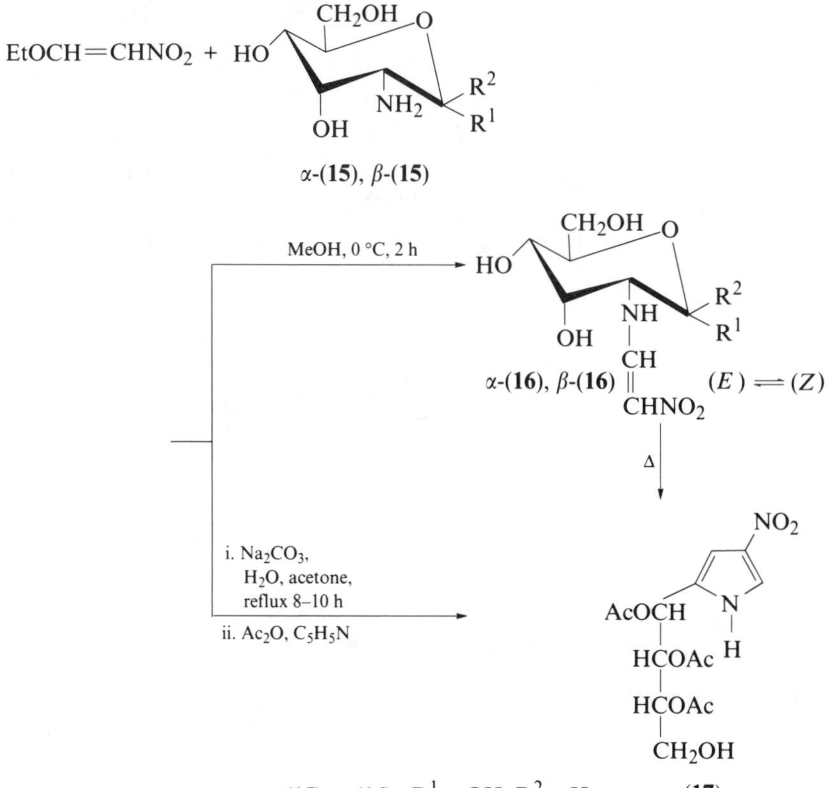

$$EtOCH=CHNO_2 \; + \; HO$$

α-(**15**), β-(**15**)

MeOH, 0 °C, 2 h

α-(**16**), β-(**16**) $(E) \rightleftharpoons (Z)$

(9)

i. Na$_2$CO$_3$,
   H$_2$O, acetone,
   reflux 8–10 h

ii. Ac$_2$O, C$_5$H$_5$N

α-(**15**), α-(**16**): R^1 = OH, R^2 = H   (**17**)

β-(**15**), β-(**16**): R^1 = H, R^2 = OH

The reaction of (**6**) with 1-amino-1-deoxy-D-fructose proceeds in a similar way. The corresponding products 2-(1-additolyl)-4-nitropyrrole and 3-(1-additolyl)-4-nitropyrrole transform into 3-nitro-4-pyrroaldehyde and 3-nitro-5-pyrroaldehyde, respectively.

β-Ethoxy-α-nitroacrylic esters (**18**; equation 10) interact with difunctional *N*,*N*-disubstituted hydrazines, urea and amides via substitution to form the derivatives of pyrazolone (**19**) and oxypyrimidine (**20, 21**) [105].

The reaction of β-ethoxy-α-nitroacrylic ester (**18**) with sulfamide (equation 11) leads to the product of heterocyclization, namely 4-nitro-1,2,6-thiadiazine 1,1-dioxide (**22**) [109, 110].

4-Nitro-1-cyclohexyl-3-ethoxy-2-oxo-3-pyrrolyne (**23**; equation 12), prepared by nitration of the corresponding carboxylic acid, is an efficient reagent for the protection of amino groups. The protecting group can be removed with

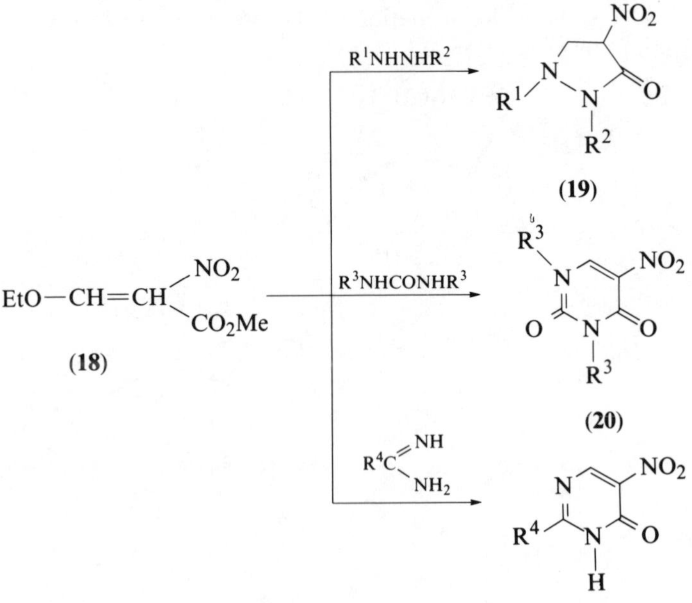

(18) (19) (20) (21)

(10)

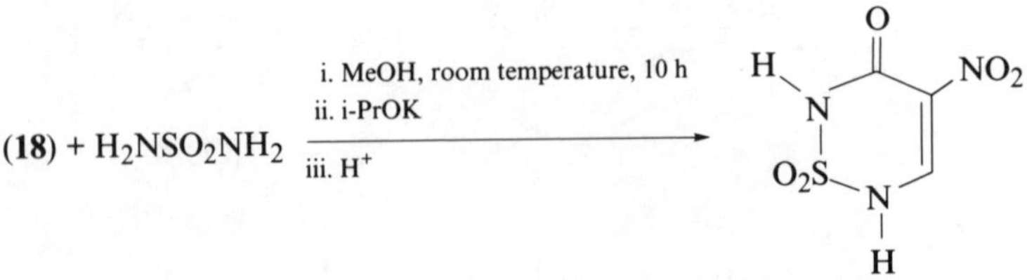

(22) 35%

(11)

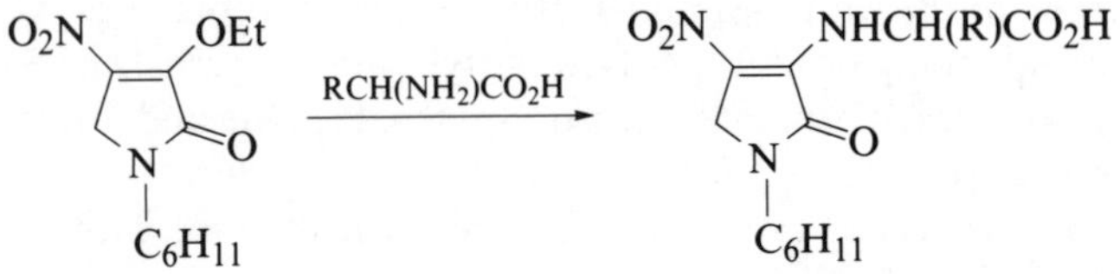

(23) (24)

(12)

ammonia. Such a method has been used for the preparation of two simple dipeptides [101].

The reactions of β-ethoxy-α-nitroacrylic ester (**18**) with indole and 6-methylindole are rather specific (equation 13). The resulting compounds (**25**) are used for the synthesis of tryptophan and its methyl derivative [111, 112]. The reaction of (**18**) with α-pyrrolecarbonic ester in THF under the action of potassium hydride results in the formation of the product of addition (26), isolated as a mixture of stereoisomers. Compound (26) is used as a starting material in the preparation of the alkaloid peramine [113, 114].

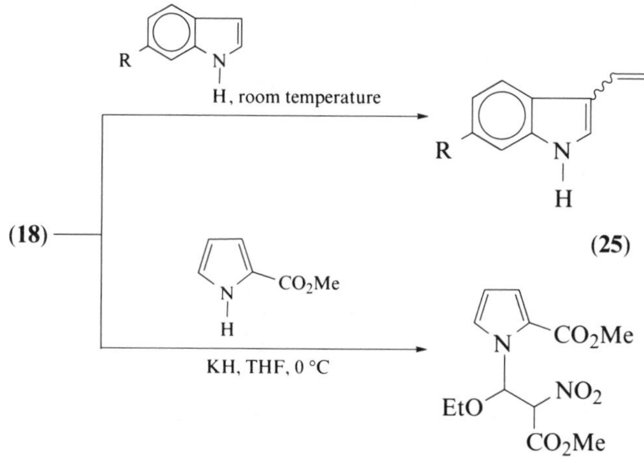

(25) R = H, Me   (13)

(26) 82%

Reactions between 2-alkoxy-1-nitrostilbenes and the thiolate anion also lead to substitution. Benasconi *et al.* [115, 116] have studied the kinetics of these reactions and demonstrated the formation of the intermediate (27; equation 14). In contrast with other β-substituted nitrostilbenes, this intermediate was detected only spectrometrically owing to the poor nucleofugicity of the methoxy group and the slowness of the last reaction step [10b, 115, 116].

Alkoxynitroalkenes undergo exchange of alkoxy groups in the reactions with alcohols. The process is catalysed by acids. For example, the reaction of ethoxy-nitroalkene (6) with 1-menthol under the action of TsOH results in the product of ethoxy group exchange (30; equation 15) in 46% yield. The non-catalysed process terminates in the formation of (31) (84%) [96].

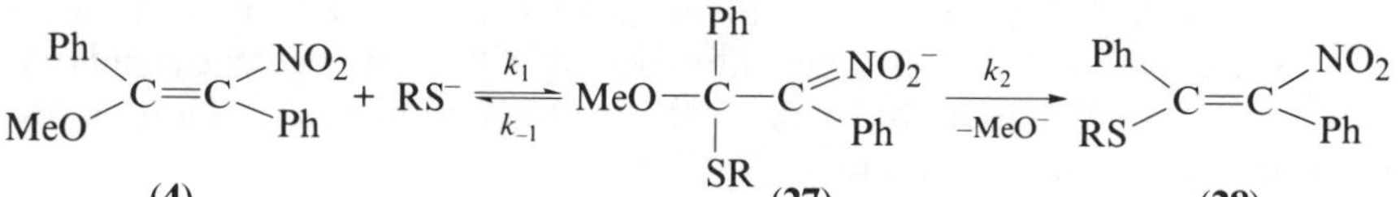

(4)   (27)   (28)   (14)

$RS^- = EtS^-, HOCH_2CH_2S^-, MeO_2CCH_2CH_2S^-$

Nitroacetals also can be produced by the action of alcoholates upon nitro-alkoxyalkenes at room temperature [92]. Alkoxynitrostilbene (4) is hydrolysed under basic catalysis to produce hydroxynitrostilbene (33a), which isomerizes readily to give the salt of the α-nitroketone (33b; equation 16) [107]. The kinetics

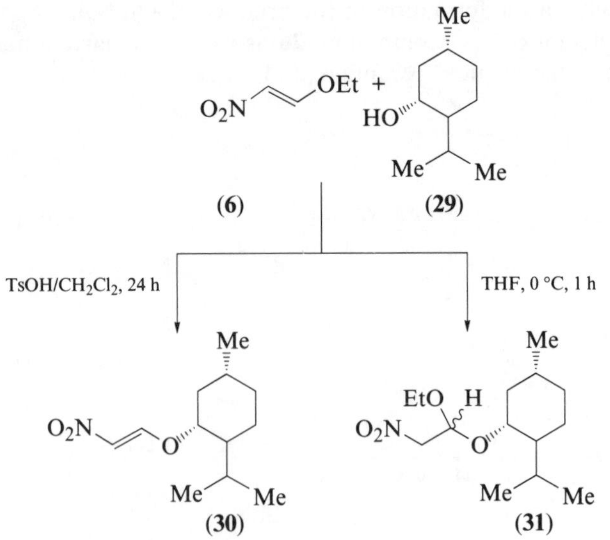

$$(15)$$

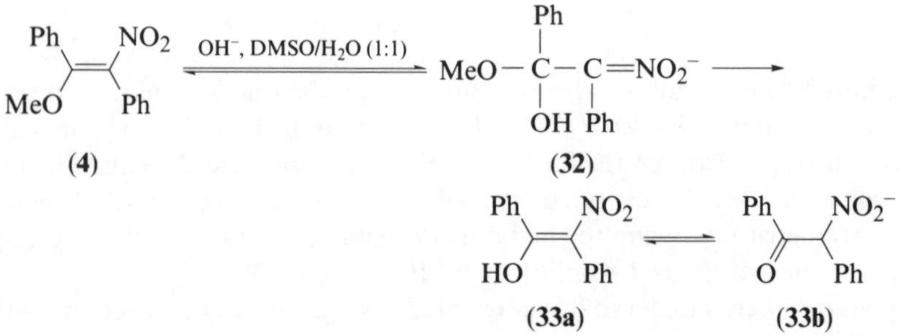

$$(16)$$

of the process are discussed by Bernsconi *et al.* [107] and the results are consistent with an addition–elimination mechanism for the reaction.

Cyclic α-nitroketones form rather stable enol structures. For example, 4-nitro-5-phenyl-3-hydroxy-2-oxo-3-pyrroline (35; equation 17), formed from the amide (34), contains the β-hydroxy-α-nitroethene fragment and shows high OH acidity [117]. The compound (35) is a useful intermediate in the synthesis of pyrrolidone derivatives. Its alkylation by diazomethane leads to the formation of the corresponding methoxy derivative (36).

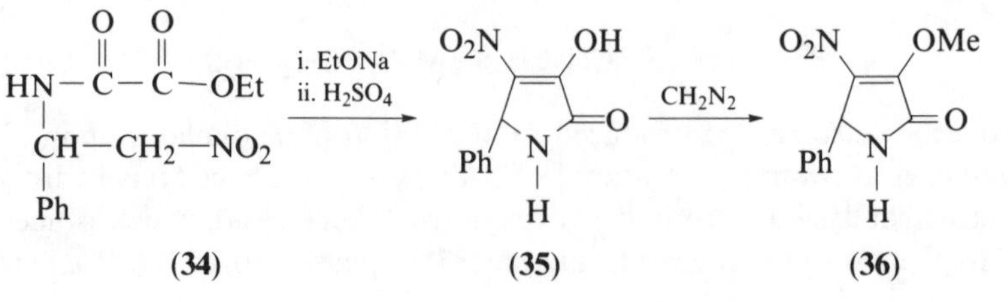

$$(17)$$

The β-hydroxy-α-nitroethene group is present in tautomers of steroids containing fragments of condensed, six-membered cyclic α-nitroketones (**37a**, **38a**) or α,α-dinitroketones (**39a**; equation 18). Such compounds are produced by steroid nitration [118–120]. The tautomeric equilibrium in the mononitro derivative is affected by the location of the nitro group and the solvent. One of the dinitrated diastereoisomers exists entirely in the enol form, while the other is a mixture with an approximate 50:50 ratio.

The synthetic approach to geminal alkoxynitroalkenes can be illustrated by the preparation of 1-benzyloxy-1-nitroalkenes (**44–46**; equation 19). Benzyloxynitromethane (**43**) condenses with aldehydes (**40–42**) that contain either the O-2-azetidinone or S-2-azetidinone fragment [1, 121].

	Y	X	R′
(**40**), (**44**)	H	O	t-Bu
(**41**), (**45**)	Me$_3$Si	S	t-Bu
(**42**), (**46**)	t-BuPh$_2$Si	S	CMe$_2$Pr-i

1-Benzyloxy-1-nitroalkenes (44–46) are useful intermediates in constructing polyfunctional, bicyclic β-lactam systems of type (47–49) (oxapenam, sulbactam, 6-aminopenicillinic acid) [1].

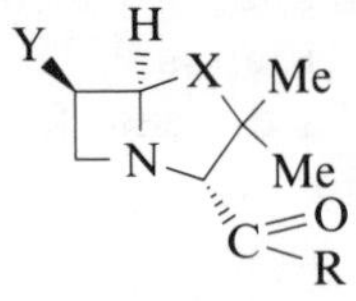

	Y	X	R
(47)	H	O	OCH₂Ph
(48)	H	SO₂	OH
(49)	H₃N⁺	S	O⁻

(47)–(49)

1-Nitroglycals may also be classified as 1-alkoxy-1-nitroalkenes and are produced by either dehydration or deacylation of the acyl derivatives of 1-deoxy-1-nitropyranose and 1-deoxy-1-nitrofuranose. For example, compounds (50) and (51) are prepared from the corresponding substituted β-D-mannopyranose and its isopropylidene derivative [1, 122].

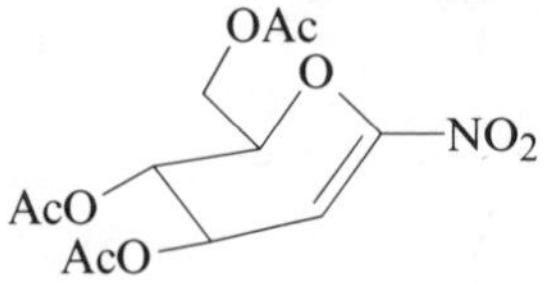

(50) 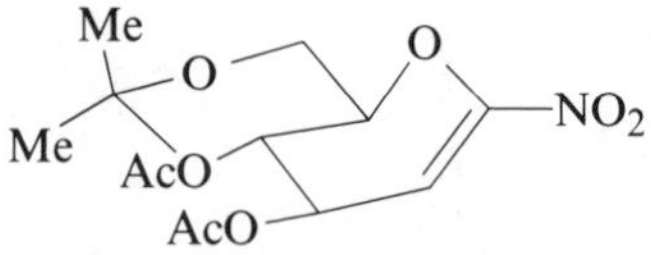   (51)

1-Nitroglycals are very useful reagents in organic synthesis owing to their ability to add nucleophiles and the possibility of transforming the nitro group into other functional groups. For example, they are used in the synthesis of some carbohydrates.

3.3 ALKENES CONTAINING THIO AND NITRO GROUPS

3.3.1 MONOTHIONITROALKENES

3.3.1.1 Synthesis of thionitroalkenes

Vicinal thionitroalkenes

A synthetic method has been developed for 2-thio-1-nitroalkenes (3) [123, 124] using α-nitroketones (1) as starting materials to produce the thioketals (2; equation 1). The thioketals undergo elimination of thiol under the action of potassium fluoride to give the products (3) (mostly (Z) isomer).

The salt mixture Hg(CF₃CO₂)₂/Li₂CO₃ is also used in this process as an eliminating agent [123]. The method has been used for cyclic α-nitroketones, i.e. nitrocyclopentanone and nitrocyclohexanone [123], but it is restricted by the instability of α-nitroaldehydes, meaning that it is impossible to prepare β-unsubstituted thioalkenes.

$$\underset{\substack{(1)}}{\underset{O}{R^1}\diagdown}C-CH_2NO_2 \xrightarrow[\substack{ZnCl_2 \text{ or} \\ BF_3, OEt_2}]{R^2SH} \underset{\substack{(2)}}{\underset{R^2S}{R^2S\diagdown}}\overset{R^1}{\underset{}{C}}-CH_2NO_2 \xrightarrow[\Delta \text{ or } AlCl_3]{KF} \underset{\substack{(3)}}{\underset{R^2S}{R^1}\diagdown}C=C\overset{H}{\underset{NO_2}{\diagup}} \quad (1)$$

$$(Z)/(E)$$

$$R^1 = Me, Et, Ar; \quad R^2 = Et, Ph$$

Another new method is based on the reactions between 1-acetoxy-2-nitro-alkanes (**4**; equation 2) and thiophenol in the presence of Et₃N, which involve substitution of the acetyl group. The high-yielded 2-thio-1-nitroalkanes (**5**) undergo chlorination and subsequent dehydrochlorination to give the thionitro-alkenes (**3**) in the form of a mixture of (E) and (Z) isomers [125, 126].

$$\underset{\substack{(4)}}{\underset{O_2N}{R^1}\diagdown}\underset{R^2}{\diagup}OAc \xrightarrow[Et_3N, 0\,°C]{PhSH} \underset{\substack{(5)}}{\underset{O_2N}{R^1}\diagdown}\underset{R^2}{\diagup}SPh \xrightarrow[\substack{ii. \, Et_3N, 0\,°C \\ 3\,h}]{i. \, SO_2Cl_2} \underset{\substack{(3) \; 83\text{–}95\%}}{\underset{O_2N}{R^1}=\underset{R^2}{SPh}} \quad (2)$$

$$(E)/(Z) \approx 1{:}1$$

$$R^1 = H, Me, Et, n\text{-}C_5H_{11}; \qquad\qquad R^1 = R^2 = H \quad (E)/(Z) = 95{:}5$$
$$R^2 = H, Me$$

Several authors [5, 10a, 106, 116, 127] have described the synthesis of 2-thio-1-nitroalkenes in the reactions of arenethiols with 2-halo-1-nitroalkenes. The structure of the starting material (**6**) does not affect the structure of the product (**3**). In all cases only one isomer is formed, presumably the (E) isomer (equation 3). The reaction is not stereospecific, consistent with the formation of an ionic intermediate that can rotate around the C(1)—C(2) bond [10b].

$$\underset{\substack{(Z) \text{ or } (E) \\ (6)}}{\underset{Ph}{Hal}\diagdown}\underset{NO_2}{R^1} \xrightarrow[\text{or } R^2SH, Et_3N, MeCN]{R^2SNa, EtOH} \underset{\substack{(3)}}{\underset{Ph}{R^2S}\diagdown}\underset{NO_2}{R^1} \quad (3)$$

$$R^1 = H, Ph; \quad R^2 = Alk, Ar$$

The main aim of many investigations [128–131] has been the synthesis of thionitroalkenes from 1,2-dinitroalkanes and H_2S, alkanethiols or arenethiols via substitution of the nitro group by the thiol. The reaction with hydrogen sulfide does not end with the formation of the nitroalkenethiol but goes further depending on the structure of the initial dinitroalkenes. Monosubstituted dinitro-alkenes give the products of disubstitution (**9**; equation 4), whereas disubstituted compounds undergo oxidation to give disulfides (**10**). The latter reaction is characteristic for thiols, particularly for thiol anions [128].

$$O_2NCR^1{=}CR^2NO_2 \xrightarrow[\text{0–18\,°C, 2–18 h}]{\text{H}_2\text{S}} [O_2NCR^1{=}CR^2SH]$$

$$\textbf{(7)} \hspace{6cm} \textbf{(8)}$$

$$R^1{=}H \hspace{4cm} R^1,R^2 \neq H \hspace{2cm} \text{(4)}$$

$$(O_2NCR^1{=}CH)_2S \hspace{5cm} (O_2NCR^1{=}CR^2S)_2$$

$$\textbf{(9)} \hspace{7cm} \textbf{(10)}$$

Ethanethiol and arenethiols combine with dinitrostyrenes and (Z)-dinitro-stilbenes, without base, to form the products of substitution, namely 2-ethyl(aryl) thio-1-nitroalkenes [129]. The less active (E) isomer, in contrast with the (Z) isomer, reacts only with arylthiolates (equation 5).

$$\begin{array}{c} R^2S \\ \diagdown \\ R^1 \diagup \end{array} C{=}C \begin{array}{c} Ph \\ \diagup \\ \diagdown NO_2 \end{array}$$

$$\textbf{(3b) } (E)$$

$$O_2NCPh{=}CR^1NO_2 \xrightarrow[\text{or } R^2S^- Na^+, 0\,°C]{R^2SH, \text{MeOH}, 20\,°C} \hspace{1cm} + \hspace{3cm} \text{(5)}$$

$$\textbf{(7)}$$
$$(Z) \text{ or } (E)$$

$$\begin{array}{c} R^2S \\ \diagdown \\ R^1 \diagup \end{array} C{=}C \begin{array}{c} NO_2 \\ \diagup \\ \diagdown Ph \end{array}$$

$$\textbf{(3a) } (Z)$$

$$R^1 = H, Ph; \quad R^2 = Et, Ar$$

This process is not stereospecific, and the stereochemical result depends on the conditions of the synthesis and the initial dinitroethene structure. Reactions with thiolates tend to produce the more stable (Z) isomer, while thiols in the absence of base lead to a mixture of isomers; the (E)/(Z) ratio does not depend upon the thiolating agent. So, the (E) isomer is the prevailing product in the reactions of dinitrostyrene (about 1:1.5) and the (Z) isomer is the prevailing product in the reactions of dinitrostilbene (3:1) [130]. (E)-thionitroalkenes, including 2-arylthio-1-nitroethenes, isomerize to the (Z) form upon heating or in the presence of a catalytic amount of thiolate [130]. The result obtained for dinitro-stilbene (equation 5) is similar to that obtained for aliphatic dinitroalkenes [131] (equation 6).

$$\begin{array}{c} R \\ \diagdown \\ O_2N \diagup \end{array} {=} \begin{array}{c} NO_2 \\ \diagup \\ \diagdown R \end{array} \xrightarrow[\text{EtOH, 25\,°C, 0.15 h}]{p\text{-MeC}_6\text{H}_4\text{SH}} \begin{array}{c} R \\ \diagdown \\ O_2N \diagup \end{array} {=} \begin{array}{c} R \\ \diagup \\ \diagdown SC_6H_4Me\text{-}p \end{array} \hspace{1cm} \text{(6)}$$

$$\textbf{(7)} \hspace{6cm} \textbf{(3)} \ 92\text{–}95\%$$

$$R = Me \quad (E)/(Z) = 2{:}3$$
$$R = Et \quad (E)/(Z) = 1{:}3$$

The reaction of (Z)-1,2-dinitrostilbene with thiols in a buffer solution can be used for the colorimetric testing of alkanethiols and arenethiols [132].

Geminal thionitroalkenes

These are prepared by condensation of aliphatic aldehydes with phenylthio-nitromethane in a methanolic solution of KOH [133–136] or a solution of t-BuOK in t-butanol [137, 138] followed by dehydration of the nitrothioalcohol (equation 7).

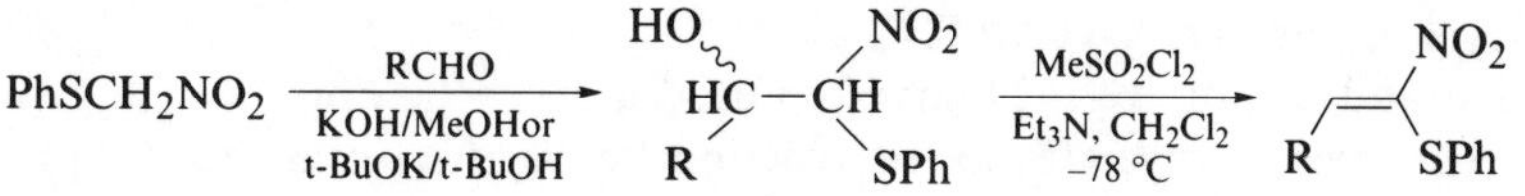

$$PhSCH_2NO_2 \xrightarrow[\substack{KOH/MeOHor \\ t\text{-}BuOK/t\text{-}BuOH}]{RCHO} \underset{\substack{R \quad SPh \\ (12)}}{\overset{\substack{HO \quad NO_2 \\ HC-CH}}{}} \xrightarrow[\substack{Et_3N, CH_2Cl_2 \\ -78\ ^\circ C}]{MeSO_2Cl_2} \underset{\substack{R \quad SPh \\ (13a)}}{\overset{NO_2}{}} \quad (7)$$

(11)

R = Me, i-Pr

This method has been used successfully in the synthesis of biologically active compounds, i.e. a wide variety of (Z)-1-phenylthio-1-nitroalkenes such as **(13b)** to **(13e)**, among others [1, 136, 138–140].

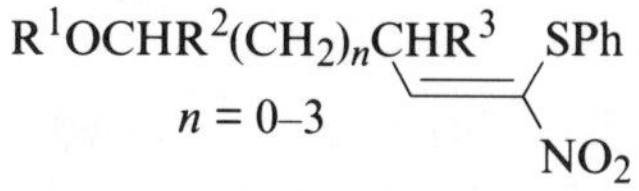

$$R^1OCHR^2(CH_2)_nCHR^3 \quad SPh$$
$$n = 0\text{-}3 \qquad\qquad NO_2$$

(13b)

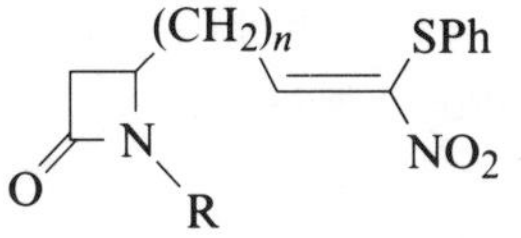

$n = 2, 3$

(13c)

R = t-BuMe$_2$Si

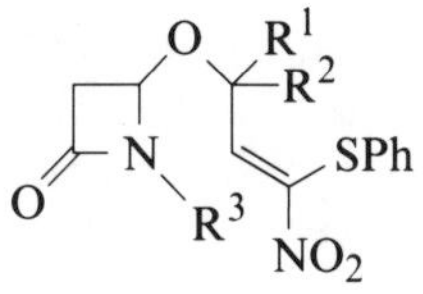

(13d)

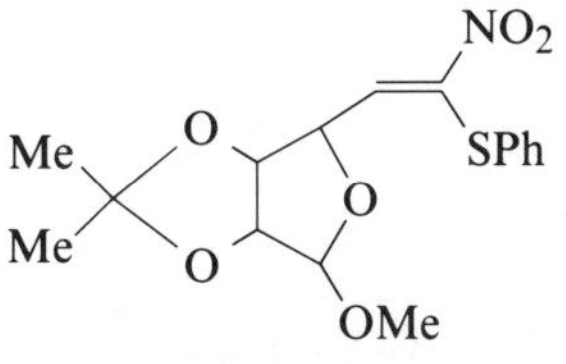

(13e)

It is possible to carry out a one-step, direct alkenylation of phenylthionitro-methane with aliphatic aldehydes and benzaldehyde to give **(13a)** in a yield of

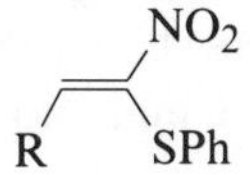

$$\underset{\substack{R \quad SPh}}{\overset{NO_2}{}}$$

(13a)

R = Me, n-Pr, i-Pr, Bu, pentyl, hexyl, Ph

56–79%. The reaction occurs in dichloromethane at $0\,°C$ in the presence of piperidinium acetate and a catalyst ($4\,\text{Å}$ molecular sieves) [137, 141].

The reaction of phenylthionitromethane with cyclohexanone in toluene at room temperature under the action of a catalytic amount of N,N-dimethylethylenediamine and $4\,\text{Å}$ molecular sieves results in the formation of 1-[nitro-(phenylthio)methyl]cyclohexene (17% yield) and 1-[nitro-(phenylthio)methylidene]cyclohexane (17% yield) [141].

Russell and Dedolph [142] have reported a synthetic approach to geminal thionitroethenes of type (15) in the reaction between 1,1-dinitro-2,2-diphenylethene and thiolate anion in DMSO (equation 8).

$$Ph_2C{=}C(NO_2)_2 + PhSK \xrightarrow{\text{DMSO, 25 °C, 1 h}} Ph_2C{=}C\overset{\displaystyle NO_2}{\underset{\displaystyle SPh}{\big\langle}} \qquad (8)$$

$$\textbf{(14)} \qquad\qquad\qquad\qquad \textbf{(15)}\ 85\%$$

3.3.1.2 Reactions of thionitroalkenes

Vicinal thionitroalkenes

2-Thio-1-nitroalkenes, like other β-functionalized nitroalkenes with nucleofugal substituents, undergo nucleophilic substitution. Reaction between compounds such as (3) and amines (equation 9) is a new synthetic approach to nitroenamines [143]. The primary amines (and aniline) produce exclusively the (Z) isomers, which are stabilized by intramolecular hydrogen bonding [128]; secondary amines give the (E) isomers.

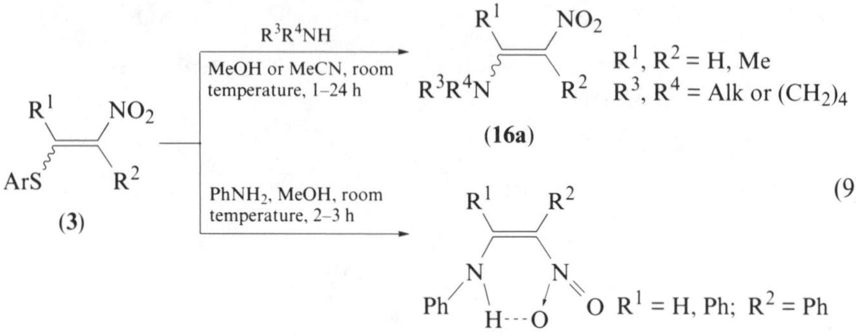

Benasconi *et al.* [106] have performed a comparative kinetic investigation of the reactions of 2-ethylthio-1-nitrostilbene and its 2-halo and 2-methoxy derivatives with piperidine. The results correlate with the assumed addition–elimination mechanism for the nucleophilic vicinal substitution with the participation of highly activated substrates.

The same authors [116] have studied in detail the reaction of the (E)-2-thio derivative of 1-nitrostilbene with aliphatic thiolate ion, and in some cases they detected the generally postulated intermediate (18) by spectrophotometric methods (equation 10).

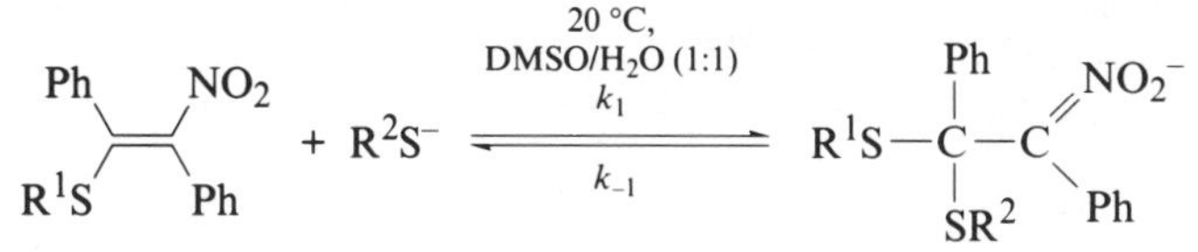

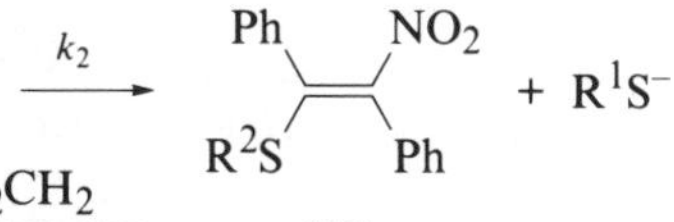

$$
R^1 = HOCH_2CH_2 \\
R^2 = Pr, HOCH_2CH_2 \quad (19)
$$

It is interesting to point out that the same reaction with aromatic thiolate ion results in substitution of the nitro group (equation 11).

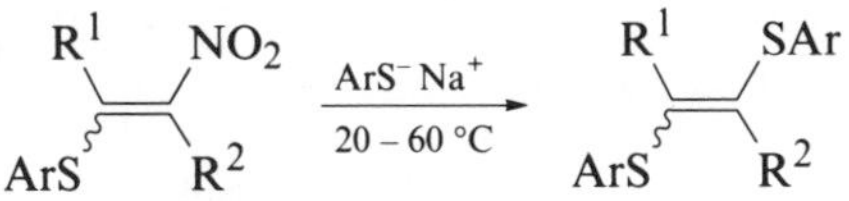

$$
R^1 = R^2 = H; \quad R^1 = H, R^2 = Ph
$$

The reactions of vicinal thionitroethenes with alkoxide anions do not result in the product of substitution. The latter undergoes the addition of a second equivalent of alkoxide anion to give the nitroaldehyde acetal (22; equation 12) [144].

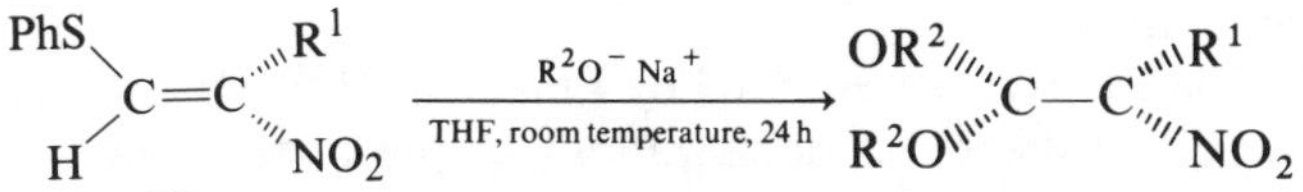

$$
R^1 = H, Me, Et, Me(CH_2)_6; \quad R^2 = Me, PhCH_2
$$

The kinetics of the reaction of β-thio-α-nitrostilbene with OH$^-$ in aqueous solution have been studied. In this case, similarly to the methoxy derivatives (see Section 3.2), the substitution product is formed, and this transforms into the α-ketonitronate in a basic medium [107]. The reaction between the organocopper–zinc reagent (24) and 2-thio-1-nitroethene (23; equation 13) is an addition–elimination process. This method can be used for the synthesis of polyfunctional nitrocyclohexenes (25) [145, 146a].

Reactions between (23) or its sulfoxide analogue and various alkyl–metal reagents give rise to a wide range of 1-nitrocycloalkenes the fragments of γ-butyrolactones, δ-valerolactones, 2-pyrrolidones and piperidines [146b].

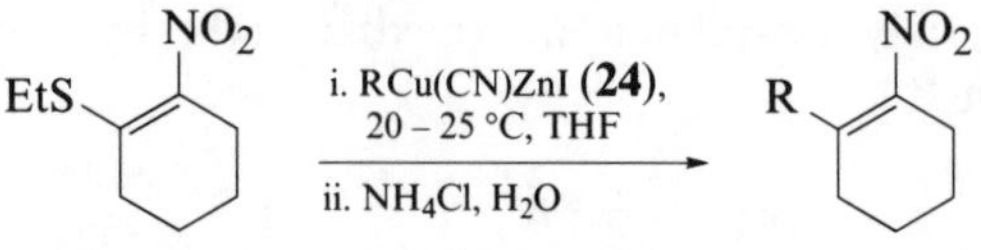

$$
\text{(23)} \xrightarrow[\text{ii. NH}_4\text{Cl, H}_2\text{O}]{\substack{\text{i. RCu(CN)ZnI (24),}\\ 20-25\ ^\circ\text{C, THF}}} \text{(25)}
\tag{13}
$$

R	Yield (%)
$(CH_2)_3CO_2Et$	89
$(CH_2)_3CN$	85
$(CH_2)_2P(O)(OEt)_2$	79
(E)-CH=CHC_6H_{13}	90

The conjugative addition of alkyl groups such as methyl to vicinal thionitro-alkenes (equation 14) differs from the conjugative addition to unsubstituted nitroalkenes and proceeds stereoselectively at low temperature [147].

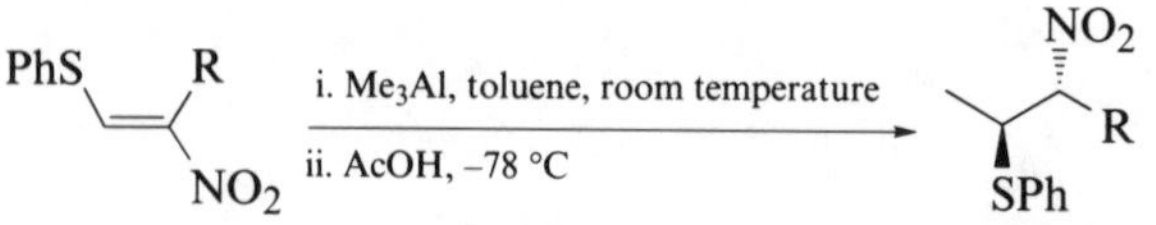

$$
\text{(3)} \xrightarrow[\text{ii. AcOH, }-78\ ^\circ\text{C}]{\text{i. Me}_3\text{Al, toluene, room temperature}} \text{(26)}\ anti
\tag{14}
$$

R	anti/syn
Me	88:12
Et	91:9

Tominaga *et al.* [148] have succeeded in the Michael reactions of 2-phenylthio-1-nitroethene (3) with specific donors, i.e. pyridinium, phthalazinium and isoquinolinium salts (equation 15). These reactions allow compound (3) to be used in the synthesis of heterocycles, particularly indolizine, pyrazole and pyridine derivatives.

Vicinal thionitroalkenes were used by Fuji *et al.* [149, 150] for regioselective and stereoselective diene functionalizations at positions 1 and 4 to give the (Z)-alkenes (31; equation 16). The conditions of the process are similar to those of the Diels–Alder reaction in the presence of acidic reagents.

Geminal thionitroalkenes

In contrast to vicinal thionitroalkenes, the geminal analogues react with nucleophilic reagents. The corresponding adducts (32; equation 17) can be reacted with ozone *in situ* to give various α-functionalized S-phenyl thioesters (33) [1, 135, 137, 138].

The reactions of benzenethiolate and t-butanethiolate with 1-phenylthio-1-nitro-2,2-diphenylethene (15) proceed via the formation of the Michael adduct to give mainly 1-phenylthio-2,2-diphenylethene and 1-t-butylthio-2,2-diphenyl-ethene (34; equation 18) [151].

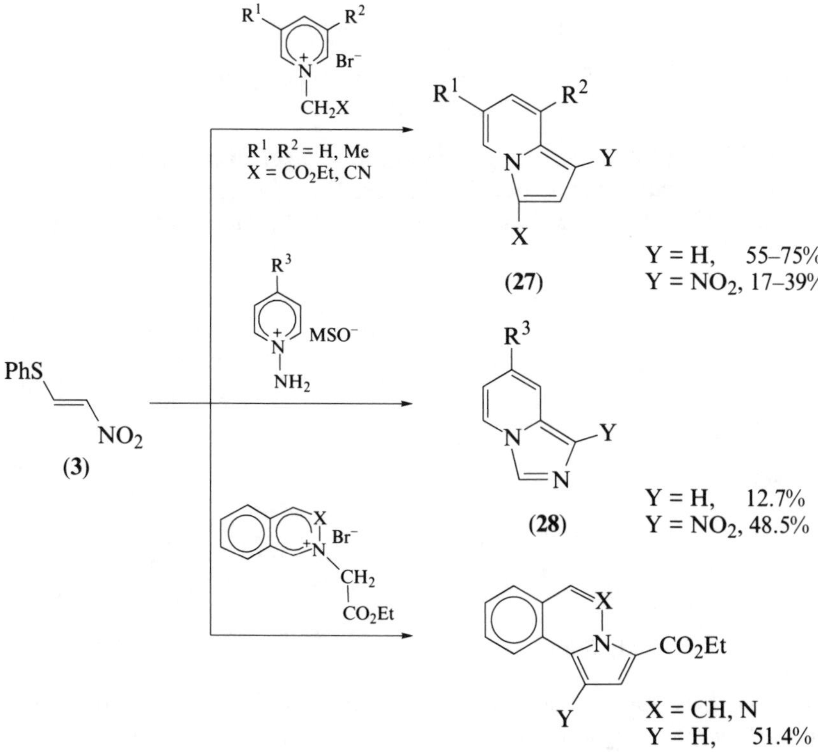

$R^1, R^2 = H, Me$
$X = CO_2Et, CN$

(27) Y = H, 55–75%
Y = NO_2, 17–39%

(15)

(28) Y = H, 12.7%
Y = NO_2, 48.5%

(29) X = CH, N
Y = H, 51.4%
Y = NO_2, 35.7%

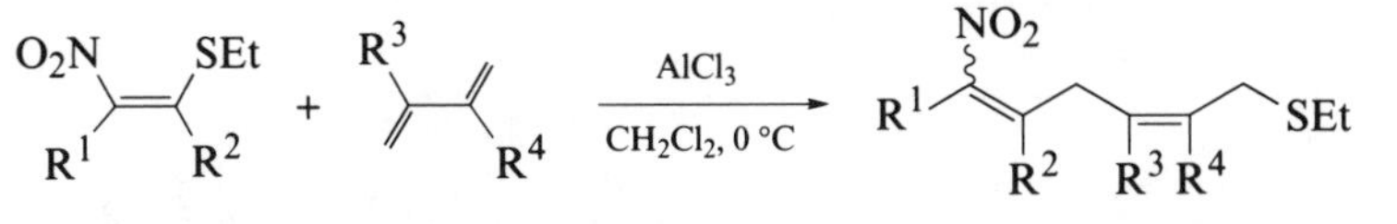

(16)

(31) 54–86%

$R^1, R^2 = H, Me, C_6H_{13}, (CH_2)_n$ (n = 4, 5)
$R^3, R^4 = H, Me$

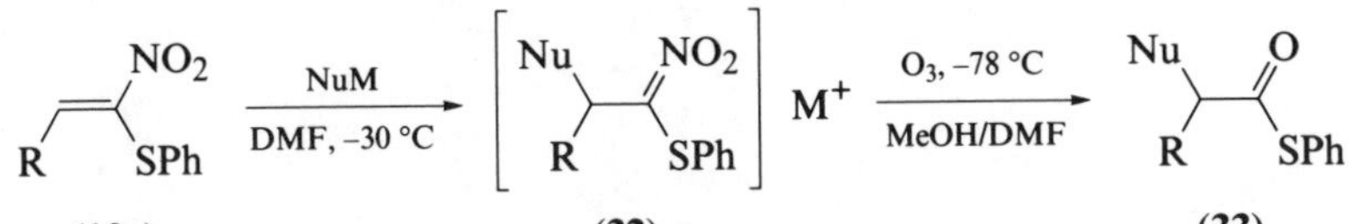

(17)

(13a) (32) (33)

R = Me, i-Pr; NuM = KOH, MeONa, i-PrONa, p-MeC$_6$H$_4$SO$_2$Na, KCH(CO$_2$Me)$_2$,

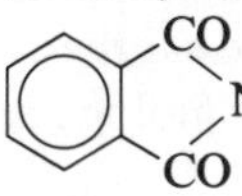

NK, CH$_2$FCONHK, PhLi (at –110 °C), PhC(OLi) =CH$_2$, Ts, PhCH$_2$NTs

$$Ph_2C{=}C(SPh)NO_2 + RS^- \xrightarrow{DMSO} Ph_2C{=}CHSR \qquad (18)$$

$$\textbf{(15)}$$

$$\textbf{(34)}\ 87{-}88\%$$

$$R = Ph, t\text{-}Bu$$

1-Phenylthio-1-nitropropene is an important reagent for the synthesis of 3-methylfuran derivatives (equation 19) that are the fragments of some natural compounds [133, 134, 152, 153] and some substituted β-lactams [136, 138].

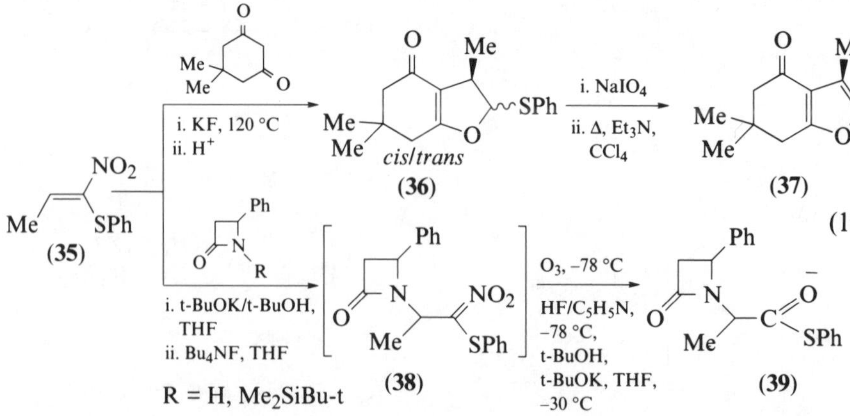

Lactams such as (13c) that contain a geminal phenylthionitroalkene fragment are the starting materials in the preparation of bicyclic β-lactams (equation 20) [138, 139].

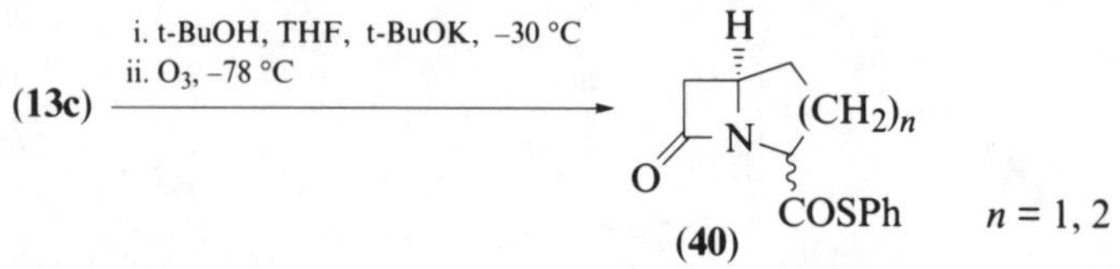

The oxygen containing analogs of oxapenam were synthesized in accordance with equation 21 [136, 139].

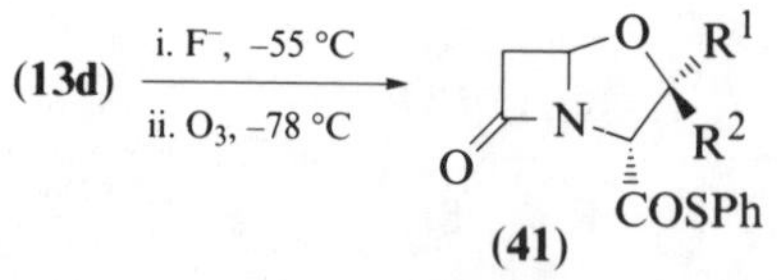

The key step in these reactions is the intramolecular nucleophilic addition of the amide nitrogen to the nitroalkene fragment. Intramolecular nucleophilic addition of the alkoxy group to the C=C bond of (13b) leads to the synthesis of some tetrahydrofuran and tetrahydropyran α-thioester homologues (equation 22) [139].

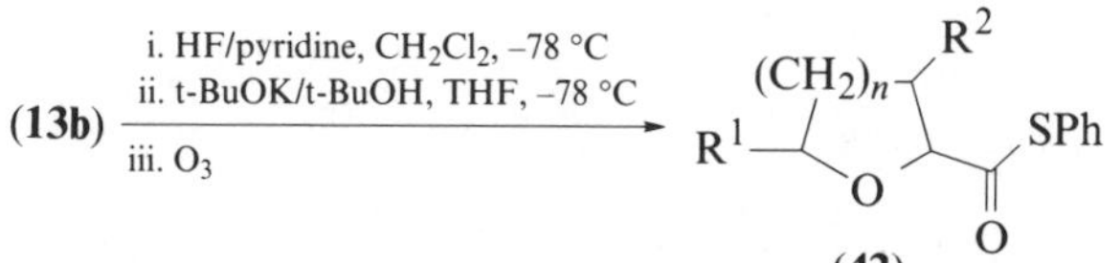

$$
\textbf{(13b)} \xrightarrow[\substack{\text{ii. t-BuOK/t-BuOH, THF, -78 °C} \\ \text{iii. O}_3}]{\text{i. HF/pyridine, CH}_2\text{Cl}_2\text{, -78 °C}} \quad R^1 \underset{O}{\overset{(CH_2)_n \overset{R^2}{\diagup}}{\diagup}} \underset{O}{\overset{}{\diagdown}} SPh \tag{22}
$$

$$
\textbf{(42)} \qquad n = 1, 2
$$

$$
R^1 = H, Me \ (n = 1), Ph, R^2 = H; \quad R^1 = H, R^2 = Ph
$$

The stereoselective addition of potassium trimethylsilanoate to the phenyl-thionitroalkene (13e) followed by ozonolysis results in the formation of the corresponding S-phenyl thioester. The latter can be transformed into antibiotics of the pyrimidine nucleoside type, e.g. polyoxin C (43) [140].

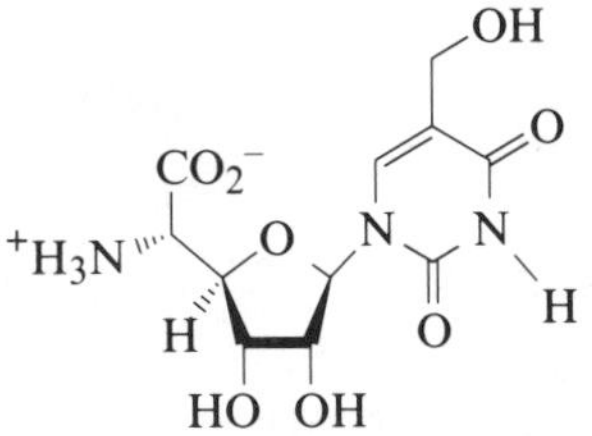

(43)

The aim of 1-phenylthio-1-nitroalkene epoxidation (equation 23) is the further transformation of the intermediate epoxide under mild conditions into the α-substituted S-phenyl thioester [141, 154].

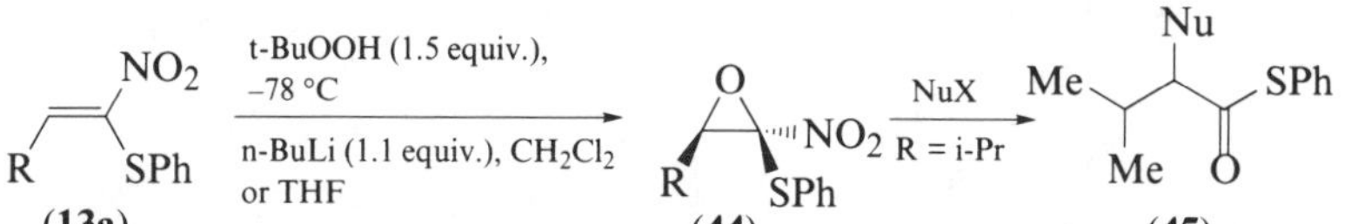

$$
R = Me, n\text{-}Pr, i\text{-}Pr, Bu, pentyl, hexyl, Ph; \quad Nu = Hal, CF_3CO_2, MeSO_3, OH;
$$
$$
X = H, Me
$$

3.3.2 DITHIONITROALKENES AND AMINOTHIONITROALKENES

Linear and cyclic nitroketene dithioacetals (2,2-dithio-1-nitroalkenes) are prepared in the reactions of dichloronitroethenes and trichloronitroethenes with monothiols and dithiols (equation 24) [13, 155].

Another synthetic method for compounds such as (50a) is based on the reaction between the salt (49) and haloalkanes (equation 25) [156–160].

2,2-Dithio-1-nitroalkenes (50b; equation 26) are the starting materials in the synthesis of 2-amino-2-thio-1-nitroalkenes (N,S-acetals of nitroketenes) (51) [157, 161–164] and 2,2-diamino-1-nitroalkenes (aminals of nitroketenes) (52) [156, 165, 166].

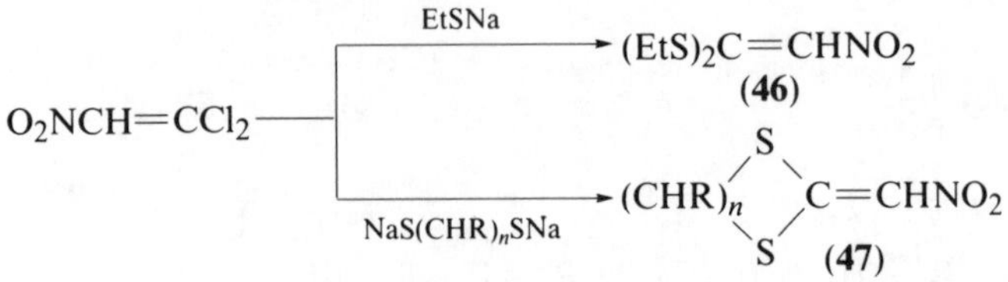

$$O_2NCH=CCl_2 \begin{cases} \xrightarrow{EtSNa} (EtS)_2C=CHNO_2 \quad \textbf{(46)} \\ \\ \xrightarrow{NaS(CHR)_nSNa} (CHR)_n \underset{S}{\overset{S}{\diagup}} C=CHNO_2 \quad \textbf{(47)} \end{cases} \qquad (24)$$

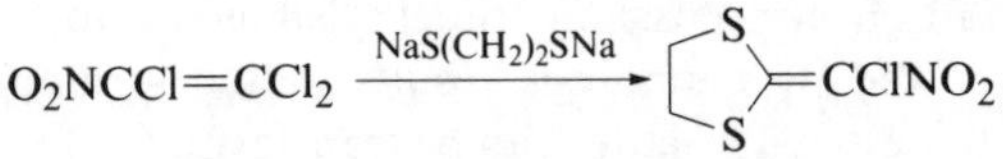

$$O_2NCCl=CCl_2 \xrightarrow{NaS(CH_2)_2SNa} \underset{S}{\overset{S}{\diagup}} C=CClNO_2$$
$$\textbf{(48)}$$

$$CS_2 + MeNO_2 \xrightarrow{KOH, EtOH} \underset{K^+S^-}{\overset{K^+S^-}{\diagup}} C=CHNO_2 \xrightarrow[-2KHal]{2R^1Hal}$$
$$\textbf{(49)}$$

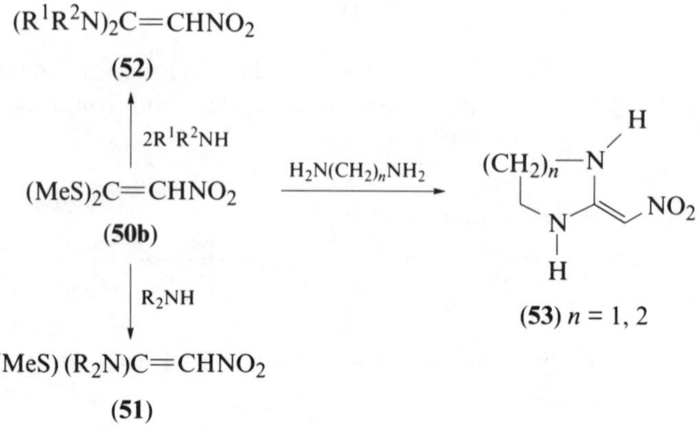

$$(R^1S)_2C=CHNO_2 \xrightarrow{R^2Hal} R^1S(R^2S)C=CHNO_2 \qquad (25)$$
$$\textbf{(50)}$$

$(R^1R^2N)_2C=CHNO_2$
$\textbf{(52)}$

$$\uparrow {\scriptstyle 2R^1R^2NH}$$

$$(MeS)_2C=CHNO_2 \xrightarrow{H_2N(CH_2)_nNH_2} (CH_2)_n \overset{H}{\underset{N}{\diagdown}} \underset{H}{\overset{}{\diagup}} C=CHNO_2 \qquad (26)$$
$$\textbf{(50b)}$$

$$\downarrow {\scriptstyle R_2NH} \qquad\qquad \textbf{(53)}\ n = 1, 2$$

$(MeS)(R_2N)C=CHNO_2$
$\textbf{(51)}$

Nitroketene and geminal benzoylnitroketene *N,S*-acetals, useful in the synthesis of nitrothiophene derivatives [167, 168] and nitrogen-containing heterocycles [169], are prepared in reactions of nitro compounds with PhNCS (equation 27) [157, 165, 167].

$$PhNCS + R^1CH_2NO_2 \xrightarrow{NaH} PhNHC=CR^1NO_2 \xrightarrow{R^2Hal} PhNHC=CR^1NO_2$$
$$\underset{S^-Na^+}{\big|} \qquad\qquad \underset{SR^2}{\big|}$$
$$\textbf{(54)} \qquad (27)$$

$$R^1 = H,\ PhCO,\ Ph;\quad R^2 = Me,\ Bn$$

Recently there was proposed a new method [170, 171] for the synthesis of
N,S-acetals of nitroketenes, illustrated by the reaction of nitromethane with
N-methylbis(methylthio)methyleneimine (56) in the presence of a zeolite catalyst
(equation 28).

$$\underset{(55)}{MeNO_2} + \underset{(56)}{MeN{=}C(SMe)_2} \xrightarrow[\text{reflux, 48 h}]{\text{zeolite catalyst}} \underset{MeS}{\overset{MeNH}{\diagup}}C{=}CHNO_2 \quad (28)$$

$$(57) \; 50\%$$

N-Arylsulfonyl derivatives of bis(methylthio)alkaneimines do not react with
nitromethane or nitroethane on zeolite, but the reaction does take place in
DMSO under the action of potash [172]. Thionitroenamines obtained in this
process exist in imine–enamine tautomeric equilibrium in various solvents. The
enamine structure (58a; equation 29) is predominant in polar solvents
(DMSO, methanol), while the imine form (58b) prevails in non-polar solvents
(CCl$_4$, chloroform).

$$\underset{\substack{(Het) \\ (58a)}}{ArSO_2NHC{=}\overset{R}{\underset{SMe}{C}}NO_2} \xrightleftharpoons[\text{DMSO, MeOH}]{CCl_4, CHCl_3} \underset{\substack{(Het) \\ (58b)}}{ArSO_2N{=}\overset{R}{\underset{SMe}{C}}CHNO_2} \quad (29)$$

R = H, Me;   Ar = Ph, p-MeC$_6$H$_4$, p-MeOC$_6$H$_4$, p-ClC$_6$H$_4$;   Het = 2-thienyl

N,S-Acetals of nitroketene hydrolyse readily under the action of mercury
salts to form N-substituted nitroacetamides [161, 172]. This method reported
has pronounced advantages over others.

The cyclic N,S-acetal of nitroketene (59) is prepared by the reaction of
nitroacetonitrile with N-methyl-o-aminothiophenol or from nitromethane and
the quaternary salt of 2-methylthiobenzthiazol (60; equation 30) [173].

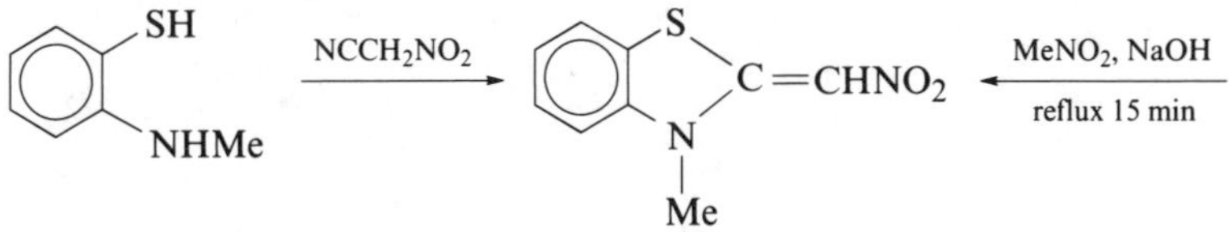

(59)

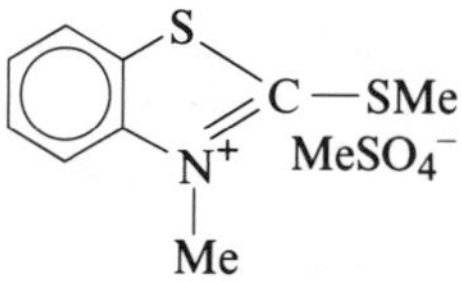

(60)

Studies of cyclic nitroketene N,S-acetals and geminal methoxycarbonyl-nitroketenes [174, 175] reveal the extraordinary stability of the (Z) form with specific orientation of the sulfur atom towards the oxygen atom of the nitro group. The (E) form is not so stable, in spite of the possible existence of intramolecular hydrogen bonding. Nitroketene dithioacetals are used as starting materials for the synthesis of various nitrogen-containing and sulfur-containing cyclic compounds [160, 162, 176, 177]. For example, 2,2-bis(methylthio)nitro-ethene (50b) is used as a two-carbon structural fragment in the synthesis of various functionalized heterocycles (equation 31) such as (61) to (64) [169, 178–180].

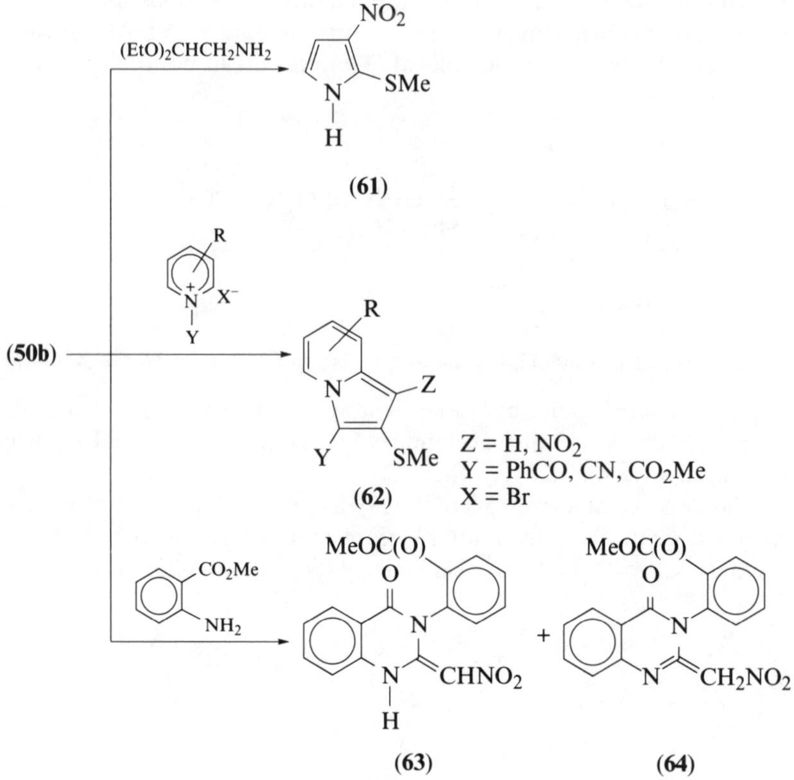

$$Z = H, NO_2$$
$$Y = PhCO, CN, CO_2Me \quad (31)$$
$$X = Br$$

$$R^1 = Ph, p\text{-}ClC_6H_4, 2\text{-thienyl}; \quad R^2 = Me$$

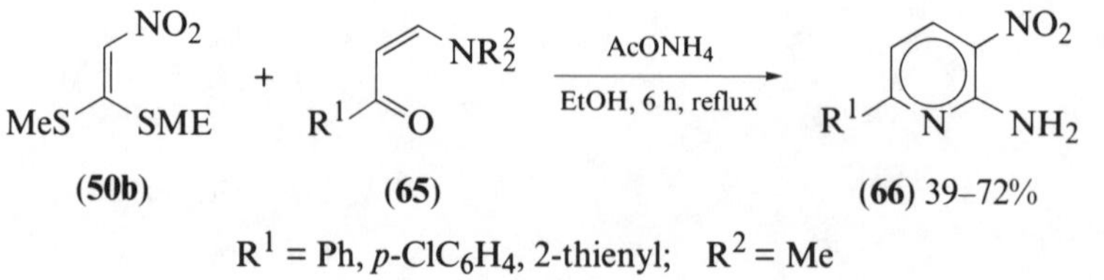

Some substituted 2-amino-3-nitropyridines (66; equation 32) are produced in the reactions of (50b) with enaminoketones (65) under the action of ammonium acetate [181].

Reactions between (50b) and N-substituted derivatives of methylpyridinium salts proceed similarly to the reactions of the latter with 2-phenylthio-1-nitroethene to give indolizine derivatives [178, 179]. Reaction of (50b) with methyl anthranilate gives the quinazolinone derivative and not the expected quinolone [169, 182]. The reaction of 2-anilino-2-methylthio-1-nitroethene with (50b) proceeds in a similar way [169]. The polysubstituted aromatic compounds, e.g. 2-amino-6-aryl(2-furyl)-4-methylthio-3-nitrobenzonitrile and its toluene analogue (68), are formed in the reactions of the nitroketene dithioacetal (50b) with aryl(furyl)alkylidenemalononitriles (67; equation 33) [183].

$$(50b) \ + \ \underset{NC}{\overset{NC}{>}}C=C\underset{CH_2R^2}{\overset{R^1}{<}} \ \xrightarrow[\text{ii. HCl}]{\text{i. } K_2CO_3, \text{ DMF}} \ \text{(68)} \tag{33}$$

(67)   (68) 69–90%

$$R^1 = Ph, \ p\text{-MeOC}_6H_4, \ 2\text{-furyl}; \quad R^2 = H, \ Me$$

The reactions of (50b) with some compounds containing activated methylene groups give the products of substitution of one MeS group; these products isomerize to give the substituted nitroallyls [162]. Manjunatha et al. [184] recommend the use of S,S-diacetals of nitroketenes as peptide synthones. The synthesis of an oligopeptide nitro precursor (69; equation 34) involves the reaction of nitroketene S,S-diacetal (50b) with an L-α-amino ester followed by hydrolysis.

$$O_2NCH=C(SMe)_2 \xrightarrow[\substack{\text{ii. HgCl}_2, \text{ MeCN/H}_2O(3:1), \\ \text{room temperature}}]{\substack{\text{i. L-R}^1\text{NHCHR}^2\text{CO}_2R^3, \\ \text{TsOH, MeCN, 30–80 °C, 10–24 h}}} O_2NCH_2C(O)NCHCO_2R^3 \tag{34}$$

(50b)   (69)

$$R^1,R^2 = (CH_2)_3, R^3 = PhCH_2, Et; \quad R^1 = H, R^2 = i\text{-Pr}, R^3 = Me; \quad R^1 = Et,$$
$$R^2 = PhCH_2, R^3 = H$$

3.3.3 SULFINYLNITROALKENES AND SULFONYLNITROALKENES

Oxidation of vicinal thionitroalkenes (equation 35) is an easy synthetic approach to a wide range of poorly investigated 2-sulfinyl-1-nitroalkenes and 2-sulfonyl-1-nitroalkenes [125, 126, 185].

$$O_2N \diagdown R^2 \diagup SR^3 \quad \underset{(3)}{} \quad \overset{\text{MCPBA (1 equiv.)}}{\underset{\text{or } H_2O_2/AcOH}{\overset{\text{MCPBA (2 equiv.)}}{\longrightarrow}}} \quad \begin{array}{c} O_2NCR^1{=}CR^2SOR^3 \\ (70) \\ O_2NCR^1{=}CR^2SO_2R^3 \\ (71) \end{array} \qquad (35)$$

The reaction is stereospecific, but a high excess of oxidizing agent and prolonged contact time lead in some cases to domination of the (Z) isomer. It is probable that such an effect can be explained by secondary isomerization into the most stable isomer [185]. Nitrovinylation of arylsulfinic acids by 1,2-dinitroalkenes [185] and 2-halo-1-nitroalkenes [50, 56] is another synthetic approach to nitroarylsulfonylalkenes (equation 36).

$$p\text{-RC}_6\text{H}_4\text{SO}_2\text{Na} + \text{ClCH}{=}\text{CHNO}_2 \longrightarrow p\text{-RC}_6\text{H}_4\text{SO}_2\text{CH}{=}\text{CHNO}_2 \quad (36)$$
$$\underset{(72)}{} \qquad\qquad\qquad\qquad\qquad\qquad \underset{(73)}{}$$
$$R = H, Cl, Br, I$$

In contrast to the oxidation reaction, the interaction between dinitroalkenes and (72) does not depend upon the initial dinitroalkene structure (equation 37). The process results entirely in the most stable isomer [185].

$$O_2N \diagdown R^2 \diagup NO_2 \quad \underset{(74)}{} \quad \overset{\text{ArSO}_2\text{Na}}{\underset{12\text{-}48 \text{ h}}{\underset{\text{MeONa, 20 °C,}}{\longrightarrow}}} \quad O_2N \diagdown R^2 \diagup SO_2Ar \quad \underset{(71)}{} \qquad (37)$$
$$R^1, R^2 = H, Me, Ph$$

The oxidation method for 1,2-dinitroalkenes has been extended by Lipina et al. [186] to the synthesis of sulfonylnitroalkenes. For example, 2-sulfonyl-1-nitroethane can be oxidized to 2-sulfonyl-1-nitroethene (equation 38). The transformation of (77) into the 1,2-dianion (79) requires a stronger base than in the case of the 1,2-dinitroalkane.

Recently it was shown that chiral 2-alkylsulfinyl-1-nitroalkenes can be used in effective asymmetric inductions [187–190a]. The synthesis of these reagents is illustrated in equation (39) [189, 190b]. The main products—chiral sulfoxides (85) – are obtained from nitrovinyl sulfoxides by exchange with (S)-2-phenyl-1-propanethiol followed by oxidation with ozone. The absolute configuration of (85a) is (S, S), determined by X-ray analysis.

2-Sulfonyl-1-nitroalkenes (71) and, less readily, 2-thio-1-nitroalkenes (3) (see Section 3.3.1.2) undergo nucleophilic substitution. For example, the reactions of (71) with amines give nitroenamines [125, 126, 128]. The reactions of vicinal sulfonylnitroalkenes with organocopper–zinc reagents are the basis for the highly selective syntheses of various polyfunctional nitroalkenes in 74–85%

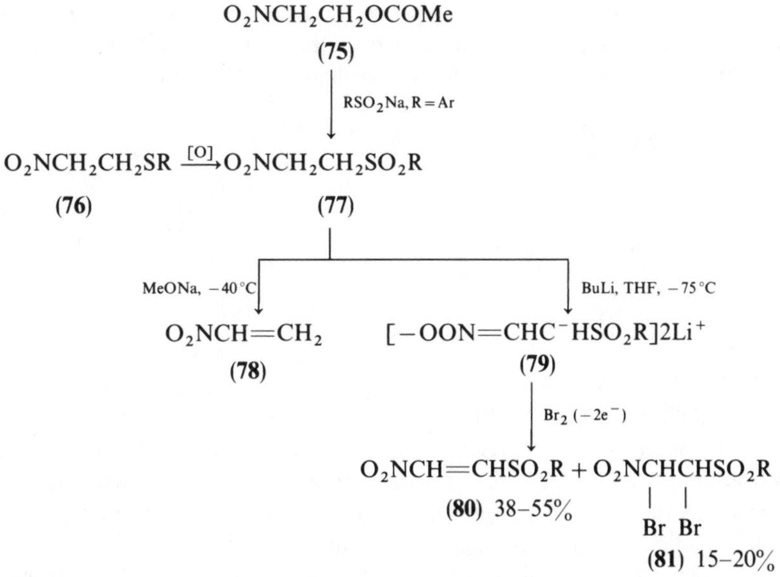

$$O_2NCH_2CH_2OCOMe$$
(75)

$$\downarrow \quad RSO_2Na, R = Ar$$

$$O_2NCH_2CH_2SR \xrightarrow{[O]} O_2NCH_2CH_2SO_2R$$
(76)   **(77)**

(38)

MeONa, −40 °C   BuLi, THF, −75 °C

$$O_2NCH{=}CH_2 \qquad [-OON{=}CHC^-HSO_2R]2Li^+$$
(78)   **(79)**

$$\downarrow Br_2 (-2e^-)$$

$$O_2NCH{=}CHSO_2R + O_2NCHCHSO_2R$$
(80) 38–55% | |
                                          Br   Br
                                      (81) 15–20%

$$R = Et, \text{ } p\text{-tolyl, } p\text{-ClC}_6H_4$$

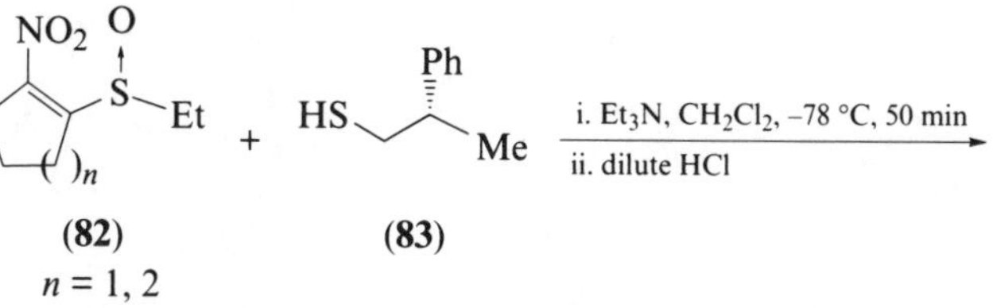

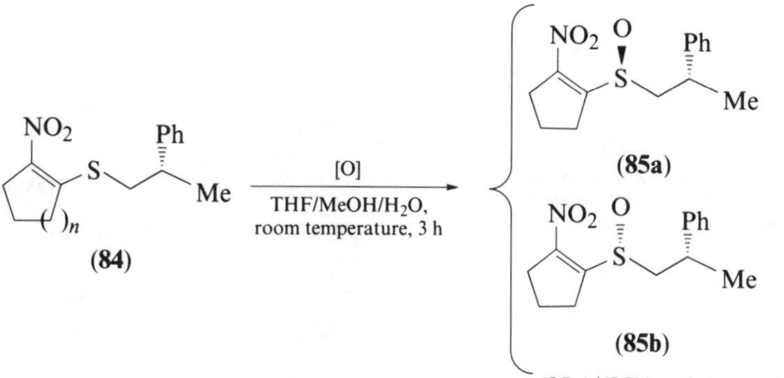

(82)   **(83)**
$n = 1, 2$

i. Et_3N, CH_2Cl_2, −78 °C, 50 min
ii. dilute HCl

(84)

[O]
THF/MeOH/H$_2$O,
room temperature, 3 h

(85a)

(39)

(85b)

(85a)/(85b) = 3:1, $n = 1$

yields. The processes take place at a lower temperature ($-78\,°C$ to $-50\,°C$)
[145, 146a] than required for the same reactions of 2-thio-1-nitrocyclohexenes.

Reactions of cyclic nitrovinyl sulfoxides with various activated methylene and methine components (equation 40) proceed through an addition–elimination process. Activated components participate in the form of their enolates. The reactions result in β-carbonylnitroalkenes in high yields [190b].

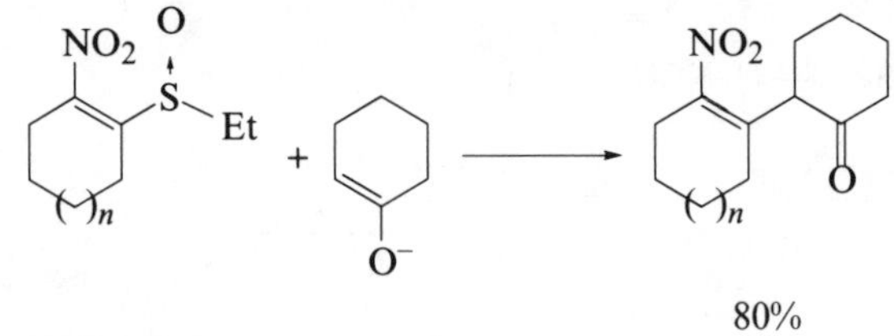

$$(40)$$

$$80\%$$

$$n = 1, 2$$

2-Arylsulfinyl-1-nitroethenes and particularly 2-arylsulfonyl-1-nitroethenes are electron-deficient reagents; thus, they behave as active dienophiles in the Diels–Alder reaction. The reactions of sulfur-containing nitroalkenes with cyclopentadiene (CPD) take place at room temperature (equations 41, 42). The reactions are stereoselective and give mostly the *endo*-nitrostereoisomers. Furan and cyclohexadiene tend to produce larger amounts of the *exo* isomers. The same nitroalkenes couple with aliphatic 1-alkadienes to give 50:50 mixtures of stereoisomers [125, 126, 191]. The processes are regioselective for asymmetric dienes and the nitro group defines the orientation of addition.

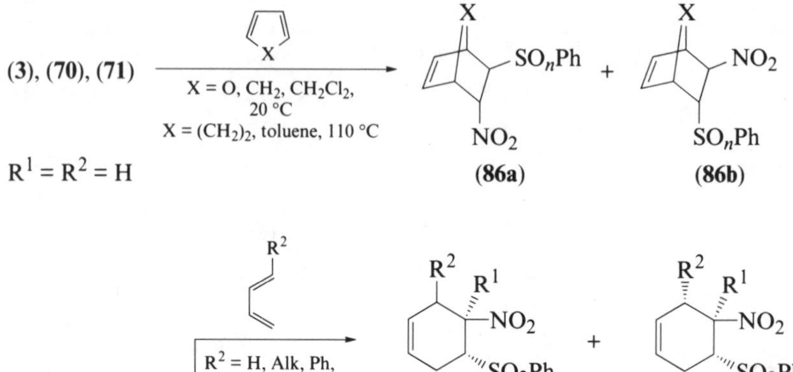

$$(3), (70), (71)$$

$$X = O, CH_2, CH_2Cl_2,$$
$$20\ ^\circ C$$
$$X = (CH_2)_2,\ \text{toluene},\ 110\ ^\circ C$$

$$R^1 = R^2 = H$$

(86a)   (86b) $$(41)$$

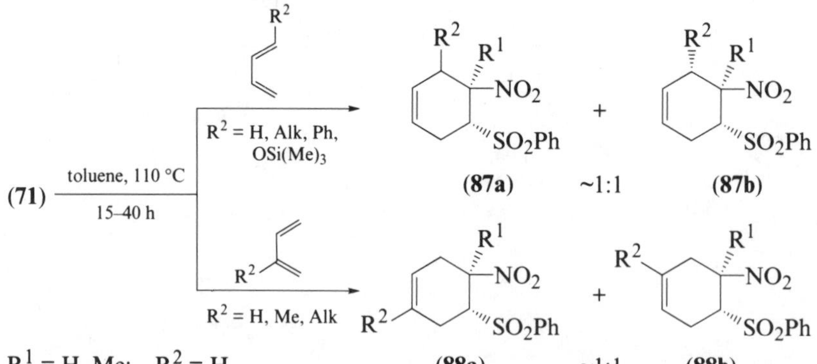

$$(71)$$

$$\text{toluene},\ 110\ ^\circ C$$
$$15\text{–}40\ h$$

$$R^2 = H, \text{Alk}, \text{Ph}, OSi(Me)_3$$

(87a)   ~1:1   (87b)

$$R^2 = H, \text{Me}, \text{Alk}$$

(88a)   ~1:1   (88b) $$(42)$$

$$R^1 = H, \text{Me};\quad R^2 = H$$

Diels–Alder adducts of phenylsulfonylnitroethenes undergo reductive elimination of the functional groups (equation 43) to form 1,4-cyclohexadienes [191].

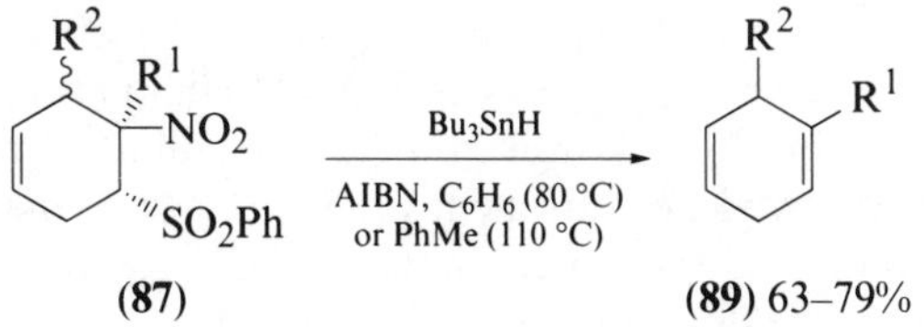

Bu3SnH
AIBN, C6H6 (80 °C) or PhMe (110 °C)
(87)
(89) 63–79%

(43)

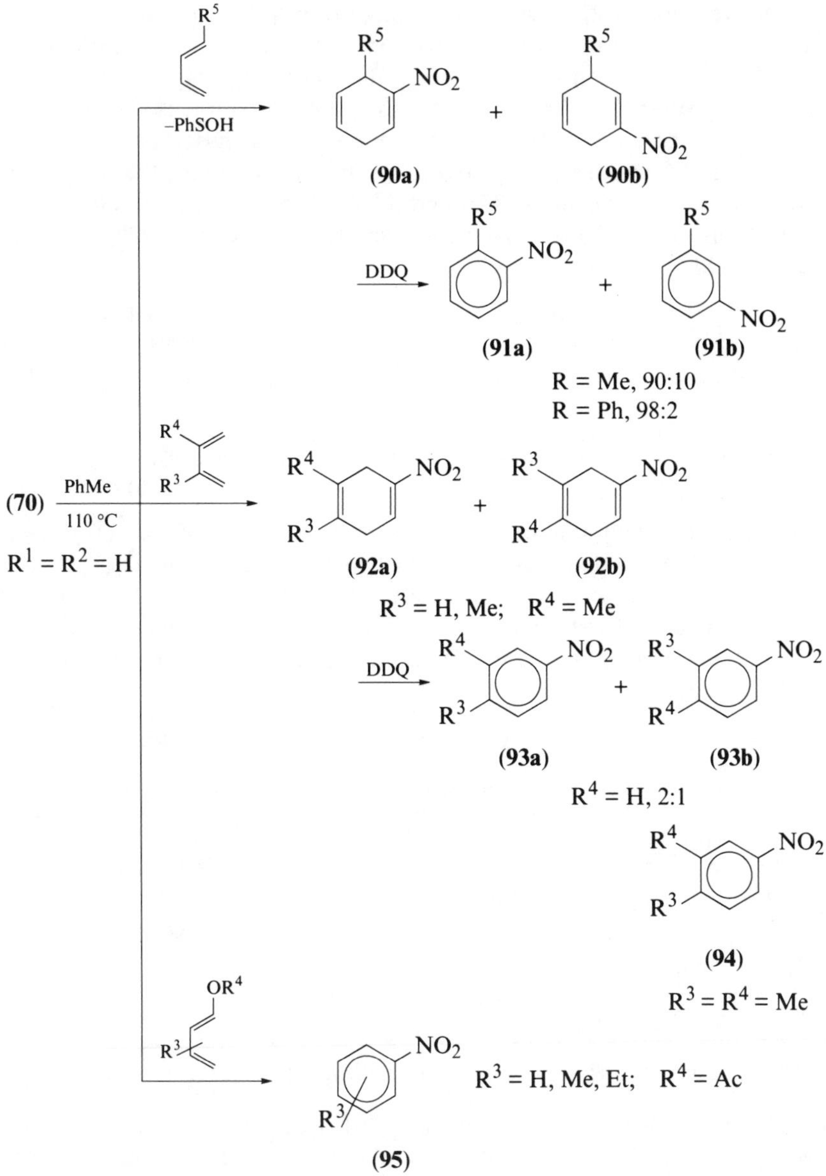

–PhSOH
(90a)
(90b)
DDQ
(91a)
(91b)
R = Me, 90:10
R = Ph, 98:2
(70)
PhMe
110 °C
R1 = R2 = H
(92a)
(92b)
R3 = H, Me; R4 = Me
DDQ
(93a)
(93b)
R4 = H, 2:1
(94)
R3 = R4 = Me
R3 = H, Me, Et; R4 = Ac
(95)

(44)

(*E*)-2-phenylsulfinyl-1-nitroethene reacts with methyl and phenyl dienes under similar conditions and readily undergoes PhSOH elimination to give 1-nitro-1,4-cyclohexadienes. In other words, the initial sulfinylnitroalkene can be considered as a nitroacetylene equivalent [125, 191]. Nitrodienes are oxidized by 2,3-dichloro-5,6-dicyano-*p*-benzoquinone (DDQ) to give the nitroaromatic compounds in high yields (equation 44). The latter are also produced in Diels–Alder reactions with oxygen-containing dienes [191], and in the case of (*Z*)-2-phenylsulfinyl-1-nitroethene and its homologues (R^2 = Me, Et, Ph), only the *o*-alkyl(phenyl)nitrobenzene derivatives are produced [124].

Chiral 2-alkylsulfinyl-1-nitroalkenes are used in the asymmetric Diels–Alder reaction for constructing optically active adducts. The latter participate as intermediates in the synthesis of biologically active compounds [188, 189]. For example, both diastereoisomers (**85a**) and (**85b**) interact with diene (**96**) under mild conditions to give diastereoisomeric, bicyclic products (**97a, 97b**; equation 45) in high yield (more than 95%).

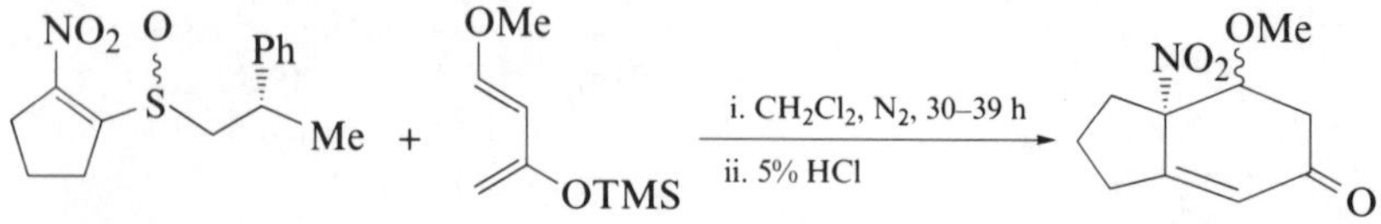

$$(45)$$

(**85a, 85b**)   (**96**)   (**97a, 97b**)
                                     (**97a**)/(**97b**) = 1:1

So, the sulfoxide group is a highly stereoselective and nucleofugal substituent in chiral dienophiles such as 2-alkylsulfonyl-1-nitroalkenes. Cycloaddition with active dienes proceeds smoothly, but such dienophiles are inert towards alkadienes and cyclopentadiene except under high pressure [190a], as shown in equation (46). The process is highly diastereoselective and produces monocyclic, bicyclic and tricyclic adducts.

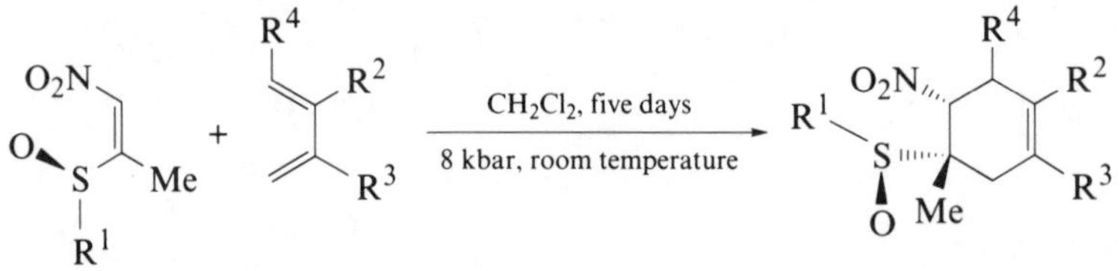

$$(46)$$

(**98**)   (**99**)   (**100**)

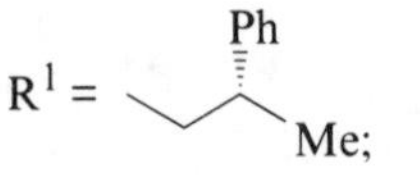

$$R^1 = \overset{Ph}{\underset{Me}{\diagdown\diagup}};$$

$R^2 = R^3 = H, R^4 = Me;$ $R^2 = R^3 = Me, R^4 = H;$ $R^2 = R^4 = H, R^3 = Me$

Optically active 2-sulfinyl-1-nitroalkenes (**101**; equation 47) undergo chiral additions with the enols of δ-lactams and γ-lactams. The reaction is an addition–elimination process and the best results are obtained for zinc enolates of δ-lactams [187].

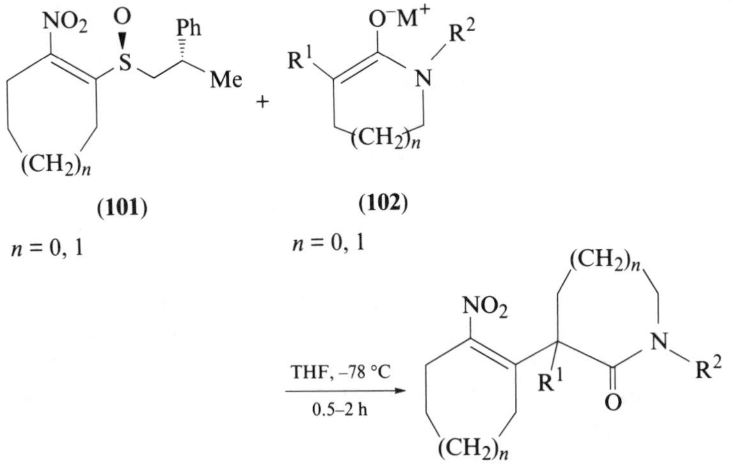

(47)

(101)

$n = 0, 1$

(102)

$n = 0, 1$

$$\xrightarrow[\text{0.5–2 h}]{\text{THF, } -78\,°C}$$

(103) 70–97%

$M = \text{Li, Zn}$

3.3.4 THIO(SULFONYL)NITRODIENES

Hydrogen sulfide reacts with 1,4-dinitrodienes in a similar manner to its reaction with dinitroalkenes to form bis(1-nitrodienyl) sulfides (equation 48) [128].

$$O_2NCH\!=\!CR^1CR^2\!=\!CHNO_2 \xrightarrow[-\text{HNO}_2]{H_2S,\,0\text{–}18\,°C} [O_2NCH\!=\!CR^1CR^2\!=\!CHSH] \longrightarrow$$

(104)

$$(O_2NCH\!=\!CR^1CR^2\!=\!CH)_2S \quad (48)$$

(105)

$$R^1 = R^2 = H, Me; \quad R^1 = Me, R^2 = H$$

The first preparation of 4-arylthio-1-nitro-1,3-butadienes was carried out in accordance with equation (49). The ^{1}H NMR data were consistent with the (E, E) structures of the products [192].

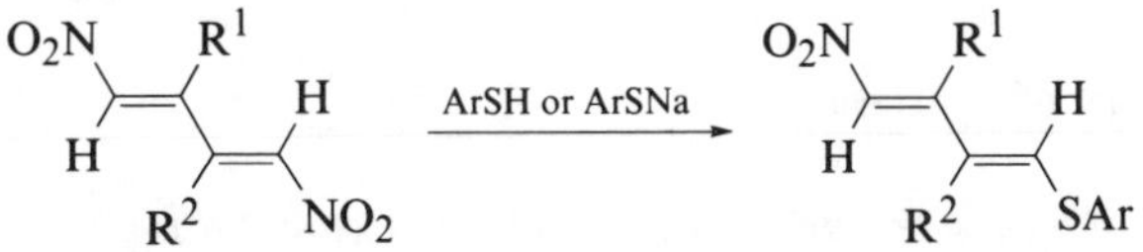

(49)

(104)

(106)

$$R^1 = H, Me, Ph; \quad R^2 = H, Me, Ph$$

Perhalo-2-nitro-1,3-butadienes react with RSH to form alkyl 2-nitro-1,3,4,4-tetrahalo-1,3-butadienyl sulfides (equation 50) via substitution of a single halogen atom [90, 193].

$$CCl_2{=}CClC(NO_2){=}CCl_2 + RSH$$

$$\textbf{(107)}$$

20–25 °C,
15–17 h

$$CCl_2{=}CClC(NO_2){=}C(Cl)SR \qquad (50)$$

$$\textbf{(108)}$$

$$R = Bn, CMe_2Et, C_8H_{17}$$

Oxidation of bis(1-nitrodienyl) sulfides or 4-arylthio-1-nitro-1,3-dienes (50 °C, 17–18 h for the former; 20 °C, two to seven days for the latter) leads to conjugated nitrodienyl sulfones (110) [185] in 40–55% or 70% yield, respectively (equation 51). Sulfonylnitrodienes can be synthesized in 20% yield by nitrovinylation of arenesulfinates with 1,4-dinitrodienes (similar to the 2-sulfonyl-1-nitroalkene synthesis) [118].

$$O_2NCH{=}CR^1CR^2{=}CHSR^3 \qquad O_2NCH{=}CR^1CR^2{=}CHNO_2$$

$$\textbf{(105, 106)}\,(E, E) \qquad\qquad \textbf{(109)}\,(E, E)$$

[O]   ArSO$_2$Na

$$(51)$$

H$_2$O/AcOH   20–60 °C, 1 h

$$O_2NCH{=}CR^1CR^2{=}CHSO_2R^3$$

$$\textbf{(110)}\ (E, E)$$

$$R^1 = H, Me; \quad R^2 = H, Me, Ph; \quad R^3 = Ar, O_2NCH{=}CR^1CR^2{=}CH$$

3.4 AMINO-CONTAINING NITROALKENES

3.4.1 MONOAMINONITROALKENES (NITROENAMINES)

Rajappa has presented a thorough review [165] on nitroenamines up to 1980. Because of this we shall discuss the publications starting from that very date and onwards, including in this division only either the principal works or those omitted from Rajappa's review but published before 1980.

3.4.1.1 Methods of synthesis

Reactions of amines with carbonyl compounds containing the nitro group

This method is restricted by the extreme instability of α-nitroaldehydes. Only the stabilized form of nitroacetic aldehyde sodium metazonate (1) reacts with primary amine salts to give the nitrovinylaryl(alkyl)amine (**2**; equation 1) [194].

$$\underset{\textbf{(1)}}{\text{HON}=\text{CHCH}=\text{NOONa}} \xrightarrow[\text{ii. ArNH}_2]{\text{i. H}^+} \underset{\textbf{(2)}}{\text{ArNHCH}=\text{CHNO}_2} \qquad (1)$$

Hurd and Nilson [195] have synthesized nitroenamines (**4**) from α-nitroketones and primary amines in acetic acid (equation 2).

$$\underset{\textbf{(3)}}{\text{R}^1\text{COCHR}^2\text{NO}_2} \xrightarrow[\text{AcOH}]{\text{R}^3\text{NH}_2} \underset{\textbf{(4)}}{\text{R}^3\text{NHCR}^1=\text{CR}^2\text{NO}_2} \qquad (2)$$

$$\text{R}^1 = \text{Alk}; \quad \text{R}^2 = \text{H, Alk, Ar}; \quad \text{R}^3 = \text{Ph}$$

The reactions of primary amine salts with the salt of nitromalonic dialdehyde (equation 3) proceed via nitroaminoacrolein derivatives (**6**) to give products such as (**7**) [196, 197].

$$\underset{\textbf{(5)}}{[\text{OHCC}^-(\text{NO}_2)\text{CHO}]\text{Na}^+} \xrightarrow{\text{R}^1\text{NH}_3^+\text{Cl}^-} \underset{\textbf{(6)}}{\text{R}^1\text{NHCH}=\text{C}(\text{NO}_2)\text{CHO}} \xrightarrow{\text{R}^2\text{NH}_2}$$

$$\underset{\textbf{(7)}}{\text{R}^1\text{NHCH}=\text{C}(\text{NO}_2)\text{CH}=\text{NR}^2} \qquad (3)$$

$$\text{R}^1, \text{R}^2 = \text{Ar, AlkCOCH}_2\text{CH}_2 \text{ NH}_2\text{CO, PhNH}$$

Reactions between (**5**) and secondary amine salts (equation 4) result in cleavage of the dialdehyde to produce two products (**8, 9**).

$$\underset{\textbf{(5)}}{} \xrightarrow{\text{Ph(R)NH}_2^+\text{Cl}^-} \underset{\textbf{(8)}}{\text{Ph(R)NCH}=\text{CHNO}_2} + \underset{\textbf{(9)}}{\text{PhNRCHO}} \qquad (4)$$

$$\text{R} = \text{Me, Et}$$

Condensation of amidoacetals with nitro compounds

One of the most common methods of nitroenamine synthesis involves the reaction of an amidoacetal with a nitro compound. For example, the simple aminonitroethene (**11**) is formed in the reaction of diethyl acetal (**10**) with MeNO$_2$ (equation 5) (85% yield) [165].

$$\underset{\textbf{(10)}}{(\text{EtO})_2\text{C(R)NMe}_2 + \text{MeNO}_2} \xrightarrow{\text{R}=\text{H}} \underset{\textbf{(11)}}{\text{Me}_2\text{NCH}=\text{CHNO}_2} \qquad (5)$$

Apart from nitromethane and dimethylformamide diethyl acetal, many other nitroalkanes and their derivatives [198–200] and various amidoacetals [200,

201] can be used (equation 6).

$$(\mathbf{10}) + H_2C(R^2)NO_2 \xrightarrow{\text{reflux}} Me_2NCR^1{=}C(R^2)NO_2 \qquad (6)$$
$$(\mathbf{12})$$
$$R^1 = H, Me, Ar; \quad R^2 = H, Et, Ph, CO_2Et$$

Nitroenamines can be prepared from arylnitroketones at room temperature in 70–72% yield [202]. The condensation of dinitromethane with dimethylformamide dimethyl acetal (equation 7) gives 2-N,N-dimethylamino-1,1-dinitroethene [203].

$$(MeO)_2CHNMe_2 + CH_2(NO_2)_2 \longrightarrow Me_2NCH{=}C(NO_2)_2 \qquad (7)$$
$$(\mathbf{13}) \qquad\qquad (\mathbf{14}) \qquad\qquad\qquad (\mathbf{15})$$

Interaction between cyclic N-methylcaprolactam diethylacetal and $MeNO_2$ under normal conditions produces a high yield of the nitroenamine (2-nitro-methylidene-1-methylhexahydroazepine) [204]. The use of amidodimethyl sulfate complexes (complexes of acid N,N-dimethylamides with dimethyl sulfate) in the reactions with CH acids containing the nitro group has made it possible to simplify the method of nitroenamine preparation owing to the generation of amidoacetals from the sulfate complexes *in situ* [200, 205, 206].

Some 2-amino-1-nitroalkenes have been prepared in a one-step synthesis (equation 8) by the reflux of a mixture of an orthoformic ester, a nitroalkane and a secondary amine [102, 103, 207, 208].

$$HC(OEt)_3 + R^1CH_2NO_2 + HNR^2R^3 \xrightarrow[\text{reflux}]{p\text{-TSA or Ac}_2O} R^2R^3NCH{=}C(R^1)NO_2$$
$$(\mathbf{16}) \qquad\qquad\qquad\qquad\qquad\qquad\qquad\qquad (\mathbf{17}) \qquad\qquad (8)$$

$$R^1 = H \,[207, 208], CO_2Et \,[102, 103], CN \,[103]; \quad R^2, R^3 = Alk, Ar, Het$$

The use of nitromethane produces nitroenamines in some cases in 56% yield. Use of other nitroalkanes decreases the yield considerably [207]. The Wolfbeis method [103] modifies [209] the process for the preparation of geminal dinitroenamines (equation 9).

$$(\mathbf{16}) + ArNH_2 \xrightarrow[\text{room temperature}]{AcOH} ArN{=}CHOEt \xrightarrow{H_2C(NO_2)_2} ArNHCH{=}C(NO_2)_2$$
$$(\mathbf{18}) \qquad\qquad\qquad (\mathbf{19}) \qquad\qquad (9)$$

The formation of the intermediate (**18**) was proven in the model synthesis. The authors point out that this method can be applied only to aromatic amines.

The chiral nitroenamine derivatives of L-pyroglutaminic acid and D-pyroglutaminic acid are synthesized in accordance with equation (10) in high yield. The products (**22**) are used as convenient synthones in the preparation of optically active piperazinones [210].

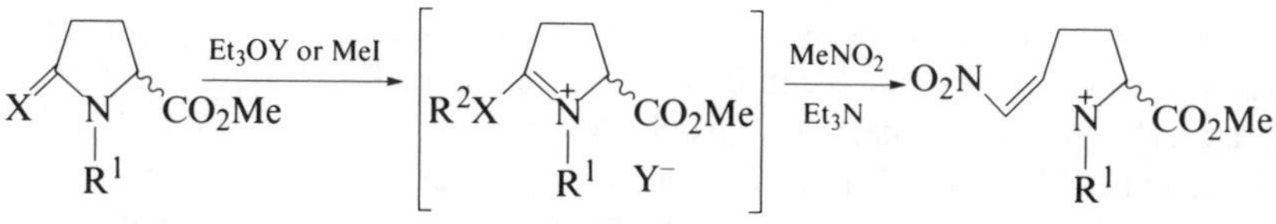

$$X = O, S; \quad R^1 = H, Br, CH_2CO_2Et; \quad R^2 = Me, Et; \quad Y = BF_4, I$$

It is probable that the process includes the formation of the lactam acetal (21). L-Nitroenamines and D-nitroenamines (23) are produced from the methyl esters of L-*N*-acetylproline and D-*N*-acetylproline in a similar way.

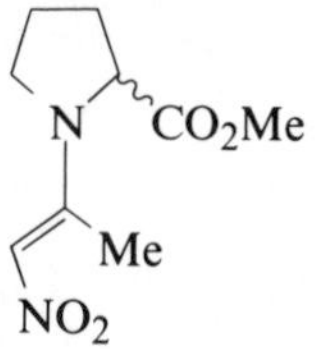

(23)

The catalytic reduction of nitroenamines (22, 23) is followed by spontaneous cyclization to give piperazinones.

Reactions of β-functionalized nitroalkenes with amines

Reactions between orthoformic esters and primary nitro compounds without amines lead to β-alkoxynitroalkenes (see Section 3.2). The latter are transformed easily into nitroenamines under the action of amines. For example, the alkoxy group in (24) is substituted by the primary or secondary amine (equation 11) to produce the (Z)-nitroenamine or the (E) nitroenamines respectively [97].

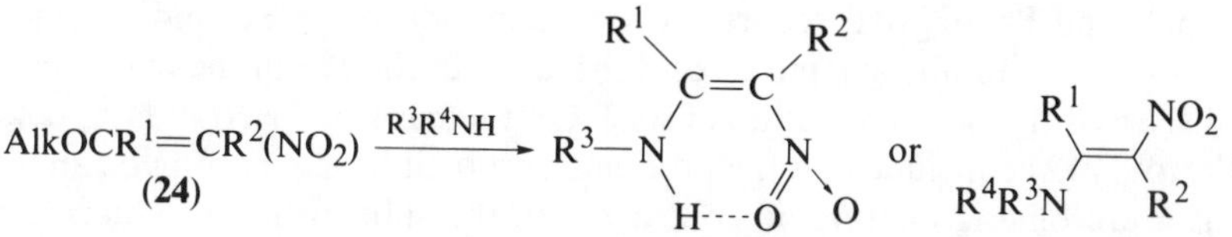

$$R^1, R^2 = H, Ph; \quad R^3 = H, Me, Ph; \quad R^3 = R^4 = Me, Et$$

For weakly basic, primary aromatic amines the nitroemamines synthesis with participation of compound (16) is preferable [102, 103].

The synthetic approach to a wide range of nitroenamine esters of substituted α-nitro-β-aminoacrylic acids is expressed in equation (12) [99, 102–104, 200].

$$R^1OCR^2{=}C(NO_2)CO_2R^1 + R^3R^4NH \longrightarrow R^3R^4NCR^2{=}C(NO_2)CO_2R^1$$

$$\textbf{(26)} \qquad\qquad\qquad \textbf{(27)} \qquad\qquad (12)$$

$$R^1 = Me, Et; \quad R^2 = H, Me; \quad R^3, R^4 = H, Alk, Ar, Het$$

Application of such amides as urea and its derivatives gives β-ureido-α-nitroacrylic esters [211] which can be used in the synthesis of 5-nitrouracil and its derivatives. A similar method is used for the preparation of nitroenamines from hydrazines [105] and 2-amino-2-deoxy-D-glucose [108] (see Section 3.2).

Substitution of the alkoxy group in 4-nitro-5-phenyl-3-methoxy-2-oxo-3-pyrroline (equation 13) takes place under the action of ammonia, primary amines or pyrrolidine to give the cyclic nitroenamine (29) [117].

$$R = H, C_6H_{11}, Ph, Br$$

Other β-substituted α-nitroalkenes such as β-halo-α-nitroalkenes or β-thio-α-nitroalkenes and α, β-dinitroalkenes can exchange the β-functional group of the amino residue. 2-Thio-1-nitroalkenes react with amines to give nitroenamines in a new synthetic method [143].

Some nitrovinylamines (31) are synthesized in accordance with equation (14) on treating chloronitroethene (30) or its precursor with various amines [13, 14].

$$ClCH{=}CHNO_2 \xrightarrow{R^1R^2NH} R^1R^2NCH{=}CHNO_2 \qquad (14)$$
$$\text{(30)} \qquad\qquad\qquad\qquad \text{(31)}$$
$$R^1 = H, Et; \quad R^2 = Et, t\text{-Bu}$$

Rappoport and Topol [10a] report that vicinal halonitrostyrenes and halonitrostilbenes give rise to nitroenamines, and the stereochemistry of the reaction has been discussed by the same authors [10a]. On the basis of spectral data, Belon and Perrot [212] conclude that the products of substitution by aromatic amines exist as a tautomeric mixture, in contrast with the substitution products from aliphatic amines, which exist in the enamine form (equation 15).

$$PhCCl{=}CHNO_2 \xrightarrow{RNH_2} RNHCPh{=}CHNO_2 \qquad (15)$$
$$\text{(32)} \qquad\qquad\qquad \text{(32a)}$$

$$ArNHCPh{=}CHNO_2 \rightleftharpoons ArN{=}CPhCH_2NO_2$$
$$\text{(33b)} \qquad\qquad\qquad \text{(33c)}$$

This synthetic approach to nitroenamines is restricted by the complicated preparation of the initial halonitroalkenes. Nitroenamines can also be prepared by the reactions of amines with vicinal dinitroalkenes (equation 16) [8, 10a,

213–218].

$$O_2NCR^1\!=\!C(R^2)NO_2 \xrightarrow{HNR^3R^4} R^3R^4NCR^1\!=\!C(R^2)NO_2 \qquad (16)$$
$$\text{(34)} \qquad\qquad\qquad\qquad \text{(25b)}$$

The reactions between 1,2-dinitrostilbene and primary and secondary aliphatic, aromatic and heterocyclic amines have been studied in detail [213, 214]. Dubois *et al.* have suggested that the reaction with 1,2-dinitrostilbene could be used as a spectroscopic, quantitative test for primary and secondary amines [213, 214]. A kinetic, stereochemical study of the reaction of (Z)-1,2-dinitrostilbene with morpholine has elucidated the two-step mechanism of the process, which involves the initial formation of a CTC between the reagents [219].

Transamination (the exchange of one amine group for another) is a possible way to new nitroenamines. The very first object of transamination was 2-dimethylamino-1-nitroethene [165, 201, 220], which under treatment with aromatic amines produced the corresponding adducts. Fetell and Feuer [221] have demonstrated the reversibility and limitations of the process. 1-t-Butylamino-2-nitro-1-propene and its butene homologue (**35**; equation 17) undergo trans-amination with pyrrolidine (76% and 23% yield, respectively) but do not react with piperidine, even though the basic and steric properties of the latter are similar to those of pyrrolidine. The reaction does not take place with secondary amines, probably because of steric factors. Compounds such as (**36**) react with primary aliphatic amines to form mixtures of isomers.

$$R^2NHCH\!=\!CR^1(NO_2) \underset{R^2NH_2}{\overset{}{\rightleftharpoons}} \quad N\!-\!CH\!=\!C(NO_2)R^1 \qquad (17)$$
$$\text{(35)} \qquad\qquad\qquad\qquad \text{(36)}$$
$$R^1 = \text{Me, Et;} \quad R^2 = \text{n-Pr, i-Pr, t-Bu}$$

A further study of transamination has promoted an effective and simple synthetic approach to primary and secondary aliphatic nitroenamines (**38**; equation 18) [208]. Other preparative methods for (**38**) are unsatisfactory.

$$R^1(O_2N)C\!=\!CHNR^2R^3 \xrightarrow[-HNR^2R^3]{R^4NH_2,\,0\text{--}20\,^\circ C} R^1(O_2N)C\!=\!CHNHR^4 \qquad (18)$$
$$\text{(37)} \qquad\qquad\qquad\qquad \text{(38) } 70\text{--}90\%$$
$$R^1 = \text{Me, Et;} \quad R^2 = \text{H, Me;} \quad R^3 = \text{Ph, Het;} \quad R^4 = \text{H, Alk}$$

The same method has been used for the synthesis of 2-t-butylamino-1-nitroalkenes [222]. The latter are produced as the (Z) and (E) isomers. The ratio of the isomers depends on the nature of the solvent. Some cyclic nitroenamines that are intermediates in the synthesis of 3-nitroquinolones are prepared via transamination of 1-aroyl-2-dimethylamino-1-nitroethenes [202].

Chiral E-nitroenamines (**40**; equation 19) can be synthesized [223] via transamination of 2-morpholyl-1-nitroalkenes (39) with chiral S-2-methoxymethyl-pyrrolidine (SMP) or (R)-2-ethylpyrrolidine.

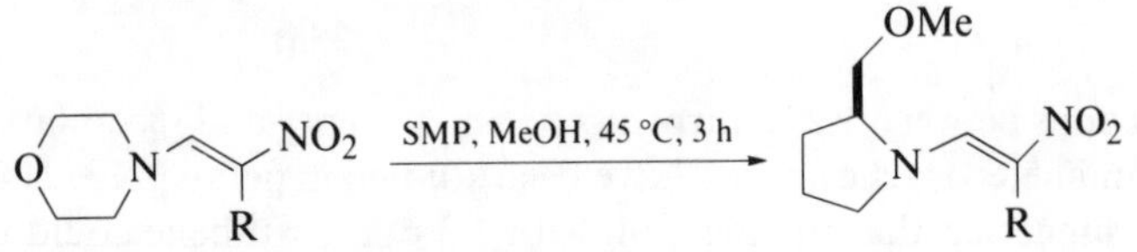

$$ \text{(39)} \qquad\qquad\qquad\qquad \text{(40)} $$

$$ R = H, Me, Et $$

The structure of aminonitropropene (40) has been elucidated by X-ray analysis. In solution, all nitroenamines studied so far have the s-(E) configuration [223].

Nitration of aldimines by alkyl nitrates

Nitration of aldimines (equation 20) is another synthetic approach to nitroenamines [224, 225].

$$ R^1CH_2CH{=}NR^2 \xrightarrow[\text{ii. } NH_4Cl]{\text{i. } KNH_2/\text{liquid } NH_3/R^3ONO_2 \ (1:2:1.5), \ -33\,°C} R^1(O_2N)C{=}CHNHR^2 $$

$$ \text{(41)} \qquad\qquad\qquad\qquad\qquad\qquad\qquad\quad \text{(42)} \tag{20} $$

$$ R^1 = n\text{-Pr}, i\text{-Pr}, t\text{-Bu}; \quad R^2 = H, Me, Et, n\text{-}C_5H_{11}; \quad R^3 = Alk $$

The yield of the product (42) depends on the nature of R^1 and varies from 50% to 67%. If the alkyl substituents in the aldimine and alkyl nitrates are different, alkyl group exchange may occur along with the nitration (equation 21).

$$ t\text{-BuN}{=}CHMe \xrightarrow[\text{ii. } NH_4Cl]{\substack{\text{i. } PrONO_2, \\ KNH_2/NH_3}} t\text{-BuNHCH}{=}CHNO_2 + t\text{-BuNHCH}{=}C(NO_2)Me $$

$$ \text{(43)} \qquad\qquad\qquad\qquad\quad \text{(44) } 21\% \qquad\qquad\qquad \text{(45) } 10\% \tag{21} $$

The optimum conditions for the preparation of aminonitrocycloalkenes (47; equation 22) have been derived by Feuer and McMillan [225] and applied in the nitration of (46).

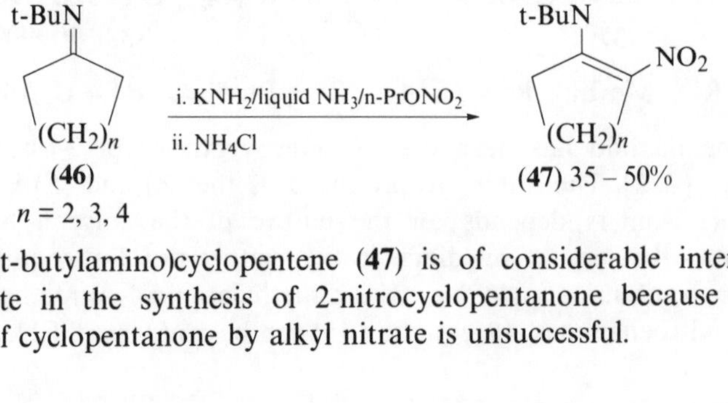

$$ \text{(46)} \qquad\qquad\qquad\qquad\qquad\qquad\qquad \text{(47) } 35-50\% $$

$$ n = 2, 3, 4 $$

1-Nitro-2-(t-butylamino)cyclopentene (47) is of considerable interest as an intermediate in the synthesis of 2-nitrocyclopentanone because the direct nitration of cyclopentanone by alkyl nitrate is unsuccessful.

Other methods of synthesis

There are some examples of nitroenamine formation as the result of nitroheterocycle cleavage. For example, nitroisoxazoles (48) transform into propenes (49) under the action of amines (equation 23). Subsequent reaction with dinucleophiles (guanidine, thiourea) produces compounds such as (50) and (51) [226].

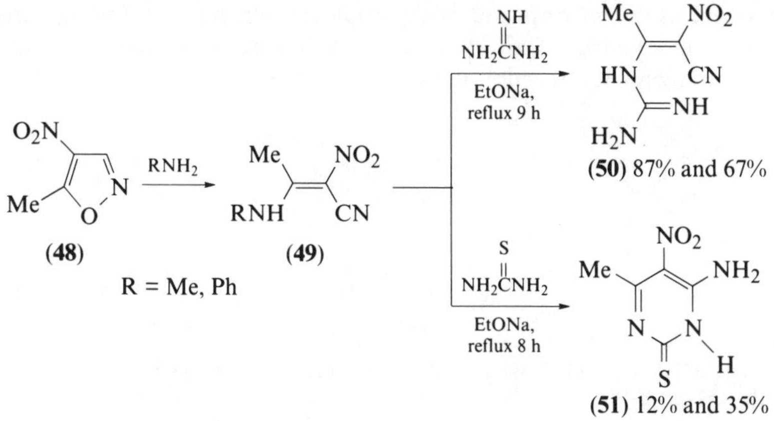

$$(23)$$

3,4-Dinitro-1-methylpyrrole (52) under reflux with pyridine converts into mostly the addition–elimination product (53; equation 24), along with 2,3-dinitro-1,4-dipiperidino-1,3-butadiene (54) [227].

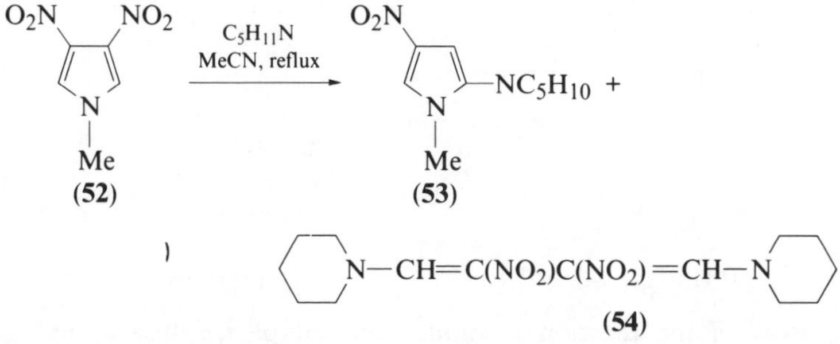

$$(24)$$

The reactions of 3,4-dinitrothiophene with aliphatic, aromatic and heterocyclic amines (equation 25) proceed entirely towards ring cleavage to give various

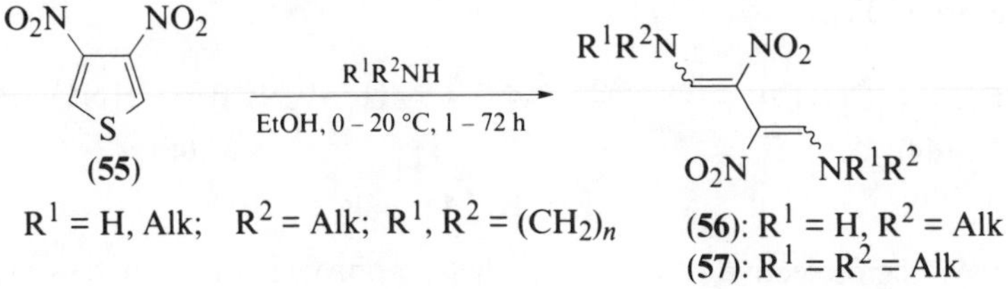

$$(25)$$

$R^1 = H, Alk; \quad R^2 = Alk; \quad R^1, R^2 = (CH_2)_n \quad$ (56): $R^1 = H, R^2 = Alk$
(57): $R^1 = R^2 = Alk$

1,4-amino derivatives of 2,3-dinitro-1,3-butadienes (56, 57) [228, 229]. Products such as (56) are isolated as mixtures of (Z, Z), (Z, E) and (E, E) isomers. The isomer ratio depends on the nature of the solvent. Compounds such as (57) are produced in the (E, E) form.

The reactions between 6-nitroazolo[1,5-a]pyrimidines and aryl(benzyl)amines are accompanied by cleavage of the pyrimidine ring and lead to 1-amino-2-nitro-3-iminopropene as the major products (equation 26). When the pyrimidine ring contains electron-donating substituents and the arylamine ring contains electron-accepting substituents the reaction does not take place [230].

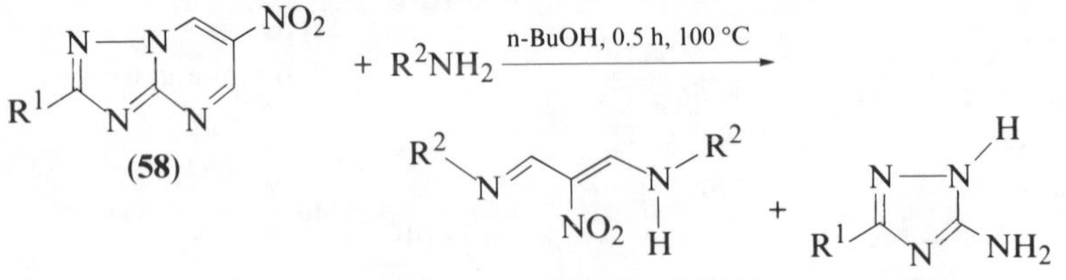

$$R^1 = H, CF_3, Me, CO_2Et, NH_2; \quad R^2 = Ph, p\text{-}MeC_6H_4, p\text{-}MeOC_6H_4, p\text{-}BrC_6H_4, CH_2Ph$$

Reaction of nitrostyrene or nitrocyclohexene with 2-vinylaziridine gives the nitroenamine [231] instead of the anticipated adduct 6-nitrotetrahydroazepine (equation 27). Formation of (63) is the result of rearrangement of the intermediate addition product.

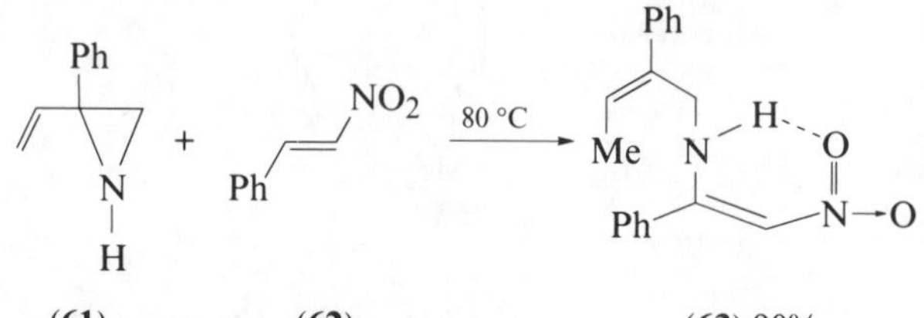

The pathway of the reaction is similar to that of α, β-unsaturated ketones and can be explained by the ready enolization of nitro and keto groups.

Two 2-amino-1-nitroalkenes (65; equation 28) were isolated from the mixture of (64) with nitromethane and piperidine after a prolonged period at room temperature [232].

$$EtOCR{=}C(CO_2Et)_2 \xrightarrow[\text{ii. MeNO}_2]{\text{i. H}_{10}\text{C}_5\text{NH, room temperature}} H_{10}C_5NCR{=}CHNO_2 \quad (28)$$
$$(64) \hspace{6cm} (65)$$

$$R = H, Me$$

The new regioselective, one-pot synthetic approach to nitroenamines (67) includes phenylacetylene nitration (equation 29) and subsequent coupling with aromatic amines [233]. Primary amines produce only (Z) isomers and secondary amines give (Z)/(E) mixtures in a ratio of 3:7.

$$
\underset{(66)}{\text{C}_6\text{H}_5-\text{C}\equiv\text{CH}} \quad
\begin{array}{c}
\text{i. HgCl}_2/\text{NaNO}_2/\text{THF, room temperature} \\
\xrightarrow{\hspace{4cm}} \\
\text{ii. } R^1\text{-C}_6\text{H}_4\text{-NHR}^2, \text{ room} \\
\text{temperature or 60 °C}
\end{array}
\quad
\underset{(67)\ 31-48\%}{
\begin{array}{c}
R^1\text{-C}_6\text{H}_4 \quad R^2 \\
\diagdown N \diagup \quad NO_2 \\
C_6H_5-C=C \\
\qquad\qquad H
\end{array}}
\quad (29)
$$

$$R^1 = \text{H, } o\text{-Me, } m\text{-Me, } p\text{-Me, } o\text{-OMe}$$

Aminonitrodienes (69) [234] and (71) [235] were obtained in accordance with equations (30) and (31).

$$\underset{(68)\ R = H,\ Me}{\text{Me}_2\text{NCH}=\text{CHCHO}} \xrightarrow{\text{RCH}_2\text{NO}_2,\ \text{EtOK}} \underset{(69)}{\text{Me}_2\text{NCH}=\text{CHCH}=\text{CRNO}_2} \quad (30)$$

$$\underset{(70)}{R^1\text{NHCH}=\text{C(NO}_2)\text{CHO}} \xrightarrow{R^2\text{CH}_2\text{X},\ \text{C}_5\text{H}_5\text{N}} \underset{(71)}{R^1\text{NHCH}=\text{C(NO}_2)\text{CH}=\text{CR}^2\text{X}} \quad (31)$$

$$R^1 = \text{Me, Ph}; \quad R^2 = \text{H, CN}; \quad X = \text{NO}_2, \text{CN}$$

3.4.1.2 Structure of nitroenamines

Three tautomeric forms (72a–72c) are probable for nitroenamines with primary and secondary amino groups.

$$
\underset{(72a)}{\overset{\displaystyle|\ \ \ |}{\underset{\overset{\displaystyle|}{H}}{-N-C}=\overset{\displaystyle|}{C}-NO_2}}
\qquad
\underset{(72b)}{-N=\overset{\displaystyle|}{C}-\overset{\displaystyle|}{C}H-NO_2}
\qquad
\underset{(72c)}{-N=\overset{\displaystyle|}{C}-\overset{\displaystyle|}{C}=N\diagup^{OH}_{\diagdown O}}
$$

There is no experimental evidence for the nitrone form (72c) [165, 218]. A study [212] of enamine–imine equilibria found no proof of (72c) through spectral investigations. Detailed NMR studies of nitroenamines and X-ray analyses of some compounds reveal structure (72a) [165, 218, 236]. According to a review [165], only two nitroenamines (74; equation 32) exist in the 2-nitroimine form owing to steric effects, whereas the more simple compounds (73) derived from the same starting materials are nitroenamines.

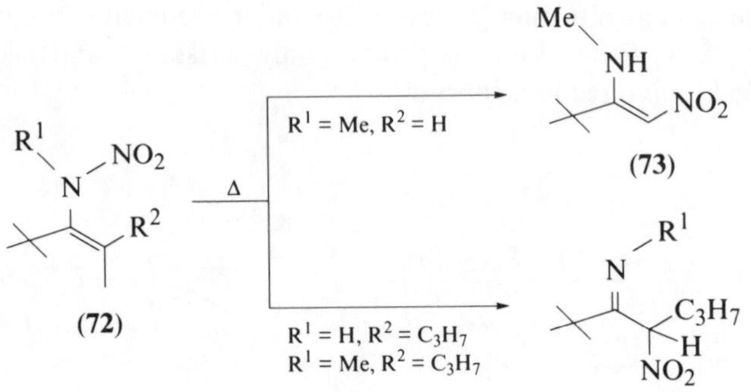

$$(32)$$

Nitroenamines are typical push–pull ethenes [165, 237], i.e. pronouncedly polarized systems. The polarization comes from the significant contributions of canonical ionic structures (equation 33) [165, 218, 222, 238].

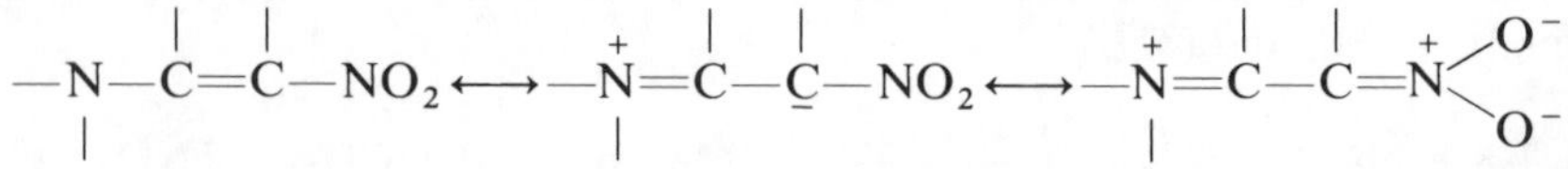

$$(33)$$

The essential delocalization of the nitrogen unshared pair has been demonstrated by PMR spectroscopy [236]. The pronounced polarization of nitroenamines means that they show distinctly different spectral characteristics from those of nitroalkenes. This is particularly true for the vibrational spectra [216, 217, 222, 238]. The 'enamine' band (1650–1550 cm^{-1}) is very strong in the IR spectrum and weak in the Raman spectrum. This is attributed to conjugation. The strong band at 1290–1230 cm^{-1} in the IR spectrum corresponds to the vibration of the polarized nitro group. The high electronic polarization correlates with the semiempirical AM1 (parametric quantum mechanical molecular model, Austin Model 1; Dewar *et al.*, *J. Am. Chem. Soc.*, 1985, **107**, 3902) method (Table 1) applied to 2-nitro-3-aminoacrylic ester (**75**) and 2-nitro-3-aminocrotonic ester (**76**) [222, 238].

The main ionic contribution to the ground state arises from the central structure in equation (33); most of the negative charge is concentrated on C(2) and the most electron-deficient group is NO$_2$. The contribution of CO$_2$Me to the electronic charge distribution in the molecules of (**75**) and (**76**) is insignificant.

Table 1 The general charges^a of the main groups of nitroenamines (**75**) and (**76**) [238]

Formula		C^{1}O$_2$Me	C^2	C^3	NH$_2$	NO$_2$
H$_2$NC^{3}H=C^2(C^{1}O$_2$Me)NO$_2$	(**75**)	−0.03	−0.37	0.17	0.20	−0.16
		(−0.07)	(−0.41)	(0.23)	(0.33)	(−0.08)
H$_2$NC^{3}Me=C^2(C^{1}O$_2$Me)NO$_2$	(**76**)	−0.03	−0.37	0.24	0.18	−0.16
		(−0.07)	(−0.23)	(0.23)	(0.31)	(−0.08)

^{a}Numbers in parentheses indicate an excess (−) or deficiency of π-electron density

The nitroenamine stereostructure has been the object of detailed study [165, 216, 236–241]. Primary and secondary nitroenamines exist in the (Z) form with intramolecular hydrogen bonding in non-polar solvents; in polar solvents, equilibration of the (Z) and (E) forms takes place because of a low energy barrier. The (E) form is characteristic for tertiary nitroenamines in the liquid and solid states because the steric hindrance is minimized. On the basis of IR and PMR data, only Dubois et al. [213] have found β-amino-γ-nitrostilbenes to exist in the (Z) configuration. According to a theoretical and spectroscopic study [241], the simplest enamine 2-amino-1-nitroethene exists as a mixture of (E) and (Z) isomers in solution. The $(E)/(Z)$ ratio increases in solvents of high polarity. A methyl group at C(2) stabilizes the (Z) form, whereas alkyl and aryl groups attached to amine nitrogen stabilize the (E) form.

A theoretical study has confirmed that the thermal (Z)–(E) isomerization proceeds via a dipolar transition state. The input of the anionic process has been defined for 2-aminonitroethene and its N-methyl derivative. For secondary and tertiary nitroaminoethenes the relative levels of (E) and (Z) isomers (at the C(1)=C(2) bond) and (E) and (Z) rotamers (at the C(2)—N bond) depend on the steric requirements of the R^1R^2N group [241]. The secondary and mixed tertiary nitroenamines of type $R^1R^2NCR^3$=CHNO$_2$ (R^3 = H, Me) can exist in four stereoisomeric forms: (Z, E), (E, E), (E, Z) and (Z, Z). The first symbol shows the configuration with respect to the C=C bond and the second symbol defines the orientation concerning the C(2)—N bond. The calculations made for 2-N-methylamino-1-nitroethene define the (Z, Z) as the least stable (equation 34).

$$\text{(34)}$$

$(C_1\text{–}C_2)\,Z,\,(C_2\text{–}N)\,E \qquad\qquad E, E \qquad\qquad E, Z$

According to the spectral and X-ray analytical data, the most stable form is the (E, Z) [236]. The isomer ratio (^{1}H NMR data) depends on the nature of the solvent. For example, only the (Z, E) isomer is found in CDCl$_3$, while in (CD$_3$)$_2$SO three forms are present–(Z, E), (E, E) and (E, Z)–and the (E, Z) form is the major component [236].

The stereostructures of nitroenamines with α-ethoxycarbonyl groups have been studied [237–239, 241–244].

Bakhmutov et al. [242–244] have devoted some dynamic NMR investigations to the mechanism of (Z)–(E) isomerization for β-amino-α-nitroacrylic ester. Compounds of type (77) that contain either a primary or secondary amino group can form structures with intramolecular hydrogen bonds involving the nitro and carbonyl groups (equation 35). Rotation around the (C(2)—COR^3 bond gives (Z, Z), (Z, E), (E, Z) and (E, E) stereoisomers for the (E) and (Z)

configurations (the second symbol defines the orientation about the $C(2)$—COR^3 bond) [237, 239].

$R^1, R^2 = H, Me \qquad R^3 = OMe, Me$ $\hspace{4cm}$ (35)

A spectroscopic investigation has revealed that 2-nitro-3-aminoacrylic ester (75) crystallizes either as a mixture of (Z, Z) and (Z, E) isomers or in the (E, Z) form [239]. All these forms exist as equilibrium mixtures in solution, the isomer ratio depending on the nature of the solvent. For example, the (Z, E) form is prevalent in DMSO. Computer analysis [240] shows that (75) can have four stable, planar structures. 2-Nitro-3-aminocrotonic ester can exist only in two forms, one of which is the planar (E, Z) form and another is the non-planar (Z, Z) form with the ester group shifted out from the plane. The corresponding stable α-nitroketones can also form non-planar isomers.

3.4.1.3 Chemical transformations of nitroenamines

The specific features of the nitroenamine structure define the chemical properties of these compounds. Enamine and push–pull resonance structures mean that these compounds can undergo such reactions as electrophilic and nucleophilic addition–elimination, 1,3-dipolar cycloaddition and reduction. The few examples of reactions with nucleophiles known before 1980 are presented in a review by Rajappa [165]. This section deals with the literature after 1980 and those works not included in the review mentioned above.

Bromination of alkylaminonitroenamines (78; equation 36) leads to the product of substitution (79) [224].

$$\text{t-BuNHCH}{=}\text{C(NO}_2)\text{R} \xrightarrow[\text{pyridine}]{\text{Br}_2, \text{CHCl}_3, 0\,°C} \left[\begin{array}{c} \text{t-BuNHCHBrC(NO}_2)\text{R} \\ | \\ \text{Br} \end{array} \right] \longrightarrow$$

$$\text{(78)}$$

$$\text{t-BuN}{=}\text{CHC(NO}_2)\text{R} \hspace{1cm} (36)$$
$$\text{(79)} \hspace{1cm} \text{Br}$$

$$\text{R = Me, Et}$$

Tokumitsu and Hayashi [245] have investigated the electrophilic reactions of primary and secondary nitroenamines with N-halosuccinimides, arylsulfenyl chlorides and thiocyanogen. In all cases, substitution at C(1) took place (equation 37).

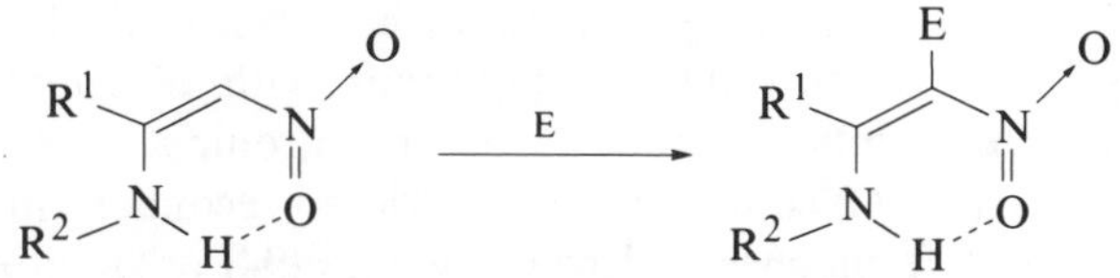

$$E = Cl, Br, I, SCN, SC_6H_4NO_2\text{-}o$$

The formation of 1-substituted nitroenamines (**83, 84**) is also the dominant direction of nitroenamine (**80**; equation 38) reactions with isocyanates and isothiocyanates [246].

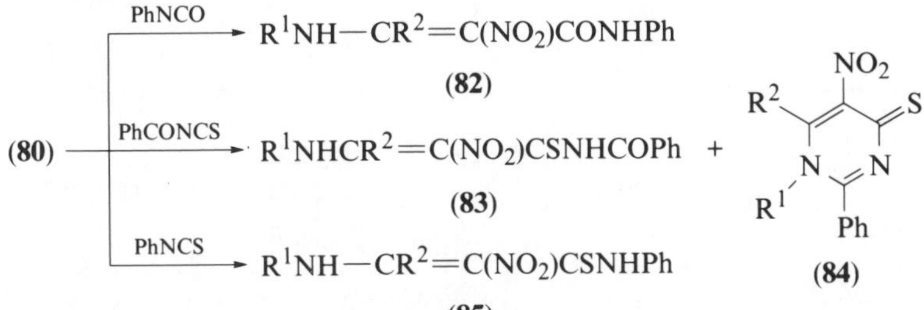

The more active benzoylisocyanate gives not only the β-substituted product but also the product of cyclization (**84**; equation 38). The spontaneous intra-molecular cyclization of the product of substitution, which probably results from the interaction between the nitroenamine (**86**) and benzoquinone (equation 39), is a good synthetic approach to new derivatives of 3-nitrobenzofuran (**87**) [247].

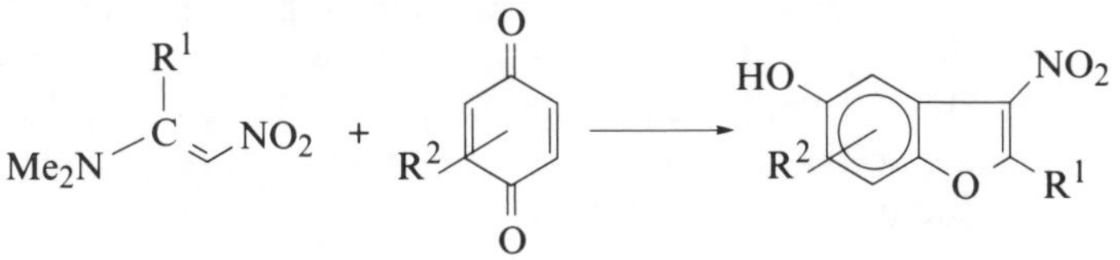

$$R^1 = H, Me; \quad R^2 = Cl, SPh, C_6H_4NO_2$$

The reactions between nitroenamines (**88**) and N-methyl-α(γ)-picolinium (or quinolinium, N-benzothiazolium) salts give rise to nitro compounds of the cyanine type (equation 40) [198].

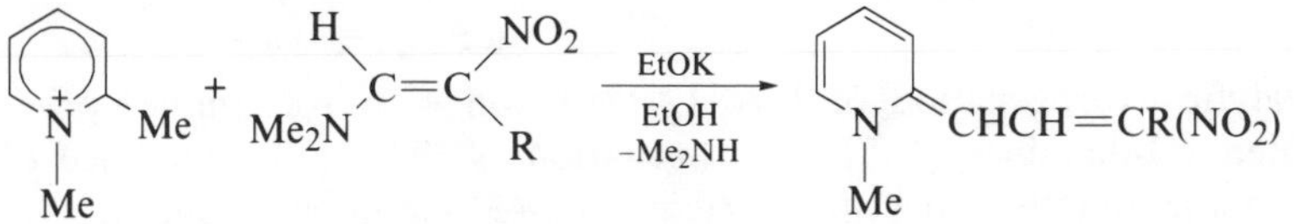

The influence of strong acids (protonation) has been studied for secondary nitroenamines. For example, those compounds with N-heteroaromatic or N-oxyaromatic substituents either rearrange via nitronic acids to give amido-oximes of α-keto acids (**92**; equation 41) or undergo secondary intramolecular cyclization to give diazaindenes and triazaindenes (**94**) or quinoxaline oxides (**93**) [248, 249]. The N-alkyl derivatives transform into dimers that are pyrazoles derivatives and 1,3-dinitro-4-aminobutadienes in trifluoroacetic acid. The type of protonation product depends on the substituents at the amine fragment and the conditions of isolation. The NMR data are consistent with nitronic acids. The mechanism of the rearrangement has been discussed [248, 249].

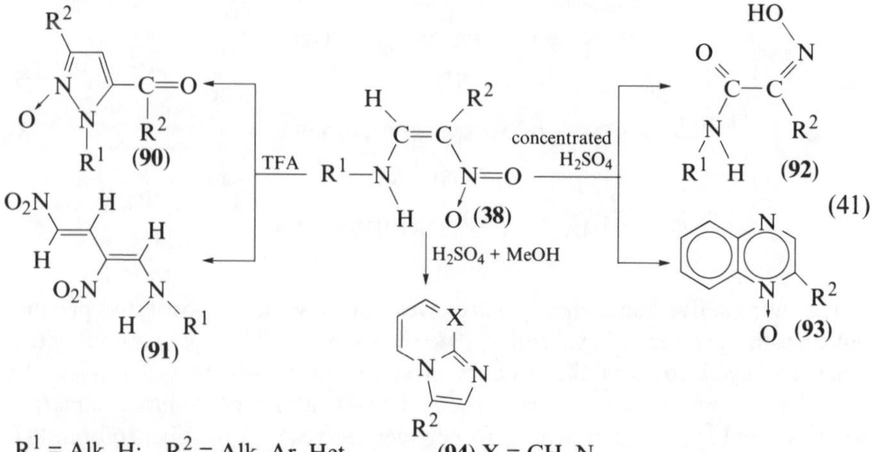

$$R^1 = \text{Alk, H}; \quad R^2 = \text{Alk, Ar, Het} \qquad (\mathbf{94})\ X = CH, N$$

Nitroenamines, like other enamines, participate in the reaction of 1,3-dipolar cycloaddition with aryl azides and nitrilimines to produce nitrotriazoles and nitropyrazoles [165]. Tokumitsu [246] has reported the [2 + 2] cycloaddition of (**86**) to dimethyl acetylenedicarboxylate (equation 42). Compound (**96**) is the product of cycloadduct (**95**) degradation.

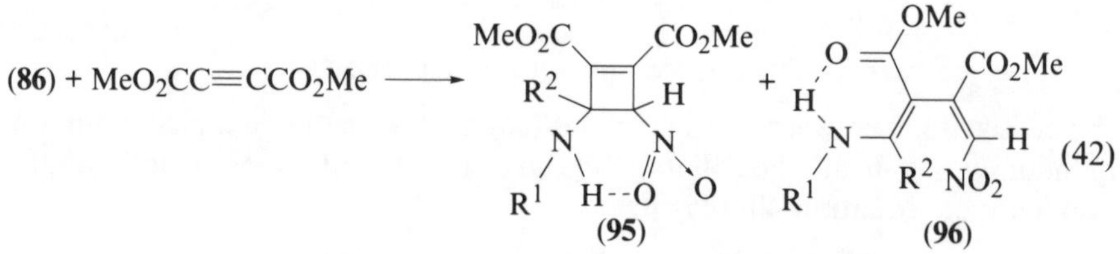

$$(\mathbf{95}) \xrightarrow{\Delta} (\mathbf{96})$$

Catalytic hydrogenation or LiAlH$_4$ reduction of nitroenamines produces saturated 1,2-diamines [165]. Hydrogenation of (**97**) in the presence of an orthoester leads to heterocyclization (equation 43). Compounds such as (**98**) are biologically active [250]. The reactions of nitroenamines with nucleophiles (for example, transamination; see Section 3.4.1.1) have been studied in more detail than other types of transformation for these compounds. For example, the first step of nitroenamine (25) hydrolysis in an alkaline medium is hydroxy addition (equation 44). Subsequent acidification of the reaction mixture leads to degradation of the intermediate (99) to give the corresponding amine and α-nitroketone salt [224].

$$
\begin{array}{c}
R^3CO \diagdown \diagup NO_2 \\
\diagup \diagdown \\
R^2 \quad NH \\
\quad | \\
(97) \text{ Ar}
\end{array}
+ R^4C(OR^1)_3 \xrightarrow[\text{Pd/C}]{H_2}
\begin{array}{c}
R^3CO \diagdown \quad N \\
\diagup \diagdown \diagdown R^4 \\
R^2 \quad N \\
\quad | \\
\text{Ar } (98)
\end{array}
\qquad (43)
$$

$$R^3 = \text{Me, OMe}$$

$$
R^1R^2NCR^3{=}C(R^4)NO_2 \; \underset{}{\overset{OH^-}{\rightleftharpoons}} \; R^1R^2NCR^3CR^4{=}NO_2{}^-
$$
$$
(25) \hspace{6cm} \underset{OH}{\overset{|}{|}}
$$
$$
(99)
$$

$$
\xrightarrow[-R^1R^2NH]{H^+} \; R^3C{-}CR^4 \qquad (44)
$$
$$
\qquad\qquad\qquad \underset{O}{\overset{\|}{\;}} \underset{NO_2{}^-}{\overset{|}{\;}}
$$
$$
\qquad\qquad\qquad (100)
$$

Nitroenamines (101) and other enamines which contain an acceptor substituent in the β position react with o-hydroxybenzyl alcohols (equation 45) to give examples of the poorly investigated, 3-functionalized 4-H-chromenes (102) [251].

$$
\begin{array}{c}
R^2 \\
\end{array}
\text{(benzene ring with OH, OH, } R^3)
+ \; \text{(}NO_2, N\text{-morpholino, }R^1\text{)} \xrightarrow[62\%]{Ac_2O, \text{ reflux 2 h}}
\begin{array}{c}
R^2 \\
\end{array}
\text{(chromene with }NO_2, O, R^1, R^3\text{)}
\qquad (45)
$$

$$
(101) \hspace{5cm} (102)
$$
$$
R^1 = \text{H, Me; } R^2, R^3 = \text{H, OMe, } CH_2CH_2CH_2CO
$$

The reactions of nitroenamines [205, 251–253] and aminonitrodienes [234] with organic metal compounds are most interesting and well studied. Reactions of the latter lead to the formation of a new C—C bond. It is worth mentioning the reactions of dimethylaminonitroalkenes with Grignard or organolithium reagents (equation 46), which give alkyl or aryl derivatives of nitroalkenes (103) [252].

$$
Me_2NCR^1{=}C(NO_2)R^2 \xrightarrow[\text{ii. } H^+, -Me_2NH]{\text{i. } R^3M} R^1CR^3{=}C(NO_2)R^2 \qquad (46)
$$
$$
(12) \hspace{5cm} (103)
$$

$$M = \text{MgBr, Li}; \quad R^1 = R^2 = \text{H, Alk, Ph}; \quad R^3 = \text{Alk, Ar, Het, RCH}{=}\text{CH, R}{\equiv}\text{C}$$

Typical CH acids react with nitroenamines (equation 47) in the presence of base to give nitronic acid salts (104) (substitution products). Upon acidification, compounds of type (104) transform into products (105) that contain the β-nitroethylidene fragment [205].

$$XCH_2Y + (12) \xrightarrow{R^1OK} XCY{=}CR^2CR^3{=}NO_2^-K^+ \xrightarrow{H^+} XCY{=}CR^2CHR^3NO_2$$

$$\qquad\qquad\qquad\qquad (104) \qquad\qquad\qquad\qquad (105)$$

$$X, Y = CO_2R^4, CN \qquad\qquad\qquad\qquad (47)$$

The salts of nitronic acid (104) can be used in the synthesis of various types of organic compound [165, 205, 206, 253, 254]. For example, the products of dimethylaminoethene condensation with CH acids (malonic esters, cyanoacetic esters and dinitriles) (104) react with arenediazonium salts to give arylhydrazones (106; equation 48). The latter are used as starting materials in the synthesis of pyridazines (107) [255].

$$(104) + p\text{-}RC_6H_4N_2BF_4 \longrightarrow X(Y)C{=}CH{-}\underset{\underset{NO_2}{|}}{C}{=}N{-}NH{-}C_6H_4R\text{-}p \xrightarrow[(-EtOH)]{H_2O, 2\,h}$$

$$(106)$$

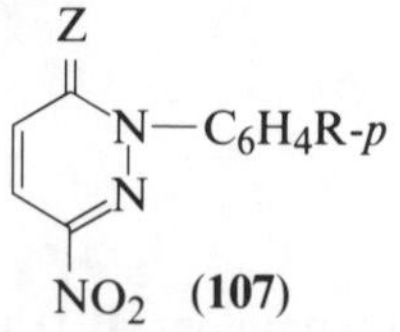

$$(48)$$

$$X, Y = CO_2Et, CN; \quad Z = O, NH; \quad R = Br, NO_2 \qquad NO_2 \quad (107)$$

Iminophosphoranepyrimidine derivative (108; equation 49) interacts with 2-dimethylamino-1-nitroethene to produce the derivative of [2,3-d]pyrimidine (109) in high yield [256].

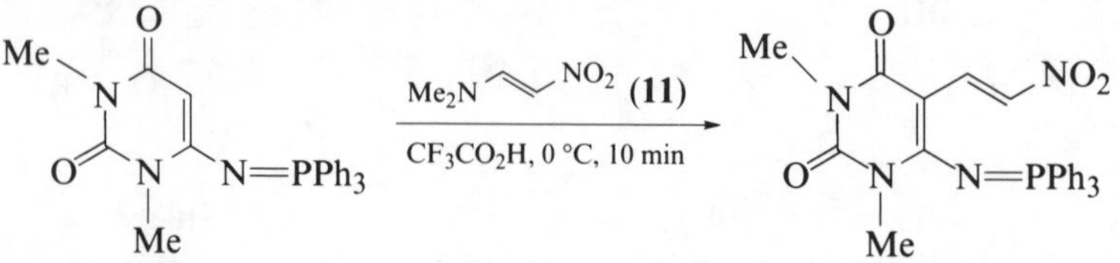

$$(49)$$

Fuji and Node have proposed a synthetic method for the preparation of nitroalkenes by the action of nitroenamines on various carbonyl compounds (110) that contain a methine carbon atom in the α position (equation 50). The reactions are highly stereoselective and produce either entirely or mostly the (E)-nitroalkenes (111) [257, 258].

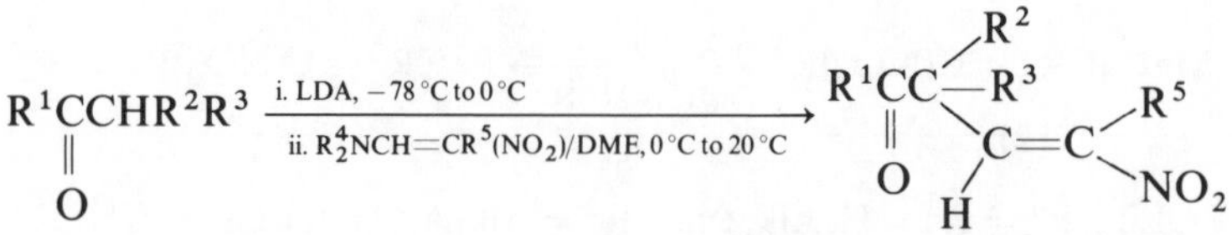

$$(50)$$

$$R^1 = H; \quad R_2^4 = NMe_2, morpholino$$

Methylphenylacetic esters, cyclohexylaldehyde and various cyclic ketones and lactones have been used as the methine-containing carbonyl compounds. These substances are introduced into the reactions with nitroenamines in the form of the Li or Zn enolates. The best results have been achieved with N-morpholino-nitroethene (55–99% yield). The use of chiral nitroenamines (112) makes it possible to carry out direct asymmetric syntheses of α,α-disubstituted δ-lactones (and γ-lactones) (114; equation 51) and cyclic ketones with chiral α carbons [223, 258, 259].

$$\qquad\qquad (51)$$

(112)   (113)   (114)

$R^1 = H, Me, Et, n = 0,1; \quad R^2 = Me, Et, CH_2 \!=\! CHCH_2;$
$M^+ = Li^+, Zn^{2+}, Cu^+; \quad X = OMe, Me$

The products are formed as a result of an addition–elimination process.

The use of substituted valerolactams as the carbonyl compounds in such reactions results in a high enantioselectivity for the process [259]. This chemical transformation has become the basis of the synthetic approach to active natural compounds such as indole alkaloids and diterpenoids.

Geminal dinitroaminoethenes (115; equations 52) react with nucleophiles to give various functionalized polynitro compounds. For example, at 80 °C compound (115) gives the salt of dinitroacetaldehyde (116) upon treatment with an alkali. At room temperature, compound (115) give the salt (117) [203].

$$R^1R^2NCH\!=\!C(NO_2)_2 \xrightarrow[R^1=R^2=Me]{KOH, 80\,°C} O\!=\!CHC^-(NO_2)_2K^+$$
$$\qquad\quad (115) \qquad\qquad\qquad\qquad\qquad\qquad (116)$$

$$\begin{array}{c} \substack{R^1=H \\ R^2=Ph} \Big\vert \substack{KOH, \\ EtOH, room temperature} \qquad\qquad \xleftarrow{H_2O,\,80\,°C} \end{array} \qquad (52)$$

$$K^+PhN^-CH\!=\!C(NO_2)_2$$
$$(117)\ 85\%$$

The stable 2-anilino-1,1-dinitroethene (115) reacts with hydroxylamine or hydrazine with substitution of the amino group (equations 53).

$$(115) \xrightarrow[ii.\,AcOK]{i.\,NH_2X,\,MeOH,\,10-20\,°C} XN\!=\!CHC^-(NO_2)_2K^+$$
$$\qquad\qquad\qquad\qquad\qquad (118)$$
$$X = OH, NH_2 \qquad\qquad (53)$$

$$(115) \xrightarrow[iii.\,AcOK]{i.\,C^-H(NO_2)_2K^+,\,DMF,\,0\,°C} HOCH[C(NO_2)_2]^{2-}2K^+$$
$$\qquad\qquad\qquad\qquad (119)$$

The product of the reaction between (115) and the salt of dinitromethane hydrolyses to give the salt (119). Compound (117) is a convenient reagent for the dinitroethylation of various nitrogen bases such as hydrazone, bishydrazines and aminotriazole (equation 54).

$$(117) \xrightarrow[\text{DMF, 0--25 °C}]{\text{RNH}_2} \text{RN}{=}\text{CH}{-}\text{C}^-(\text{NO}_2)_2\text{K}^+ \qquad (54)$$
$$(120)$$

$$R = \text{PhNH}, \; \text{N}{=}\text{CH}{-}\text{CH}{=}\text{N}, \; \begin{array}{c} \text{N} {-}\!\!\diagdown \\ \diagup\diagup \quad\diagdown\text{N} \\ \text{N} \quad \diagup \\ | \\ \text{H} \end{array}$$

3.4.2 DIAMINONITROALKENES (NITROKETENE AMINALS)

The synthetic approach to symmetrical 2,2-diamino-1-nitroalkenes (nitroketene aminals) (equation 55) is the interaction between 2,2-dithio-1-nitroalkenes (121) and amines (2 moles) or diamines (1 mole) [157, 165, 260, 261].

$$(\text{MeS})_2\text{C}{=}\text{CHNO}_2$$
$$(121)$$

$$\begin{array}{cc} \text{H} & \text{H} \\ | & | \\ \text{H}{-}\text{N}{-}(\text{CH}_2)_n{-}\text{N}{-}\text{H} \end{array} \qquad\qquad 2\text{R}^1\text{R}^2\text{NH} \qquad (55)$$

$$\begin{array}{c} \text{H} \\ | \\ \text{CH}_2{-}\text{N} \\ | \qquad\qquad \diagdown \text{C}{=}\text{CHNO}_2 \quad (\text{R}^1\text{R}^2\text{N})_2\text{C}{=}\text{CHNO}_2 \\ (\text{CH}_2)_{n-1}{-}\text{N} \\ \qquad | \\ \qquad \text{H} \quad (122) \end{array} \qquad\qquad (123a)$$

$$n = 1, 2$$

Asymmetrical nitroketene aminals (125; equation 56) are prepared either from nitroketene S,N-acetals [165, 166] (see Section 3.3) or in the condensation of S-methylthiourea (124) with nitromethane [156, 165].

$$\begin{array}{c} \text{MeN}{=}\text{CNR}^1\text{R}^2 + \text{MeNO}_2 \xrightarrow{\text{reflux 16 h}} \begin{array}{c} \text{MeNH} \\ \diagdown \text{C}{=}\text{CHNO}_2 \\ \text{R}^1\text{R}^2\text{N} \diagup \end{array} \qquad (56) \\ | \\ \text{SMe} \end{array}$$

$$(124) \qquad\qquad\qquad\qquad (125)$$

$$R^1 = H, R^2 = Ph; \quad R^1, R^2 = (CH_2)_4$$

Some α-functionalized aminals of nitroketene (128; equation 57) with one primary amino group have been synthesized from benzoylcyanamide (126) and nitro compounds (127) under the action of an equimolar quantity of $Ni(OAc)_2$ [262].

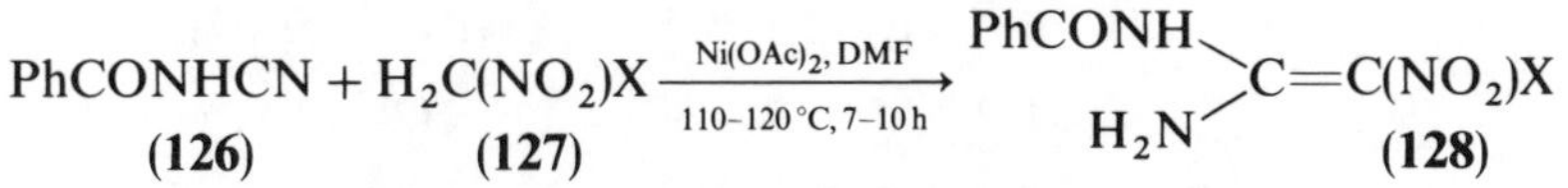

$$PhCONHCN + H_2C(NO_2)X \xrightarrow[110-120\,°C,\,7-10\,h]{Ni(OAc)_2,\,DMF} \begin{array}{c} PhCONH \\ \diagdown \\ H_2N \diagup \end{array} C=C(NO_2)X \quad (57)$$

$$\text{(126)} \qquad\quad \text{(127)} \qquad\qquad\qquad\qquad\qquad\qquad \text{(128)}$$

$$X = CO_2Et,\ COPh,\ NO_2$$

A study of nitroketene aminal dipole moments has revealed high polarity with a pronounced contribution from zwitterionic structures (123b) in aqueous solution [263].

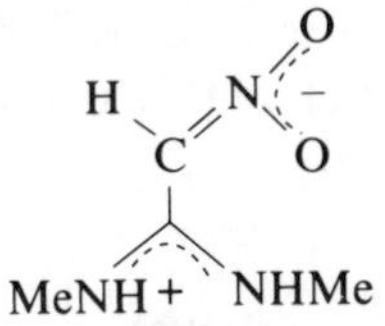

(123b)

Nitroketene aminals exhibit moderate reactivity of the enamine structure [165]. That nitroketene aminals react with electrophiles is an example of their high activity in comparison with nitroenamines (equation 58).

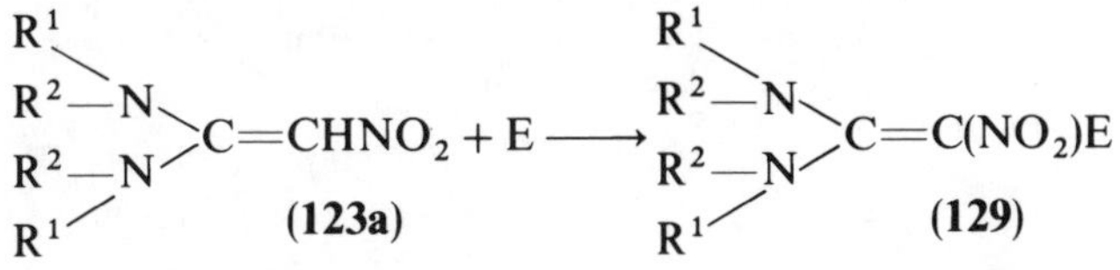

$$\begin{array}{c} R^1 \\ \diagdown \\ R^2{-}N \\ \diagdown \\ R^2{-}N \\ \diagup \\ R^1 \diagup \end{array} C=CHNO_2 + E \longrightarrow \begin{array}{c} R^1 \\ \diagdown \\ R^2{-}N \\ \diagdown \\ R^2{-}N \\ \diagup \\ R^1 \diagup \end{array} C=C(NO_2)E \quad (58)$$

$$\text{(123a)} \qquad\qquad\qquad\qquad\qquad \text{(129)}$$

This statement can be illustrated by the reactions with isothiocyanates [264], phenylsulfenyl chlorides [165] and by coupling with diazonium salts to give 2-arylazo-2-nitroketene aminals (131; equation 59) [265].

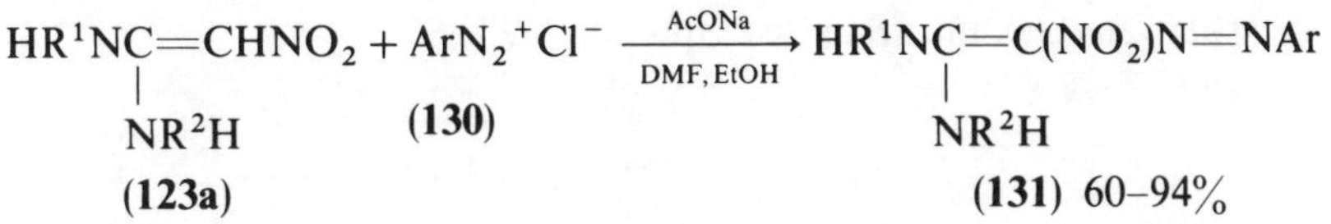

$$HR^1NC=CHNO_2 + ArN_2{}^+Cl^- \xrightarrow[DMF,\,EtOH]{AcONa} HR^1NC=C(NO_2)N=NAr \quad (59)$$
$$\quad\ \ |\qquad\qquad\qquad \text{(130)} \qquad\qquad\qquad\qquad\qquad\qquad | $$
$$\ \ NR^2H \qquad\qquad\qquad\qquad\qquad\qquad\qquad\qquad\qquad\quad NR^2H$$
$$\text{(123a)} \qquad\qquad\qquad\qquad\qquad\qquad\qquad\qquad\qquad \text{(131)}\ \ 60-94\%$$

$$R^1, R^2 = Alk, Ar$$

Some functionalized electrophiles such as α-bromoketones, bromonitromethane and acylisothiocyanates participate in this process to give the products of heterocyclization [165]. The five-membered rings are prepared from dipolarophiles [165]. The use of enaminoketones (133; equation 60) as dielectrophiles provides a route to the derivatives of 2-amino-3-nitropyridines (134) [181].

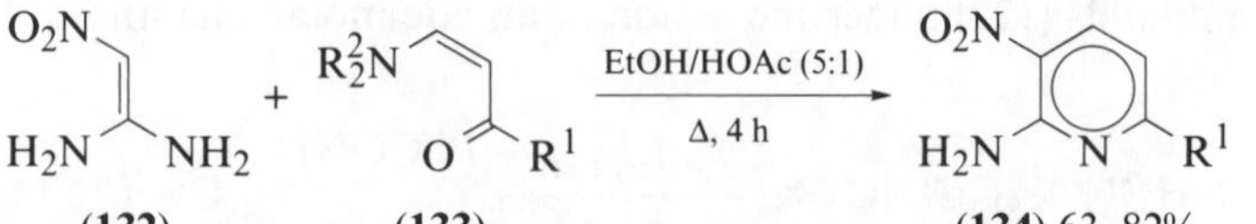

$$
\text{(132)} + \text{(133)} \xrightarrow[\Delta,\,4\,h]{\text{EtOH/HOAc (5:1)}} \text{(134) } 63\text{–}82\%
\tag{60}
$$

R^1 = H, Me, Ph, p-ClC$_6$H$_4$, 2-thienyl; R^2 = H, Me, Et

Mertens *et al.* [181] have suggested a scheme for this process according to which the reaction proceeds via a monoelectrophilic addition step followed by interaction with the second electrophilic centre to give heteroarylation. Some substituted 4-aryl-1,4-dihydronicotine and 4-aryl-4,5-dihydronicotine esters are synthesized from nitroketene aminals and substituted α-acetylcinnamic esters or diazoketones [261]. Pyridine derivatives (135, 136; equation 61) are prepared by the reactions between compound (123a) and 1-ethoxy-2-cyanoethene [266] or diphenylcyclopropenone [267].

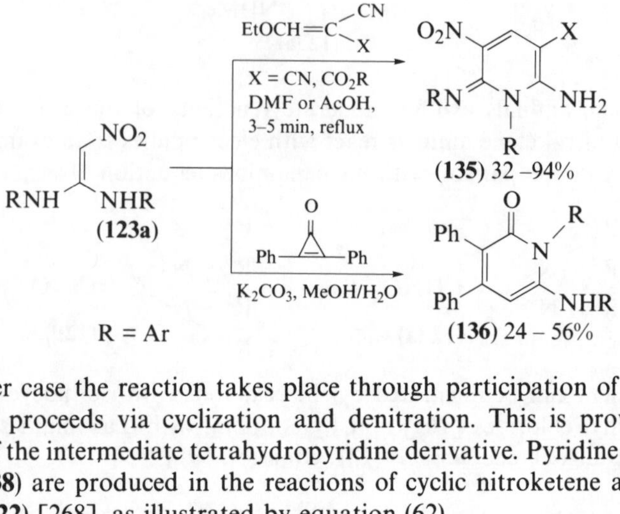

$$
\tag{61}
$$

In the latter case the reaction takes place through participation of one amino group and proceeds via cyclization and denitration. This is proven by the isolation of the intermediate tetrahydropyridine derivative. Pyridine derivatives such as (138) are produced in the reactions of cyclic nitroketene aminals, for example (122) [268], as illustrated by equation (62).

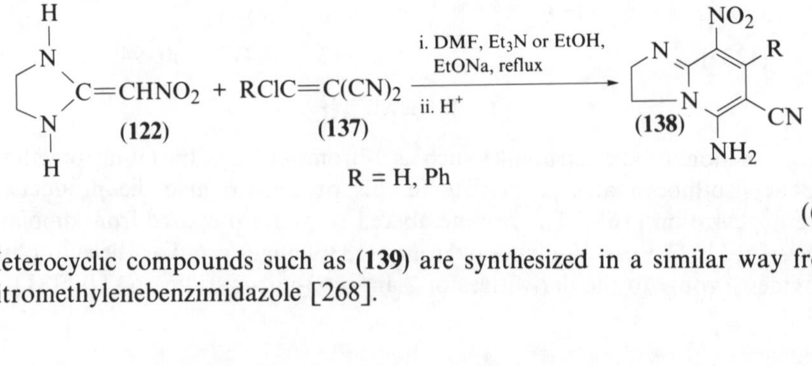

$$
\text{(122)} + \text{RClC=C(CN)}_2 \xrightarrow[\text{ii. H}^+]{\substack{\text{i. DMF, Et}_3\text{N or EtOH,}\\ \text{EtONa, reflux}}} \text{(138)}
\tag{62}
$$

R = H, Ph

Heterocyclic compounds such as (139) are synthesized in a similar way from nitromethylenebenzimidazole [268].

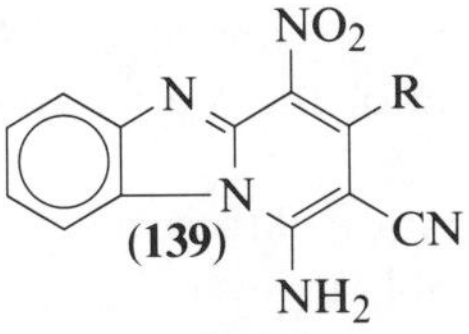

$$R = H, Ph$$

Nitroketene aminals, in contrast with nitroenamines, do not react with nucleophiles. The nitration of cyclic dinitroketene aminals (**140**; equation 63) gives rise to a number of tetranitro compounds of type (**141**) [269].

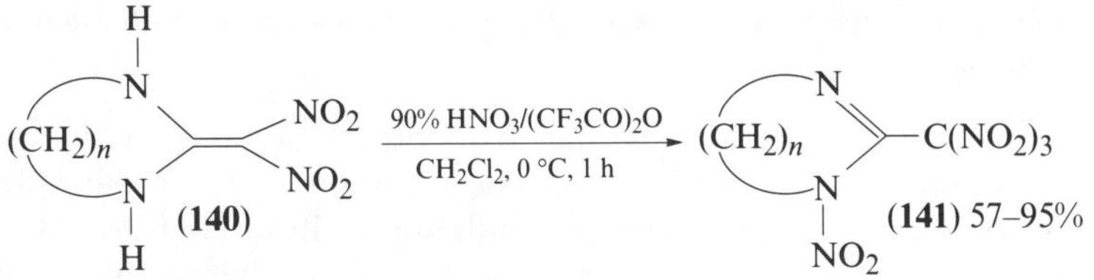

$$n = 2, 3, 4$$

3.5 HETEROCYCLIC NITROALKENES (SULFUR-CONTAINING, PHOSPHORUS-CONTAINING AND SILICON-CONTAINING NITROHETEROCYCLOPENTENES)

A systematic study devoted to the synthesis of heterocycles containing the nitroethene fragment was performed recently for the structural analogues of the heterocyclopentene type, e.g. thiolene 1,1-dioxides, phospholene-1-alkoxy(phenyl) 1-oxides and silolenes.

The significance of these heterocyclenes is defined by the wide range of their chemical transformations, high biological activity and some other valuable properties [270–275]. This statement can be illustrated by the psychotropic, antiinflammatory, antiastigmatic, antidrug, nematocidal, insecticidal, herbicidal and fungicidal activities of thiolane and thiolene 1,1-dioxide derivatives [270, 276]. Phospholene 1-oxides show anticonvulsive and pesticidal activity [277] and are used as catalysts in the synthesis of carbodiimides and their polymers [278, 279]. Silolenes are of certain interest as starting monomers for the production of high molecular weight compounds with good heat resistance.

3.5.1 METHODS OF SYNTHESIS

The simplest way to introduce the nitro group into the heterocyclenes under discussion is through direct nitration of the double bond. A recent review by

Berestovitskaya and Perekalin [280] presents a detailed analysis of the problem; therefore, only a short résumé of the most important results is given in this section.

The investigation of the nitration of these heterocyclopentenes under comparable conditions makes it possible to:

(i)   define the correlation between the structure of the heterocyclene, the nature of the nitrating agent and the conditions of nitration;

(ii) find out the specific influences of the heteroatomic fragments (sulfonyl, phosphoryl and dialkylsilyl) upon the process pathways; and

(iii) compare the properties of synthesized nitro thiolene 1,1-dioxides, pospholene 1-oxides and silolenes among themselves and with their linear analogues [281–285].

The primary experiments showed that silacyclopentenes have a tendency to undergo a smooth hydrolysis in the acidic nitration medium. Phospholene oxides give the corresponding complexes with dilute nitric acid [281, 282]. The sulfolene cycle is stable in both oxidizing and acidic media, so that it can be nitrated by various nitrating agents.

The most pronounced characteristic features of the heteroatomic groups are exposed by nitration of such structural analogues as the Δ^2-heterocyclenes (equations 1). 2-Thiolene 1,1-dioxide nitration requires severe conditions, such as reflux in fuming nitric acid. The simplest unsubstituted thiolene 1,1-dioxide does not undergo nitration at all, despite the application of a large variety of nitrating agents (N_2O_4, dilute and fuming HNO_3, HNO_3 and AcOH mixture, nitrogen oxides) and experimental conditions (various solvents, time of contact, temperature). Phenylthiolene 1,1-dioxide gives the product of nitration of the benzene ring. Only the electron-donating effect of the methyl group brings about the successful nitration to give the product (2) [283a]. Methyl substituents in 1-phenyl-2-phospholene 1-oxide do not direct the process to the desired result, and the aromatic fragment is nitrated instead [283b].

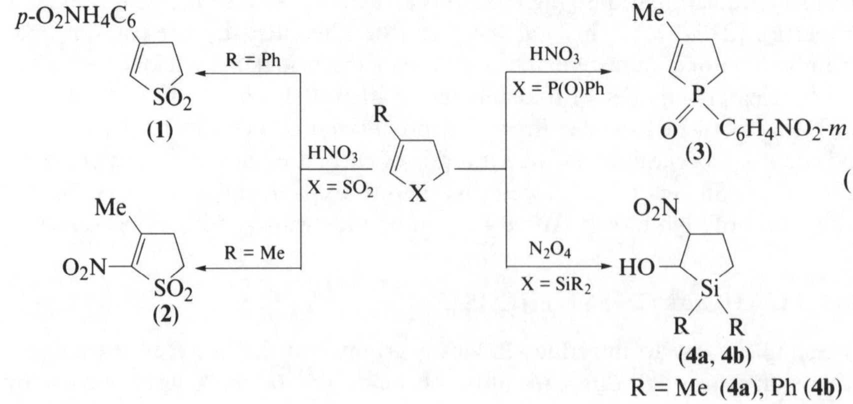

p-$O_2NH_4C_6$

(1)

R = Ph

$\xrightarrow[X = SO_2]{HNO_3}$

Me

O_2N—

SO_2

(2)

R = Me

HNO₃
X = P(O)Ph

Me

P
O $C_6H_4NO_2$-m

(3)

N_2O_4
X = SiR₂

O_2N

HO—Si

R R

(4a, 4b)

R = Me (4a), Ph (4b)

(1)

The exchange of strongly electron-accepting, protophilic groups (SO_2, $P{=}O$) in the heterocycle for the dimethyl(diaryl)silyl group directs the process along a different pathway. Even non-substituted 2-silolenes interact with dinitrogen tetroxide readily to form 3-nitro-2-hydroxy-1,1-dimethyl(diphenyl)silolanes (**4**) [283c].

Dinitrogen tetroxide is the most convenient nitrating agent for all sulfur-containing, phosphorus-containing and silicon-containing Δ^3-heterocyclopentenes, allowing one to vary the experimental conditions within a wide range. In all cases, structurally similar products are isolated (equation 2), i.e. dinitro heterocyclopentanes (**5**) and hydroxynitroheterocyclopentanes (**6**) and nitroheterocyclopentenes with nitroallyl (**7**) and nitrovinyl (**8**) fragments [286–293].

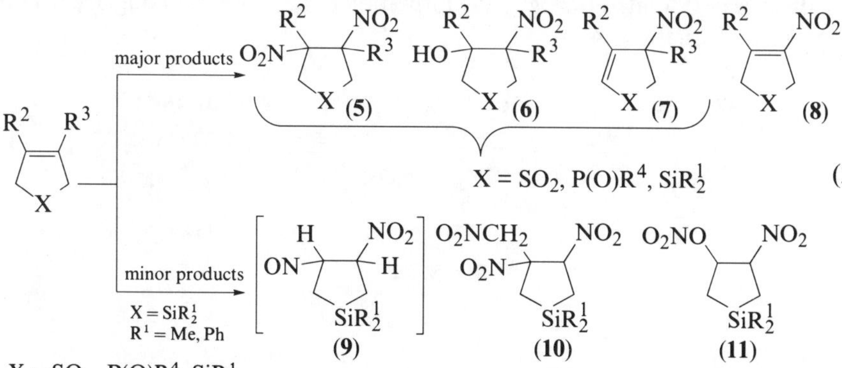

$$X = SO_2, P(O)R^4, SiR_2^1 \qquad (2)$$

The nitration of Δ^3-silolenes is the least selective process. The major products are compounds of type (**5–7**) and the minor products are of type (**9–11**).

Under more rigid conditions (heating of the Δ^3-heterocyclopentene in an excess of dinitrogen tetroxide), oximes of nitrophospholenone 1-oxides and nitrothiolenone 1,1-dioxides (**12**; equation 3) are formed unexpectedly [294–296]. Such oximes can also be prepared by nitrosation of 4-nitro-3-thiolene 1,1-dioxides. The structures of compounds (**12**) have been proven by spectroscopic methods [295, 296].

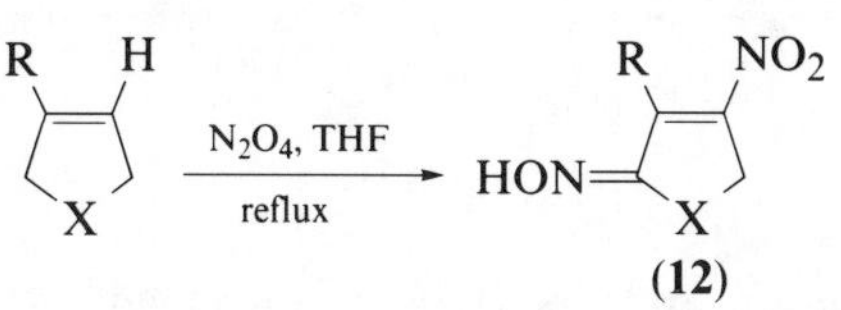

$$\xrightarrow[\text{reflux}]{N_2O_4,\ THF} \qquad (3)$$

$$X = SO_2,\ R = H,\ Me,\ Cl;\quad X = P(O)OAlk,\ R = H,\ Me$$

The X-ray analysis of compounds (**12**) has shown that these substances exist in the form of NH nitrones in the solid state. This is the very first evidence of oxime–nitrone tautomerism [297–299]. The NH nitrone stabilization has been discussed in terms of the specific stereostructure and electronic structure of (12), i.e. the presence of nitro and sulfonyl (or phosphoryl) groups and the conjugated endocyclic and exocyclic double bonds that are fixed in the heterocycle molecule [297–299]. The specific packing of molecules in the crystal must also be taken into consideration.

A second endocyclic or exocyclic double bond introduced into the sulfolene ring changes the result of nitration completely: only dinitro compounds (13, 15, 17–19) are formed in this case [300–302]. The regioselectivity of the reaction depends upon the conditions. The elimination of HNO_2 takes place under extremely mild conditions, upon dissolving dinitrothiolene 1,1-dioxides in alcohol or through chromatography on silica. The chlorine-containing nitro-sulfodiene is not isolated owing to the formation of the corresponding acetal (14) in methanol (equations 4–6).

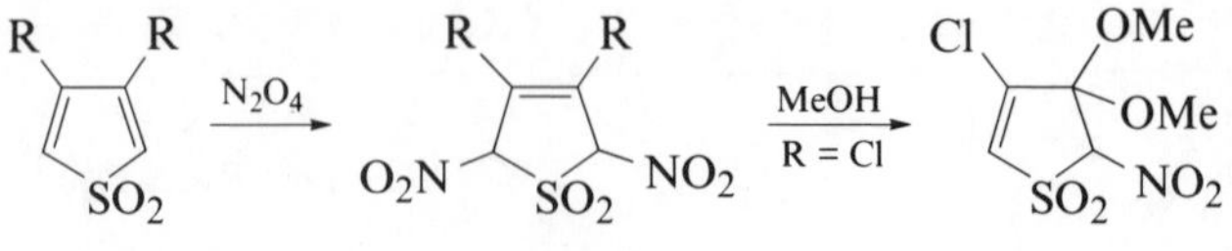

(4)

(13)   (14)

R = Me, Cl

(5)

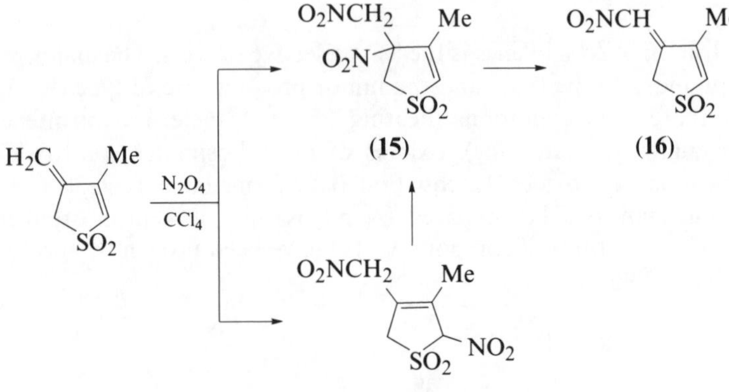

(15)   (16)

(17)

The elimination reaction appears to be a convenient synthetic method for mononitroheterocyclopentenes. The common methods are dehydration, denitration, deacylation and debenzoylation of nitrohydroxy, dinitro, nitroacetoxy and nitrobenzoyloxy derivatives [282, 289, 290, 303]. For example, 2-hydroxy-3-nitrosilolanes eliminate water readily under the action of PCl_5 and transform into 3-nitro-2-silolenes (21; equation 7) [283c].

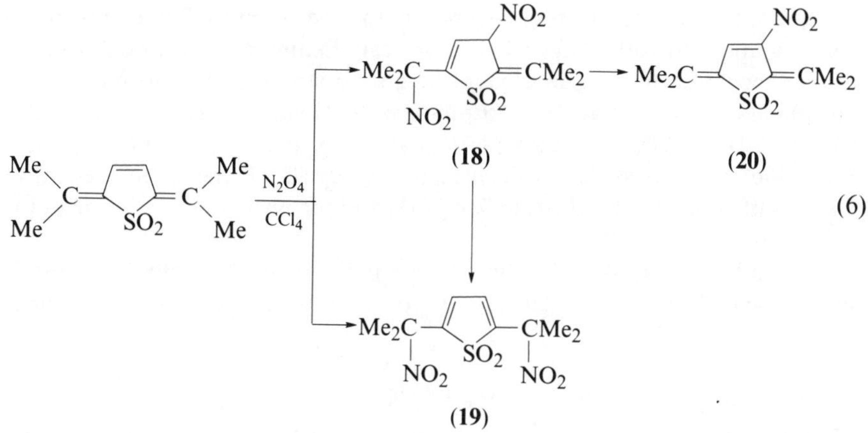

$$\text{Me}_2\text{C}=\overset{\overset{\displaystyle NO_2}{|}}{\underset{\underset{\displaystyle NO_2}{|}}{\text{CMe}_2}} \quad (18) \quad \longrightarrow \quad \text{Me}_2\text{C}=\overset{\overset{\displaystyle NO_2}{|}}{\underset{\underset{\displaystyle SO_2}{|}}{\text{CMe}_2}} \quad (20) \tag{6}$$

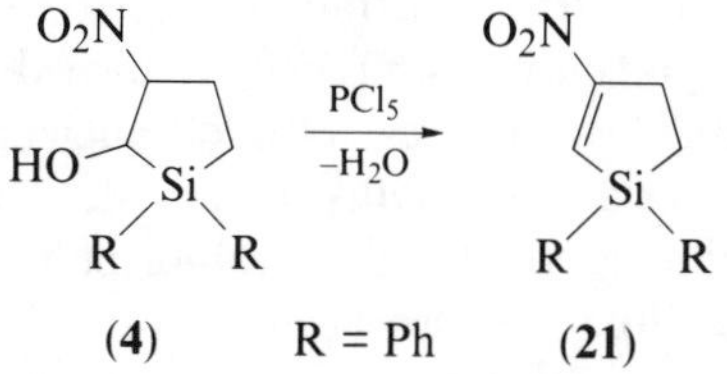

Me₂C=C(Me)–SO₂–C(Me)=CMe₂ →(N₂O₄, CCl₄)→ (18), (20), (19)

$$\text{Me}_2\text{C}\!\!-\!\!\overset{|}{\underset{NO_2}{\text{SO}_2}}\!\!-\!\!\text{CMe}_2 \quad (19)$$

$$\underset{R \quad R}{\overset{O_2N}{\underset{HO}{\bigcirc}\text{Si}}} \quad \xrightarrow[-H_2O]{PCl_5} \quad \underset{R \quad R}{\overset{O_2N}{\text{Si}}} \tag{7}$$

(4) R = Ph (21)

The conditions and the results of denitration for sulfonyl-containing, phosphoryl-containing and silyl-containing dinitroheterocyclopentanes essentially differ from those for the aliphatic and carbocyclic analogues and are structure dependent. For example, dinitrosilolanes eliminate HNO₂ in the presence of

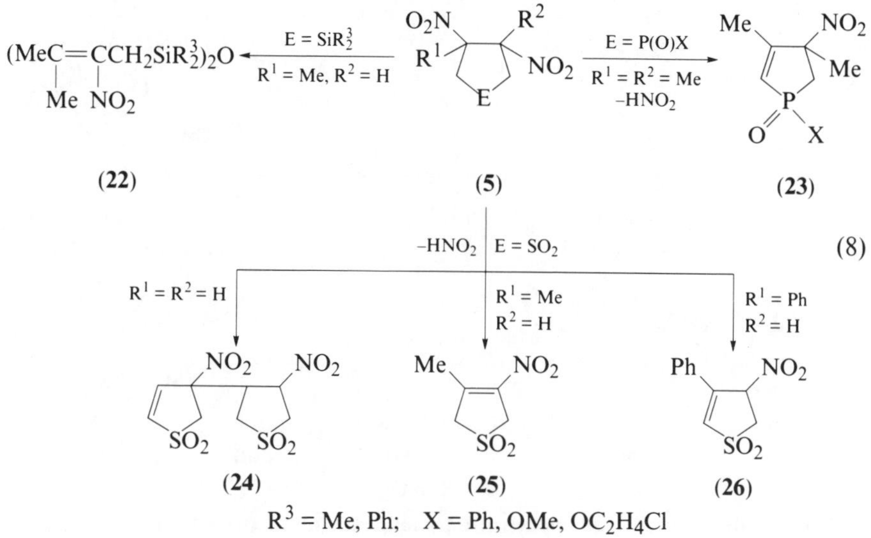

$$(\text{Me}\overset{|}{\underset{\text{Me}}{\text{C}}}=\text{C}\overset{|}{\underset{NO_2}{\text{CH}_2\text{SiR}_2^3}})_2\text{O} \xleftarrow[R^1 = Me,\, R^2 = H]{E = SiR_2^3} \underset{E}{\overset{O_2N \quad R^2}{R^1 \diagup\!\!\!\diagup NO_2}} \xrightarrow[R^1 = R^2 = Me \\ -HNO_2]{E = P(O)X} \underset{O \quad X}{\overset{Me \quad NO_2}{\underset{P}{\diagup\!\!\diagup Me}}} \tag{8}$$

(22) (5) (23)

−HNO₂ │ E = SO₂

R¹ = R² = H

$$\underset{SO_2 \quad SO_2}{\overset{NO_2 \quad NO_2}{\diagup\!\!\!\diagup}} \quad (24)$$

R¹ = Me, R² = H

$$\underset{SO_2}{\overset{Me \quad NO_2}{\diagup\!\!\!\diagup}} \quad (25)$$

R¹ = Ph, R² = H

$$\underset{SO_2}{\overset{Ph \quad NO_2}{\diagup\!\!\!\diagup}} \quad (26)$$

R³ = Me, Ph; X = Ph, OMe, OC₂H₄Cl

base and undergo ring opening to form compounds such as (22). Dinitrophos-
pholanes and dinitrothiolane 1,1-dioxides can eliminate HNO_2 even without
base, either under heating in alcohol or during storage. The influence of the
substituents upon the reaction pathway is most obvious in the case of dinitrothiolane 1,1-dioxides. For example (equation 8), the simplest unsubstituted
dinitrothiolane 1,1-dioxide transforms into the bicycle (24), the methyl derivative
gives Δ^3-nitrothiolene 1,1-dioxide (25) [303] and the phenyl derivative gives the
Δ^2 isomer (26) [304, 305].

Potentially, nitrothiolene 1,1-dioxides are the most synthetically useful nitro-
heterocyclenes owing to their ability to form nitroethene and nitrodiene
fragments.

3.5.2 CHEMICAL TRANSFORMATIONS
OF Δ^2-NITROTHIOLENE AND Δ^3-NITROTHIOLENE 1,1-DIOXIDES

Nitrothiolene 1,1-dioxide, with its geminal, electron-withdrawing nitro and
sulfonyl groups (3-methyl-2-nitro-2-thiolene 1,1-dioxide), is a highly electron-
deficient system. For this reason it easily adds nucleophilic reagents such as
CH acids, aromatic amines and phenylhydrazine. Besides the common adducts
(27–29; equation 9), the products of intramolecular heterocyclization, i.e.
quinoxaline derivatives (30), have also been obtained [283a, 306].

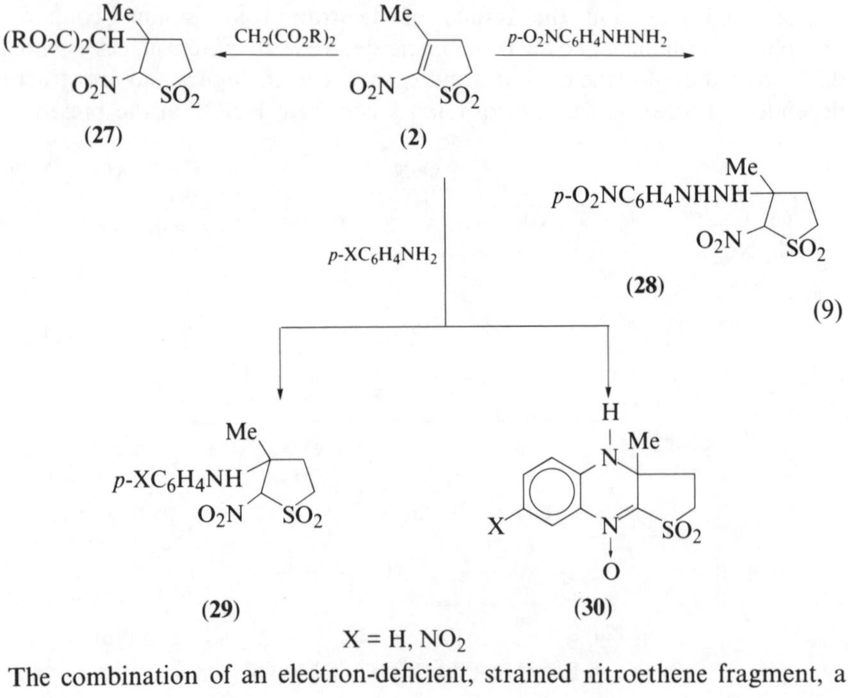

(9)

X = H, NO_2

The combination of an electron-deficient, strained nitroethene fragment, a

CH acid and a sulfonyl group in the same molecule explains the tendency of 4-nitro-3-thiolene 1,1-dioxides towards allyl-vinyl isomerization and some other reactions.

The equilibrium transformations of Δ^2 and Δ^3 in nitrothiolenedioxides (equation 10) proceed via prototropic isomerizations and essentially differ from those of nitroalkenes and thiolene 1,1-dioxides [305, 307–309].

$$\text{(8)} \qquad \qquad \qquad \qquad \qquad \qquad \qquad \qquad \qquad \text{(7)}$$

The equilibrium is affected by the nature of the substituent at C(3) in the heterocycle [303, 307–309]. When small substituents (R = H, Cl, Me) are attached to the heterocycle the isomerization takes place under very mild conditions, usually upon dissolution of the compound in a polar solvent. The high lability of the double bond results from the tendency of nitro and sulfonyl groups to couple with the unsaturated fragment. The negative inductive effect of the chlorine atom and the participation of its p electrons in conjugation lower the isomerization energy barrier. Large substituents stabilize the Δ^2 form. Arylamino groups fix the Δ^3 form in 3-substituted 4-nitrothiolene 1,1-dioxides owing to effective conjugation in the β-aminonitroethene fragment.

Competitive reactions of 4-nitro-3-thiolene 1,1-dioxides with nucleophiles are directed along the pathways: conjugative addition and deprotonation. Mild nucleophiles (alcohols, aromatic amines, thiophenols) in the absence of steric effects react with the double bond with formation of either the products of addition (32; equation 11) (R = H, Me) or addition–elimination (33, 34) (R = Cl). The strong bases such as MeONa, Et_3N, Et_2NH and others initiate deprotonation of the methylene group to give salts (31) [310–312].

Nitrothiolene 1,1-dioxides are rather active in homolytic reactions. Formation of an intermediate conjugated allyl radical is demonstrated by the formation of the same products from Δ^2 and Δ^3 forms owing to the reactions of the allyl radical at both reactive centres. Such a process extends the synthetic potential of nitrothiolene 1,1-dioxides and presents a new, simple method of synthesis for geminal dinitro compounds (equation 12) [313].

Δ^3-Nitrosulfolenes undergo heliotropic reactions that are of high preparative significance owing to the production of sterically identical isomers. Thiolene 1,1-dioxides are used in such reactions for the synthesis of pheromones.

High-temperature elimination of SO_2 from Δ^3-nitrosulfolenes leads to mono-nitrobutadienes or the dinitrobutadienes [314, 315]. This was the original preparative method for 2-nitro-1,3-butadienes and 1,4-dinitro-1,3-butadienes. This very method presents the possibility of introducing the conjugated nitrodiene at the very moment of its formation, i.e. in the most active s-cis form, into

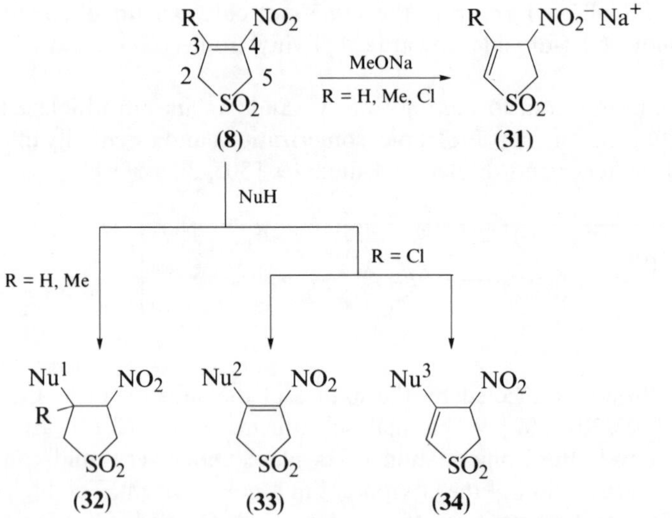

$$(11)$$

Nu^1 = OMe, OEt, NHPh; Nu^2 = SC_6H_4Cl-p, NHPh, NHC_6H_4Me-p,

NHC_6H_4OMe-p, $NHC_6H_4NMe_2$-p; Nu^3 = NH(Ph)CH$_2$Ph,

NHNHPh, N⟨ ⟩, N⟨ ⟩O

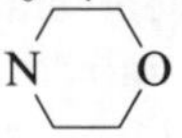

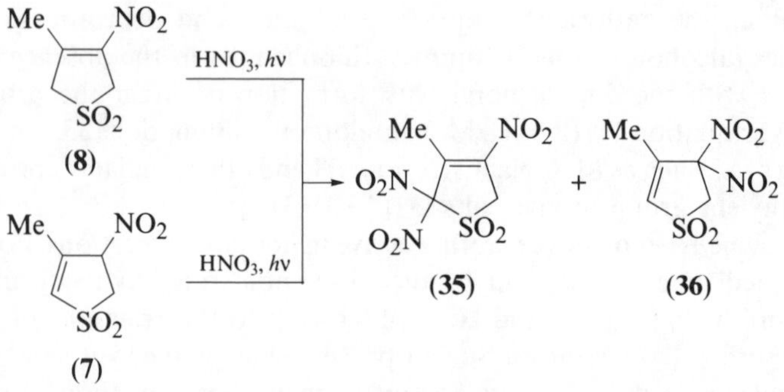

$$(12)$$

reactions of direct and reversible diene synthesis. The catchers of the arising dienes are such dienophiles as maleic anhydride and p-benzoquinone (equations 13).

Desulfonylations of dinitrothiolene-1,1-dioxides (13; equation 14) in the presence of styrene (donor-dienophile) give electrophilic dienes (acceptors) which can participate in the diene synthesis with a reversal electronic character [315].

The combination of the four electron-withdrawing substituents, the activated CH_2 group and the double bond in 2,2,4-trinitro-3-thiolene 1,1-dioxide explains the extremely smooth desulfonylation of (35) under the action of an aromatic

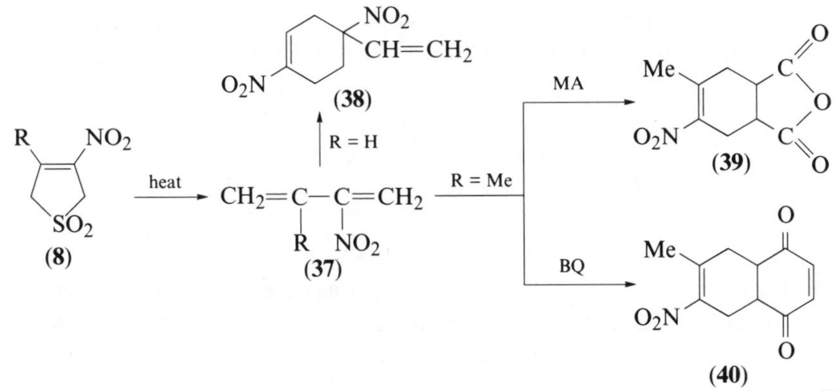

(13)

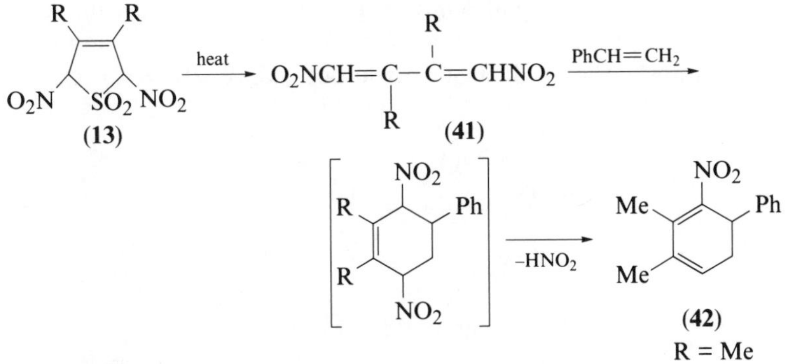

(14)

amine (equation 15). The process takes place at room temperature and leads to the aminodinitrobutadiene (**43**) [316].

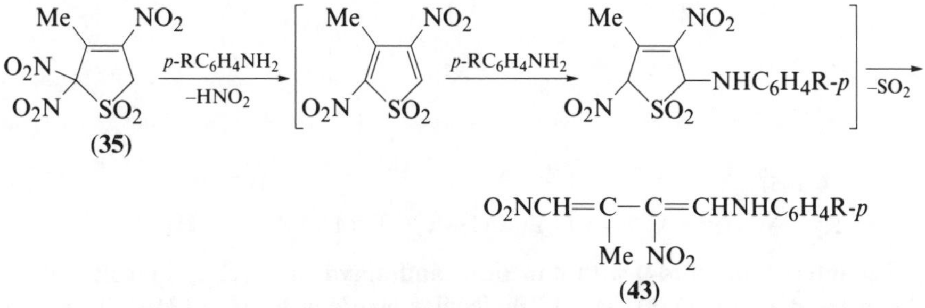

$$O_2NCH=\underset{\underset{Me}{|}}{C}-\underset{\underset{NO_2}{|}}{C}=CHNHC_6H_4R\text{-}p$$

(**43**)

R = H, Me, OMe (15)

Further development of nitrothiolene 1,1-dioxide chemistry, particularly the condensation of 4-nitro-3-thiolene 1,1-dioxides with aromatic and heterocyclic aldehydes under acidic catalysis, has led to the synthesis of 2-benzylidene(heteryllidene)-4-nitro-3-thiolene 1,1-dioxides (**44**; equation 16). The synthesized compounds contain the *s-trans*-nitrosulfodiene system [303, 317].

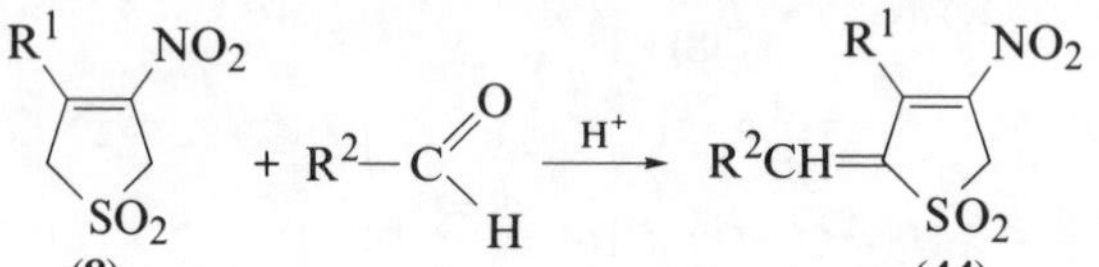

$$R^1 \diagup NO_2 \qquad + \; R^2{-}C\diagup_{H}^{O} \; \xrightarrow{\;H^+\;} \; R^2CH{=}\diagdown SO_2 \qquad (16)$$

(**8**)   (**44**)

R^1 = H, Me, Cl; R^2 = Ar, Het

The high electrophilic properties of nitrosulfodienes result in low polarographic reduction potentials [318]. These potentials are lower than those for β-nitrostyrene, but correlate with the data for dinitrobutadienes.

The quantum-chemical calculations performed for nitrosulfodienes have been aimed at forecasting the direction of nucleophilic addition. The data reveal that attack at C(1) dominates over attack at C(3) when the difference in bulkiness of these two centres is not significant [318]. This is illustrated by the reactions of (**44**) with malonic esters, acetoacetic esters and acetylacetone, loading in each case to formation of the product of 1,4 addition (**45**; equation 17).

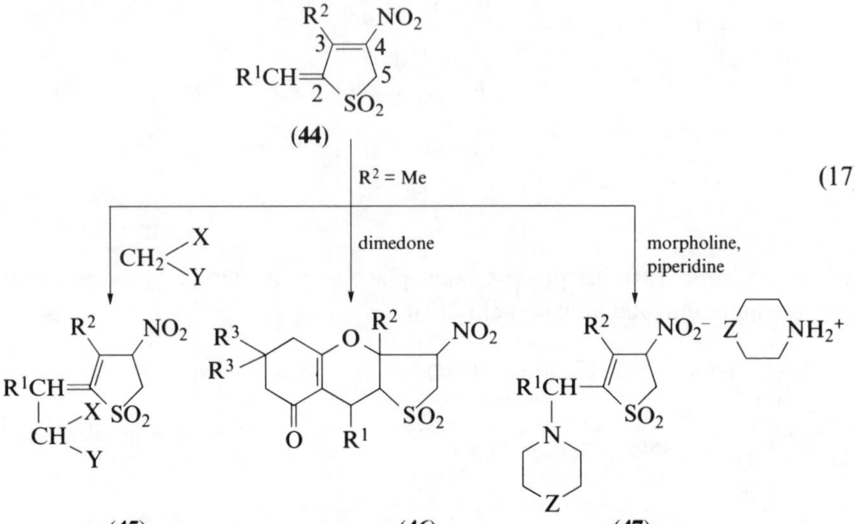

(**44**)

R^2 = Me   (17)

(**45**)   (**46**)   (**47**)

R^1 = Ar, Het;   X, Y = CO_2Et, CO_2Me, $C(O)Me$;   Z = O, CH_2

The interaction of (**44**) with dimedone and dihydroresorchinol results in the chromene derivative (**46**) via intramolecular heterocyclization [319]. The high basicities of morpholine and piperidine define the same order of addition in the formation of nitronates (**47**) [320].

Biological tests of nitroheterocycles have revealed high fungicidal and herbicidal activities matched with low toxicity [321, 322].

3.6 REFERENCES

[1]   Barrett, A. G. M. *Chem. Soc. Rev.*, 1991, **20**, 95.
[2]   Bartashevich, Yu. E., Kopranenkov, V. N., Burdelev, O. T., and Unkovskii, B. V. *Khim. Pharm. Zh.*, 1972, **6**, 13.
[3]   Campbell, R. D. and Schultz, F. J. *J. Org. Chem.*, 1960, **25**, 1877.
[4]   Owai, I., Tomita, K., and Ide, J. *Chem. Pharm. Bull. (Tokyo)*, 1965, **13**, 118; *Chem. Abstr.*, 1965, **62**, 14 541e.
[5]   Perrot, R. and Berger, R. *C. R. Hebd. Seances Acad. Sci.*, *Ser. C*, 1952, **235**, 185.
[6]   Koremura, M. and Tomita, K. *Nippon Nogei Kagaku Kaishi*, 1962, **36**(6), 479; *Chem. Abstr.*, 1965, **62**, 3962b.
[7]   Schlubach, H. H. and Braun, A. *Justus Liebigs Ann. Chem.*, 1959, **627**, 28.
[8]   Freeman, J. P. and Emmons, W. D. *J. Am. Chem. Soc.*, 1957, **79**, 1712.
[9]   Stevens, T. E. and Emmons, W. D. *J. Am. Chem. Soc.*, 1958, **80**, 338; *J. Org. Chem.*, 1958, **23**, 136.
[10] (a) Rappoport, Z. and Topol, A. *J. Org. Chem.*, 1989, **54**, 5967.
       (b) Rappoport, Z. *Acc. Chem. Res.*, 1992, 474.
[11] Vasilyev, S. V., Burdelev, O. T., and Kopranenkov, V. N. *Zh. Org. Khim.*, 1969, **5**, 434; *Chem. Abstr.*, 1969, **71**, 2906q.
[12] Yakubovich, A. Ya. and Lemke, A. L. *Zh. Obshch. Khim.*, 1949, **19**, 649.
[13] Francotte, E., Verbruggen, R., Viehe, H. G., Van Meersche, M., Germain, G., and Declercq, J. P. *Bull. Soc. Chim. Belg.*, 1978, **87**, 693.
[14] Ogloblin, K. A. and Semenov, V. P. *Zh. Obshch. Khim.*, 1964, **34**, 1522; *Chem. Abstr.*, 1964, **61**, 5500c.
[15] Shin, C., Yonezawa, Y., Suzuki, K., and Yoshimura, J. *Bull. Chem. Soc. Jpn*, 1978, **51**, 2614.
[16] Park, K. P., Ha, H. J., and Williard, P. G. *J. Org. Chem.*, 1991, **56**, 6725.
[17] Sopova, A. S., Perekalin, V. V., Lebednova, V. M., and Yurchenko, O. I. *Zh. Obshch. Khim.*, 1964, **34**, 1185.
[18] Eremenko, L. T. and Oreshko, G. V. *Izv. Akad. Nauk SSSR, Ser. Khim.*, 1969, 724.
[19] Wilkendorf, R. and Trenel, M. *Chem. Ber.*, 1924, **57**, 306.
[20] Schmidt, E. and Rutz, G. *Chem. Ber.*, 1928, **61**, 2142.
[21] Loevenich, J., Koch, J., and Pucknat, U. *Chem. Ber.*, 1930, **63**, 636.
[22] Reichert, B. and Koch, W. *Chem. Ber.*, 1935, **68**, 445.
[23] Baker, J. W. *J. Chem. Soc.*, 1931, 2416.
[24] Worrall, D. and Tatilbaum, A. *J. Am. Chem. Soc.*, 1942, **64**, 1739; Worrall, D. and Cohen, L. *J. Am. Chem. Soc.*, 1944, **66**, 842.
[25] Nakasawa, A. *Nangaoka Koguo Tanki Daigaku Koto Semmon Gakko Kenkynkiyo*, 1937, **3**, 305; *Chem. Abstr.*, 1968, **69**, 2398j.
[26] Edasery, J. P. and Cromwell, N. H. *J. Heterocycl. Chem.*, 1979, **16**, 831.
[27] Carroll, F. I. and Kepler, J. A. *Can. J. Chem.*, 1966, **44**, 2909.
[28] Tronchet, J. M. J., Bonenfant, A. P., Pallic, K. D., and Habashi, F. *Helv. Chim. Acta*, 1979, **62**, 1622.
[29] Ruggli, P. and Schetty, O. *Helv. Chim. Acta*, 1940, **23**, 718.
[30] Kono, H., Shiga, M., Motoyama, I., and Hata, K. *Bull. Chem. Soc. Jpn*, 1969, **42**, 3270.
[31] Nasarova, Z. N. *Zh. Obshch. Khim.*, 1954, **24**, 575.
[32] Tzukervanik, I. P. and Potyemkin, G. F. *Dokl. Akad. Nauk Uz. SSR*, 1951, 26.
[33] Dauzonne, D. and Royer, R. *Synthesis*, 1987, 1020.
[34] Vasilyev, S. V. and Burdelev, O. T. *Izv. Vyssh. Ucheb. Zaved. Khim. Tekhnol.*, 1970, **13**, 73; *Chem. Abstr.*, 1970, **72**, 132 199t.

[35] Ger. Pat. 1 103 907, 1961; Chem. Abstr., 1962, 56, 1345g; Ott, E. and Bossaller, W. Chem. Ber., 1943, 76, 88.
[36] Zakharkin, L. I. Izv. Akad. Nauk SSSR, Ser. Khim., 1957, 1064.
[37] Martynov, I. V., Khromova, Z. I., and Kruglyak, Yu. L. Probl. Org. Sint., Akad. Nauk SSSR, 1965, 60.
[38] Kaberdin, R. V., Nikolaeva, E. E., Potkin, V. I., and Oldekop, Yu. A. Zh. Org. Khim., 1990, 26, 2223.
[39] Johnston, H. US Pat 3 054 828, 1962; Chem. Abstr., 1963, 58, 3315e.
[40] Buevich, V. A., Grineva, V. S., and Perekalin, V. V. Zh. Org. Khim., 1971, 7, 2624.
[41] Baum, K., Bigelow, S. S., Nguyen, N. Y., Archibald, T. G., Gilardi, R., Flippen-Anderson, J. L., and George, C. J. Org. Chem., 1992, 57, 235.
[42] Lipina, E. S. and Perekalin, V. V. Zh. Org. Obshch. Khim., 1964, 34, 3644.
[43] Carroll, F. I., Kerbow, S. C., and Wall, M. E. Can. J. Chem., 1966, 44, 2115.
[44] Rowley, G. L. and Frankel, M. B. J. Org. Chem., 1969, 34, 1512.
[45] Nekrasova, G. V., Lipina, E. S., Boldysh, E. E., and Perekalin, V. V. Zh. Org. Khim., 1988, 24, 1144.
[46] Braye, E. H. Bull. Soc. Chim. Belg., 1963, 72, 699.
[47] Miller, D. B., Flanagan, P. W., and Shechter, H. J. Org. Chem., 1976, 41, 2112.
[48] Oldekop, Yu. A. and Kaberdin, R. V. Zh. Org. Khim., 1976, 12, 2039; 1979, 15, 1321.
[49] Potkin, V. I., Kaberdin, R. V., and Oldkop, Yu. A. Zh. Org. Khim., 1991, 27, 56.
[50] (a) Aleksiev, D. Zh. Org. Khim., 1976, 12, 2038.
     (b) Drosd, V. N., Knyazev, V. N., Nam, N. L., Yufit, D. S., Struchkov, Y. T., Stankevich, I. V., Chistyakov, A. L., Lezina, V. P., Mozhaeva, T. Ya., and Savelyev, V. V. Tetrahedron, 1992, 48, 469; Knyazev, V. N., Drosd, N. M., Mozhaeva, T. Ya., and Savelyev, V. L. Zh. Org. Khim., 1989, 25, 669.
[51] Yurchenko, O. I., Komarov, N. V., and Dibova, T. N. Zh. Org. Khim., 1976, 12, 230.
[52] Volynskii, V. E., Perekalin, V. V., and Sopova, A. S. Zh. Org. Khim., 1967, 3, 1345; 1970, 6, 938.
[53] Buevich, V. A., Deiko, L. I., and Volynskii, V. E. Zh. Org. Khim., 1980, 16, 2399.
[54] Armand, J. and Convert, O. Collect. Czech. Chem. Commun., 1971, 36, 351.
[55] Campbell, M. M., Cosford, N., Zongli, L., and Sainsbury, M. Tetrahedron, 1987, 43, 1117.
[56] Aleksiev, D. Zh. Org. Khim., 1975, 11, 211, 908; Vestsi Akad. Navuk B. SSR, Ser. Khim. Navuk, 1976, 121; Chem. Abstr., 1976, 85, 5321v.
[57] Sopova, A. S., Perekalin, V. V., and Yurchenko, O. I. Zh. Obshch. Khim., 1963, 33, 2140; 1964, 34, 1188.
[58] Sopova, A. S., Perekalin, V. V., Yurchenko, O. I., and Arnautova, G. M. Zh. Org. Khim., 1969, 5, 858.
[59] Sopova, A. S., Yurchenko, O. I., Perekalin, V. V., and Tkhor, T. G. Zh. Org. Khim., 1971, 7, 820.
[60] Sopova, A. S., Perekalin, V. V., and Semenova, G. K. Zh. Org. Khim., 1967, 3, 1900.
[61] Yurchenko, O. I., Sopova, A. S., Perekalin, V. V., Berestovitskaya, V. M., Polyanskaya, A. S., and Aboskalova, N. I. Dokl. Akad. Nauk SSSR, 1966, 171, 1123.
[62] Sopova, A. S. and Bakova, O. V. Zh. Org. Khim., 1970, 6, 1339; 1972, 8, 934.
[63] Tkhor, T. G., Sopova, A. S., and Ionin, B. I. Zh. Org. Khim., 1976, 12, 648.
[64] Metelkina, E. L., Sopova, A. S., and Ionin, B. I. Zh. Org. Khim., 1972, 8, 2082; 1973, 9, 2204.
[65] Berestovitskaya, V. M., Sopova, A. S., and Perekalin, V. V. Khim. Geterotsikl. Soedin., 1967, 396; Berestovitskaya, V. M., Sopova, A. S., and Perekalin, V. V. Zh. Org. Khim., 1966, 2, 1123; 1967, 3, 1703; Zh. Obshch. Khim., 1963, 33, 2143.
[66] Sopova, A. S., Tkhor, T. G., Perekalin, V. V., and Ionin, B. I. Zh. Org. Khim., 1972, 8, 2301; Tkhor, T. G., Sopova, A. S., and Ionin, B. I. Zh. Org. Khim., 1977, 13, 402, 851.

[67] Dauzonne, D., Josien, H., and Demerseman, P. *Tetrahedron*, 1990, **46**, 7359.
[68] (a) Dauzonne, D. and Demerseman, P. *Synthesis*, 1990, 66.
(b) Dauzonne, D. and Adam-Launay, A. *Tetrahedron*, 1992, **48**, 3069.
[69] Aboskalova, N. I., Polyanskaya, A. S., Perekalin, V. V., Golubkova, N. K., and Paperno, T. Ya. *Zh. Org. Khim.*, 1966, **2**, 2132.
[70] Metelkina, E. L., Sopova, A. S., and Perekalin, V. V. *Zh. Org. Khim.*, 1974, **10**, 209.
[71] Magdesieva, N. N., Sergeeva, T. A., and Kyandzhetsian, R. A. *Zh. Org. Khim.*, 1985, **21**, 1980.
[72] Nasakin, O. E., Lukin, P. M., Zilberg, S. P., Terentyev, P. B., Bulai, A. Kh., Dyachenko, O. A., Zolotoi, A. B., Konovalikhin, S. V., and Atovmyan, L. O. *Zh. Org. Khim.*, 1988, **24**, 997.
[73] Bedford, C. D. and Nielsen, A. T. *J. Org. Chem.*, 1978, **43**, 2460.
[74] Neuman, H. and Seebach, D. *Chem. Ber.*, 1978, **111**, 2785.
[75] Khisamutdinov, G. Kh., Bondarenko, O. A., Kupriyanova, L. A., Klimenko, V. G., and Demina, L. A. *Zh. Org. Khim.*, 1979, **15**, 1307; 1975, **11**, 2445.
[76] Parham, W. E. and Bleasdale, J. L. *J. Am. Chem. Soc.*, 1951, **73**, 4664.
[77] Buevich, V. A. and Nakova, N. *Zh. Org. Khim.*, 1977, **13**, 2619; 1978, **14**, 2229; 1979, **15**, 1473.
[78] Buevich, V. A., Rudchenko, V. V., and Perekalin, V. V. *Zh. Org. Khim.*, 1976, **12**, 910; 1977, **13**, 1383.
[79] Buevich, V. A., Rudchenko, V. V., and Perekalin, V. V. *Khim. Geterotsikl. Soedin.*, 1976, 1429.
[80] Rudchenko, V. V., Buevich, V. A., Grineva, V. S., and Perekalin, V. V. *Khim. Geterotsikl. Soedin.*, 1975, 1576; Buevich, V. A., Grineva, V. S., Rudchenko, V. V., and Perekalin, V. V. *Zh. Org. Khim.*, 1975, **11**, 2620; Buevich, V. A., Rudchenko, V. V., and Perekalin, V. V. *Zh. Org. Khim.*, 1979, **15**, 411.
[81] Buevich, V. A., Nakova, N., Kempter, G., and Perekalin, V. V. *Zh. Org. Khim.*, 1977, **13**, 2618.
[82] Grineva, V. S., Buevich, V. A., and Rudchenko, V. V. *Khim. Geterotsikl. Soedin.*, 1975, 855; Buevich, V. A., Grineva, V. S., and Rudchenko, V. V. *Zh. Org. Khim.*, 1976, **11**, 1768; Buevich, V. A., Rudchenko, V. V., Grineva, V. S., and Perekalin, V. V. *Zh. Org. Khim.*, 1978, **14**, 2179.
[83] Baum, K., Bigelow, S. S., and Nguyen, N. V. *Tetrahedron Lett.*, 1992, **33**, 2141.
[84] Buevich, V. A., Deiko, L. I., and Perekalin, V. V. *Zh. Org. Khim.*, 1977, **13**, 972; *Khim. Geterotsikl. Soedin.*, 1977, 311; Deiko, L. I., Buevich, V. A., Grineva, V. S., and Perekalin, V. V. *Khim. Geterotsikl. Soedin.*, 1975, 1148; *Zh. Org. Khim.*, 1975, **11**, 653.
[85] Buevich, V. A., Deiko, L. I., and Perekalin, V. V. *Zh. Org. Khim.*, 1981, **17**, 1324.
[86] Potkin, V. I., Kaberdin, R. V., and Oldekop, Yu. A. *Vestsi Akad. Navuk B. SSR, Ser. Khim. Navuk*, 1987, 114; *Chem. Abstr.*, 1988, **108**, 130970a.
[87] Oldekop, Yu. A., Kaberdin, R. V., Potkin, V. I., and Shingel, I. A. *Zh. Org. Khim.*, 1979, **15**, 46; 1979, **15**, 1099.
[88] Oldekop, Yu. A., Kaberdin, R. V., Potkin, V. I., and Shingel, I. A. *Zh. Org. Khim.*, 1979, **15**, 276.
[89] Oldekop, Yu. A., Kaberdin, R. V., and Potkin, V. I. *Zh. Org. Khim.*, 1978, **14**, 1594.
[90] Potkin, V. I., Kaberdin, R. V., and Oldekop, Yu. A. *Zh. Org. Khim.*, 1986, **22**, 1389.
[91] Kaberdin, R. V., Potkin, V. I., and Oldekop, Yu. A. *Dokl. Akad. Nauk SSSR*, 1988, **300**, 1153.
[92] Pavlova, Z. F. and Lipina, E. S. *Gertsenovskie Chteniya, Khim. Nauchn. Dokl.*, 1973, **2**, 63; *Chem. Abstr.*, 1974, **81**, 135591m.
[93] Rappoport, Z. and Gazit, A. *J. Org. Chem.*, 1985, **50**, 3184.
[94] Rene, L. and Royer, R. *Synthesis*, 1981, 878.
[95] Kabusz, S. and Tritschler, W. *Synthesis*, 1971, 312.

[96] Kogan, T. P. and Gaeta, F. C. A. *Synthesis*, 1988, 706.
[97] Gariou, M. *C. R. Hebd. Seances Acad. Sci., Ser. C*, 1979, **289**, 255.
[98] Shiga, M., Tsunashima, M., Kono, H., Motoyama, I., and Hata, K. *Bull. Chem. Soc. Jpn*, 1970, **43**, 841.
[99] Kamlet, M. J. *J. Org. Chem.*, 1959, **24**, 714.
[100] Babievskii, K. K. and Belikov, V. M. *Synth. Commun.*, 1977, **7**, 269.
[101] Southwick, P. L., Dufresne, R. F., and Lindsey, J. J. *J. Org. Chem.*, 1974, **39**, 3351.
[102] Babievskii, K. K., Tikhonova, N. A., and Belikov, V. M. *Izv. Akad. Nauk SSSR, Ser. Khim.*, 1969, 2755.
[103] Wolfbeis, O. S. *Chem, Ber.*, 1977, **110**, 2480.
[104] Knippel, E., Knippel, M., Michalik, M., Kelling, H., and Kristen, H. *Z. Chem.*, 1975, **15**, 446.
[105] De La Cuesta, E. and Avendano, C. *J. Heterocycl. Chem.*, 1985, **22**, 337.
[106] Bernasconi, C. F., Fassberg, J., Killion, R. B., and Rappoport, Z. *J. Org. Chem.*, 1990, **55**, 4568.
[107] Bernasconi, C. F., Fassberg, J., Killion, R. B., Schuck, D. F., and Rappoport, Z. *J. Am. Chem. Soc.*, 1991, **113**, 4937.
[108] Gomez-Sanchez, A., Hidalgo, F.-J., and Chiara, J. L. *Carbohydr. Res.*, 1987, **167**, 55.
[109] Goya, P. and Stud, M. *J. Heterocycl. Chem.*, 1978, **15**, 253.
[110] Goya, P., Martines, P., Ochoa, C., and Stud, M. *J. Heterocycl. Chem.*, 1981, **18**, 459.
[111] Babievskii, K. K., Belikov, V. M., and Tikhonova, N. A. *Khim. Geterotsikl. Soedin. Sb. 1, Azotsoderzh. Heterocykli*, 1967, 46; *Chem. Abstr.*, 1969, **70**, 78 343d.
[112] Hengartner, U., Valentine, D., Johnson, K. K., Larscheid, M. E., Pigott, F., Sheide, F., Sott, J. W., Sun, R. C., Townsend, J. M., and Williams, T. H. *J. Org. Chem.*, 1979, **44**, 3741.
[113] Brimble, M. A. and Rowan, D. D. *J. Chem. Soc., Chem. Commun.*, 1988, 978.
[114] Brimble, M. A. and Rowan, D. D. *J. Chem. Soc., Perkin Trans. 1*, 1990, 311.
[115] Bernasconi, C. F., Killion, R. B., Fassberg, J., and Rappoport, Z. *J. Am. Chem. Soc.*, 1989, **111**, 6862.
[116] Bernasconi, C. F., Fassberg, J., Killion, R. B., and Rappoport, Z. *J. Am. Chem. Soc.*, 1990, **112**, 3169.
[117] Southwick, P. L., Fitzgerald, J. A., Malhav, R., and Welsh, D. K. *J. Org. Chem.*, 1969, **34**, 3279.
[118] Suginome, H., Kurokawa, Y., and Orito, K. *Bull. Chem. Soc. Jpn*, 1988, **61**, 4005.
[119] Suginome, H. and Kurokawa, Y. *Bull. Chem. Soc. Jpn*, 1989, **62**, 1107.
[120] Rank, W. *Tetrahedron Lett.*, 1991, **32**, 5353.
[121] Barrett, A. G. M., Cheng, M.-C., Spilling, C. D., and Taylor, S. J. *J. Org. Chem.*, 1989, **54**, 992.
[122] Brade, W. and Vasella, A. *Helv. Chim. Acta*, 1989, **72**, 1649.
[123] Node, M., Kawabata, T., Fujimoto, M., and Fuji, K. *Synthesis*, 1984, 234.
[124] Jung, M. and Grove, D. *J. Chem. Soc. Chem. Commun.*, 1987, 753.
[125] Ono, N., Kamimura, A., and Kaji, A. *J. Org. Chem.*, 1986, **51**, 2139.
[126] Ono, N., Kamimura, A., and Kaji, A. *Tetrahedron Lett.*, 1986, **27**, 1595.
[127] Rappoport, Z. and Topol, A. *J. Am. Chem. Soc.*, 1980, **102**, 406.
[128] Mukhina, E. S., Pavlova, Z. F., Lipina, E. S., and Perekalin, V. V. in *Paper Summaries of the Butlerov Memorial Conference*, Kazan, 1986, Vol. 1, p. 153.
[129] Pavlova, Z. F., Kasem, Ya. A., Lipina, E. S., and Perekalin, V. V. *Zh. Org. Khim.*, 1982, **18**, 2524.
[130] Mukhina, E. S., Pavlova, Z. F., Berkova, G. A., Lipina, E. S., and Perekalin, V. V. *Zh. Org. Khim.*, 1990, **26**, 1447.
[131] Park, K. P. and Ha, H. J. *Bull. Chem. Soc. Jpn*, 1990, **63**, 3006.

[132] Dubois, P. D., Levillain, P., and Viel, C. *C. R. Hebd. Seances Acad. Sci.*, 1980, **290**, 21.

[133] Miyashita, M., Kumazawa, T., and Yoshikoshi, A. *J. Org. Chem.*, 1980, **45**, 2945.

[134] Miyashita, M., Kumazawa, T., and Yoshikoshi, A. *J. Chem. Soc. Chem. Commun.*, 1978, 362.

[135] Banks, B. J., Barrett, A. G. M., and Russell, M. A. *J. Chem. Soc. Chem. Commun.*, 1984, 670.

[136] Barrett, A. G. M., Graboski, G. G., Sabat, M., and Taylor, S. J. *J. Org. Chem.*, 1987, **52**, 4693.

[137] Barrett, A. G. M., Graboski, G. G., and Russell, M. A. *J. Org. Chem.*, 1986, **51**, 1012.

[138] Barrett, A. G. M., Graboski, G. G., and Russell, M. A. *J. Org. Chem.*, 1985, **50**, 2603.

[139] Barrett, A. G. M., Flygare, J. A., and Spilling, Ch. D. *J. Org. Chem.*, 1989, **54**, 4723.

[140] Barrett, A. G. M. and Lebold, S. A. *J. Org. Chem.*, 1990, **55**, 3853.

[141] Ashwell, M., Jackson, R. F. W., and Kirk, J. M. *Tetrahedron*, 1990, **46**, 7429.

[142] Russell, G. and Dedolph, D. *J. Org. Chem.*, 1985, **50**, 3878.

[143] Kamimura, A. and Ono, N. *Synthesis*, 1988, 921.

[144] Kamimura, A. and Nagashima, T. *Synthesis*, 1990, 694.

[145] Retherford, C. and Knochel, P. *Tetrahedron Lett.*, 1991, **32**, 441.

[146] (a) Jubert, C. and Knochel, P. *J. Org. Chem.*, 1992, **57**, 5431.
     (b) Node, M., Itoh, A., Nishide, K., Abe, H., Kawabata, T., Masaki, Y., and Fuji, K. *Synthesis*, 1992, 1119.

[147] Kamimura, A., Sasatani, H., Hashimoto, T., Kawai, T., Hori, K., and Ono, N. *J. Org. Chem.*, 1990, **55**, 2437.

[148] Tominaga, Y., Ychichara, Y., and Hosomi, A. *Heterocycles*, 1988, **27**, 2345.

[149] Fuji, K., Khanapure, S. P., Node, M., Kawabata, T., and Itoh, A. *Tetrahedron Lett.*, 1985, **26**, 779.

[150] Fuji, K., Khanapure, S. P., Node, M., Kawabata, T., Itoh, A., and Masaki, Y. *Tetrahedron*, 1990, **46**, 7393.

[151] Russell, G. R., Yao, C.-F., Tashtoush, H. I., Russell, J. E., and Dedolph, D. F. *J. Org. Chem.*, 1991, **56**, 663.

[152] Miyashita, M., Kumasawa, T., and Yoshikoshi, A. *J. Org. Chem.*, 1984, **49**, 3728.

[153] Yoshikoshi, A. and Miyashita, M. *Acc. Chem. Res.*, 1985, **18**, 284.

[154] Ashwell, M. and Jackson, R. F. W. *J. Chem. Soc., Chem. Commun.*, 1988, 282.

[155] Drozd, V. N., Komarova, E. N., and Dmitriev, L. V. *Zh. Org. Khim.*, 1989, **25**, 2171.

[156] Rajappa, S., Sreenivasan, R., Advani, B. G., Summerville, R. H., and Hoffmann, R. *Indian J. Chem., Sect. B*, 1977, **15**, 297.

[157] Gompper, R. and Schaefer, H. *Chem. Ber.*, 1967, **100**, 591.

[158] Ysaksson, G. and Sandstrom, J. *Acta Chem. Scand.*, 1973, **27**, 1183.

[159] Yensen, K. A., Buchardt, O., and Lohse, Ch. *Acta Chem. Scand.*, 1967, **21**, 2797.

[160] Henriksen, L. and Autrup, H. *Acta Chem. Scand.*, 1970, **24**, 2629.

[161] Manjunatha, S. G., Reddy, K. V., and Rajappa, S. *Tetrahedron Lett.*, 1990, **31**, 1327.

[162] Tominaga, Y. and Matsuda, Y. *J. Heterocycl. Chem.*, 1985, **22**, 937.

[163] *Belg. Pat. 888 727* (Cl. CO7D), 1981; *Chem. Abstr.*, 1982, **96**, 181 127x.

[164] *UK Pat. 2 621 092*, 1976; *Chem. Abstr.*, 1977, **86**, 72 655r.

[165] Rajappa, S. *Tetrahedron*, 1981, **37**, 1453.

[166] Price, B. J. and Jack, D. *Eur. Pat. 58 492* (Cl. CO7 D 295/12), 1982; *Chem. Abstr.*, 1983, **98**, 16 574z.

[167] Rudorf, W. D. *Z. Chem.*, 1979, **19**, 100.

[168] Schäfer, H. and Gewald, K. *Z. Chem.*, 1975, **15**, 100.

[169] Rajappa, S. and Sreenivasan, R. *Indian J. Chem., Sect. B*, 1985, **24**, 795.

[170] Deshmukh, A. R. A. S., Bhawal, B. M., Shiralkar, V. P., and Rajappa, S. *US Pat.* 4 967 007 (Cl. 564–501, CO7 C 321/00), 1990; *Chem. Abstr.*, 1991, **114**, 142 666t.
[171] Deshmukh, A. R. A. S., Reddy, T. I., Bhawal, B. M., Shiralkar, V. P., and Rajappa, S. *J. Chem. Soc., Perkin Trans. 1*, 1990, 1217.
[172] Maybhate, S. P., Deshmukh, A. R. A. S., and Rajappa, S. *Tetrahedron*, 1991, **47**, 3887.
[173] Kiprianov, A. I. and Verbovskaya, T. M. *Zh. Obshch. Khim.*, 1962, **32**, 3703.
[174] Rajappa, S., Nagarajan, K., Venkatesan, K., Kamath, N., Padmanabhan, V. M., Philipsborn, V., Chen Ban Chin, and Muller, R. *Helv. Chim. Acta*, 1984, **67**, 1669.
[175] Rajappa, S., Bhawal, B. M., Deshmukh, A. R. A. S., Manjunatha, S. G., and Chandrasekhar, J. *J. Chem. Soc., Chem. Commun.*, 1989, 1729.
[176] Börner, A., Krister, H., Peseke, K., and Michalik, M. *J. Prakt. Chem.*, 1986, **328**, 21.
[177] Schäfer, H., Bartho, B., and Gewald, K. *Z. Chem.*, 1973, **13**, 294; Schäfer, H. and Gewald, K. *Z. Chem.*, 1976, **16**, 272.
[178] Kolb, M. *Synthesis*, 1990, 171.
[179] Tominaga, Y., Shiroshita, Y., and Hosomi, A. *J. Heterocycl. Chem.*, 1988, **25**, 1745.
[180] Kumar, A., Ila, H., and Junjappa, H. *J. Chem. Soc., Chem. Commun.*, 1976, 593.
[181] Mertens, H., Troschütz, R., and Roth, H. J. *Liebigs Ann. Chem.*, 1986, 380.
[182] Schäfer, H., Geward, K., and Seifert, M. *J. Prakt Chem.*, 1976, **318**, 39.
[183] Peseke, K. and Suarez, J. Q. *Z. Chem.*, 1981, **21**, 405.
[184] Manjunatha, S. G., Chittari, P., and Rajappa, S. *Helv. Chim. Acta*, 1991, **74**, 1071.
[185] Mukhina, E. S., Pavlova, Z. F., Berkova, G. A., Lipina, E. S., Mostyaeva, L. V., and Perekalin, V. V. *Zh. Org. Khim.*, 1991, **27**, 910.
[186] Mukhina, E. S., Pavlova, Z. F., Nekrasova, G. V., Lipina, E. S., and Perekalin, V. V. *Zh. Org. Khim.*, 1990, **26**, 2285.
[187] Fuji, K., Node, M., Abe, H., Itoh, A., Masaki, Y., and Shiro, M. *Tetrahedron Lett.*, 1990, **31**, 2419.
[188] Fuji, K., Tanaka, K., Abe, H., Itoh, A., Node, M., Taga, T., Miwa, Y., and Shiro, M. *Tetrahedron Asymm.*, 1991, **2**, 179.
[189] Fuji, K., Tanaka, K., Abe, H., Itoh, A., Node, M., Taga, T., Miwa, Y., and Shiro, M. *Tetrahedron Asymm.*, 1991, **2**, 1319.
[190] (a) Fuji, K., Tanaka, K., Abe, H., Matsumoto, K., Taga, T., and Miwa, Y. *Tetrahedron Asymm.*, 1992, **3**, 609.
(b) Fuji, K. and Node, M. *Synlett*, 1991, 601.
[191] Ono, N., Kamimura, A., and Kaji, A. *J. Org. Chem.*, 1988, **53**, 251.
[192] Pavlova, Z. F., Kasem, Ya. A., Lipina, E. S., Berkova, G. A., and Perekalin, V. V. *Zh. Org. Khim.*, 1985, **21**, 2300.
[193] Oldekop, Yu. A., Kaberdin, R. V., and Potkin, V. I. *Zh. Org. Khim.*, 1980, **16**, 543.
[194] Freeman, J. F. and Emmons, W. D. *J. Am. Chem. Soc.*, 1956, **78**, 3405.
[195] Hurd, C. D. and Nilson, M. E. *J. Org. Chem.*, 1955, **20**, 927.
[196] Ulbricht, T. L. V. and Price, C. C. *J. Org. Chem.*, 1957, **22**, 235.
[197] Freeman, J. P. and Parker, C. O. *J. Org. Chem.*, 1956, **21**, 579.
[198] Severin, T. and Böhme, H. J. *Chem. Ber.*, 1968, **101**, 2925.
[199] Glushkov, R. G., Maslova, M. M., Marchenko, N. B., and Bolshakov, V. I. *Khim. Pharm. Zh.*, 1991, **25**, 61.
[200] Babievskii, K. K., Belikov, V. M., and Tikhonova, N. A. *Izv. Akad. Nauk SSSR, Ser. Khim.*, 1970, 1161.
[201] Bredereck, H., Kantehner, W., and Scweiser, D. *Chem. Ber.*, 1971, **104**, 3475.
[202] Maslova, M. M., Marchenko, N. B., and Glushkov, R. G. *Khim. Pharm. Zh.*, 1991, **25**, 62.
[203] Laikhter, A. L., Cherkasova, T. I., Melnikova, L. G., Ugrak, B. I., Fainzilberg, A. A., and Semenov, V. V. *Izv. Akad. Nauk SSSR, Ser. Khim.* 1991, 1849.

[204] Granik, V. G. and Glushkov, R. G. *Zh. Org. Khim.*, 1971, **7**, 1146.
[205] Severin, T. and Brûck, B. *Chem. Ber.*, 1965, **98**, 3847.
[206] Severin, T. and Kullmer, H. *Chem. Ber.*, 1971, **104**, 440.
[207] Faulgues, M., Rene, L., and Royev, R. *Synthesis*, 1982, 260.
[208] Krowczynski, A. and Koserski, L. *Synthesis*, 1983, 489.
[209] Bedford, C. D. and Nielsen A. T. *J. Org. Chem.*, 1979, **44**, 633.
[210] Jain, S., Sujatha, K., Rama Krishna, K. V., Roy, R., Singh, J., and Anand, N, *Tetrahedron*, 1992, **48**, 4985.
[211] Prystas, M. and Gut, J. *Collect. Czech. Chem. Commun.*, 1963, **28**, 2501.
[212] Belon, J.-P. and Perrot, R. *Bull. Soc. Chim. Fr.*, 1977, 329.
[213] Dubois, P., Levillain, P., and Viel, C. *C. R. Hebd. Seances Acad. Sci., Ser. C*, 1979, **288**, 311.
[214] Dubois, P., Levillain, P., and Viel, C. *Talanta*, 1981, **28**, 843.
[215] Dubois, P., Levillain, P., and Viel, C. *Bull. Soc. Chim. Fr.*, 1986, 297.
[216] Todres, Z. V., Diusengaliev, K. I., and Garbusova, I. A. *Zh. Org. Khim.*, 1986, **23**, 370.
[217] Lipina, E. S., Pavlova, Z. F., and Prikhodko, L. V. *Zh. Org. Khim.*, 1970, **6**, 1123.
[218] Allade, I., Dubois, P., Levillain, P., and Viel, C. *Bull. Soc. Chim. Fr.*, 1983, 339.
[219] Korotaev, S. V., Ermolov, D. S., Todres, Z. V., and Malievskii, A. D. *Izv. Akad. Nauk SSSR, Ser. Khim.*, 1992, 78.
[220] Marchetti, L. and Passalacqua, V. *Ann. Chim. (Rome)*, 1967, **57**, 1266; *Chem. Abstr.*, 1968, **68**, 95 420.
[221] Fetell, A. and Feuer, H. *J. Org. Chem.*, 1978, **43**, 1238.
[222] Ostercamp, D. L. and Taylor, P. *J. Chem. Soc., Perkin Trans. 2*, 1985, 1021.
[223] Fuji, K., Node, M., Nagasawa, H., Naniwa, Y., Taga, T., Machida, K., and Snatzke, C. *J. Am. Chem. Soc.*, 1989, **111**, 7921.
[224] Fetell, A. I. and Feuer, H. *J. Org. Chem.*, 1978, **43**, 497.
[225] Feuer, H. and McMillan, R. *J. Org. Chem.*, 1979, **44**, 3410.
[226] Alberola, A., Antolin, L. F., Gonzalez, A., Laguna, M. A., and Pulido, F. J. *Heterocycles*, 1987, **25**, 393.
[227] Devincenzis, G., Mencarelli, P., and Stegel, F. *J. Org. Chem.*, 1983, **48**, 162.
[228] Dell'Erba, C., Novi, M., Petrillo, G., and Stagnaro, P. *Tetrahedron Lett.*, 1990, **31**, 4933.
[229] Dell'Erba, C., Mele, A., Novi, M., Petrillo, G., and Stagnaro, P. *Tetrahedron*, 1992, **48**, 4407.
[230] Rusinov, V. L., Timashov, A. A., Pilicheva, T. L., and Chupakhin, O. I. *Zh. Org. Khim.*, 1991, **27**, 1100.
[231] Hassner, A. and Chau, W. *Tetrahedron Lett.*, 1982, **23**, 1989.
[232] Hurd, C. D. and Sherwood, L. T. *J. Org. Chem.*, 1948, **13**, 471.
[233] Bavluenga, J., Aznav, F., Lir, R., and Bayod, M. *Synthesis*, 1983, 159.
[234] Severin, T. and Ipach, I. *Chem. Ber.*, 1978, **111**, 692.
[235] Kvitko, S. M. and Perekalin, V. V. *Dokl. Akad. Nauk SSSR*, 1962, **143**, 345.
[236] Cato, E. N., Meek, M. A., Schwalbe, C. H., Stevens, M. F. G., and Threadgill, M. D. *J. Chem. Soc., Perkin Trans. 1*, 1985, 251.
[237] Marcos, E. S., Maraver, J. J., Chiara, J. L., and Gomez-Sanchez, A. *J. Chem. Soc., Perkin Trans. 2*, 1988, 2059.
[238] Chiara, J. L., Gomez-Sanchez, A., Marcos, E. S., and Bellanato, J. *J. Chem. Soc., Perkin Trans. 2*, 1990, 385.
[239] Chiara, J. L., Gomez-Sanchez, A., and Hidalgo, F.-J. *J. Chem. Soc., Perkin Trans. 2*, 1988, 1691.
[240] Pappalardo, R. R. and Marcos, E. S. *J. Chem. Soc., Faraday Trans.*, 1991, **87**, 1719.

[241] Chiara, J. L., Gomez-Sanchez, A., and Bellanato, J. *J. Chem. Soc., Perkin Trans.* 2, 1992, 787.

[242] Bakhmutov, V. I., Burmistrov, V. A., Babievskii, K. K., Belikov, V. M., and Fedin, E. I. *Izv. Akad. Nauk SSSR, Ser. Khim.*, 1977, 2820.

[243] Bakhmutov, V. I., Babievskii, K. K., Burmistrov, V. A., Fedin, E. I., and Belikov, V. M. *Izv. Akad. Nauk SSSR, Ser. Khim.*, 1978, 2719.

[244] Borisov, Yu. A., Babievskii, K. K., Bakhmutov, V. I., Struchkov, Yu. T., and Fedin, E. I. *Izv. Akad. Nauk SSSR, Ser. Khim.*, 1982, 123.

[245] Tokumitsu, T. and Hayashi, T. *J. Org. Chem.*, 1985, **50**, 1547.

[246] Tokumitsu, T. *Bull. Chem. Soc. Jpn*, 1986, **59**, 3871.

[247] Lyubchanskaya, V. M. and Granik, V. G. *Khim. Geterotsikl. Soedin.*, 1990, 597.

[248] Krowozynski, A. and Kozerski, L. *Heterocycles*, 1986, **24**, 1209.

[249] Krowozynski, A. and Kozerski, L. *Bull. Soc. Chim. Polon.*, 1986, **34**, 341.

[250] Gomez-Sanchez, A., Hidalgo, F.-J., and Chiara, J. L. *J. Heterocycl. Chem.*, 1987, **24**, 1757.

[251] Rene, L. *Synthesis*, 1989, 69.

[252] Severin, T., Scheel, D., and Adhikary, P. *Chem. Ber.*, 1969, **102**, 2966.

[253] Severin, T., Bräutigam, I., and Bräutigam, K.-H. *Chem. Ber.*, 1977, **110**, 1669.

[254] Schäfer, H. and Gewald, K. *Z. Chem.*, 1983, **5**, 179.

[255] Genkina, N. K., Ipatkin, V. V., and Kurkovskaya, L. N. *Zh. Org. Khim.*, 1991, **5**, 1105.

[256] Molina, P. and Vilaplana, M. J. *Synthesis*, 1990, 474.

[257] Node, M., Nagasawa, H., Naniwa, Y., and Fuji, K. *Synthesis*, 1987, 729.

[258] Fuji, K., Node, M., Nagasawa, H., Naniwa, Y., and Terada, S. *J. Am. Chem. Soc.*, 1986, **108**, 3855.

[259] Fuji, K. and Node, M. *Synlett*, 1991, 603.

[260] Schäfer, H., Bartho, B., and Gewald, K. *J. Prakt. Chem.*, 1977, **B319**, 149.

[261] Troschuetz, R. and Lueckel, A. *Arch. Pharm.*, 1991, **324**, 73.

[262] Dorokhov, V. A., Gordeev, M. F., Bogdanov, V. S., Laikhter, A. L., Kislyi, V. P., and Semenov, V. V. *Izv. Akad. Nauk SSSR, Ser. Khim.*, 1990, 2660.

[263] Young, R. C., Ganellin, C. R., Graham, M. J., and Grant, E. H. *Tetrahedron*, 1982, **38**, 1493.

[264] Rajappa, S. and Sreenivasan, R. *Indian J. Chem., Sect. B*, 1977, **15**, 301.

[265] Schäfer, H. and Gewald, K. *J. Prakt. Chem.*, 1980, **322**, 87.

[266] Schäfer, H. and gewald, K. *Z. Chem.*, 1978, **18**, 335.

[267] Takahashi, M., Nazaki, C., and Shibazaki, Y. *Chem. Lett.*, 1987, 1229.

[268] Schäfer, H., Gruner, M., Grossmann, G., and Gewald, K. *Monatsh. Chem.*, 1991, **122**, 959.

[269] Baum, K., Nguyen, N. V., Gilardi, R., Flippen-Anderson, J. L., and George, C. *J. Org. Chem.*, 1992, **57**, 3026.

[270] Bezmenova, T. E. *Fiz. Aktiv. Veshch.*, 1985, **17**, 3.

[271] Arbuzov, B. A., Vizel, A. O., and Giniiatullin, R. S. *Khim. Pharm. Zh.*, 1973, **11**, 27.

[272] Quin, L. D. and Parket, J. P. *J. Chem. Soc., Chem. Commun.*, 1967, 914.

[273] Monagle, J. J. *J. Org. Chem.*, 1962, **27**, 3851; Campbell, T. W. and Monagle, J. J. *J. Am. Chem. Soc.*, 1962, **84**, 1493.

[274] Mathey, F. *Tetrahedron*, 1974, **30**, 3127.

[275] Chernyshev, E. A., Komalenkova, S. A., and Bashkirova, S. A. *Usp. Khim.*, 1976, **45**, 1482.

[276] Arbuzov, B. A., Vizel, A. O., Zvereva, M. A., Studentsova, I. A., and Gareev, R. S. *Izv. Akad. Nauk SSSR, Ser. Khim.*, 1966, 1482.

[277] Monagle, J. J. *J. Org. Chem.*, 1962, **27**, 3851.

[278] Guberman, F. S., Bakhitov, M. I., and Kuznetsova, E. V. *Zh. Obshch. Khim.*, 1974, **44**, 757.

[279] Hunger, K. *Tetrahedron Lett.*, 1966, **47**, 5929.

[280] Berestovitskaya, V. M. and Perekalin, V. V. *Zh. Obshch. Khim.*, 1993, **63**, 261.

[281] Efremov, D. A., Berestovitskaya, V. M., and Perekalin, V. V. *Mezhvuz. Sborn. Nauchn. Tr. Leningr. Tekhn. Inst., Leningrad*, 1980, 98; *Chem. Abstr.*, 1981, 157 082s.

[282] Trukhin, E. V., Berestovitskaya, V. M., and Perekalin, V. V. in *Siliconorganic Compounds and Materials on Their Base*, Nauka, Leningrad, 1984, p. 28.

[283] (a) Titova, M. V., Berestovitskaya, V. M., and Perekalin, V. V. *Zh. Org. Khim.*, 1981, **17**, 1322.

(b) Efremova, I. E., Berestovitskaya, V. M., and Berkova, G. A. *Zh. Obshch. Khim.*, 1984, **53**, 530.

(c) Perekalin, V. V., Berestovitskaya, V. M., Trukhin, E. V., Komalenkova, N. G., Bashkirova, S. A., and Chernyshov, E. A. *Zh. Obshch. Khim.*, 1982, **52**, 922.

[284] Perekalin, V. V., Berestovitskaya, V. M., Speranskii, E. M., and Sulimov, I. G. *Zh Org. Khim.*, 1971, **7**, 2630; Berestovitskaya, V. M., Speranskii, E. M., and Perekalin, V. V. *Zh. Org. Khim.*, 1976, **12**, 2256.

[285] Berestovitskaya, V. M., Speranskii, E. M., and Perekalin, V. V. *Zh. Org. Khim.*, 1977, **13**, 1934.

[286] Speranskii, E. M., Berestovitskaya, V. M., and Perekalin, V. V. *Zh. Org. Khim.*, 1974, **10**, 875.

[287] Berestovitskaya, V. M., Lyamina, T. N., and Efremova, I. E. *Zh. Org. Khim.*, 1983, **19**, 1767.

[288] Berestovitskaya, V. M., Trukhin, E. V., and Perekalin, V. V. *Zh. Org. Khim.*, 1976, **12**, 911.

[289] Efremov, D. A., Berestovitskaya, V. M., and Perekalin, V. V. *Zh. Org. Khim.*, 1976, **12**, 912.

[290] Efremov, D. A., Berestovitskaya, V. M., and Perekalin, V. V. *Zh. Obshch. Khim.*, 1979, **49**, 946.

[291] Berestovitskaya, V. M., Efremov, D. A., Berkova, G. A., Perekalin, V. V., and Zakharov, V. I. *Zh. Obshch. Khim.*, 1980, **50**, 2680.

[292] Trukhin, E. V., Berestovitskaya, V. M., and Perekalin, V. V. *Zh. Obshch. Khim.*, 1981, **51**, 708.

[293] Trukhin, E. V., Berestovitskaya, V. M., and Perekalin, V. V. *Zh. Obshch. Khim.*, 1982, **52**, 1167.

[294] Berestovitskaya, V. M., Efremov, D. A., and Perekalin, V. V. *Zh. Obshch. Khim.*, 1979, **49**, 2390.

[295] Berestovitskaya, V. M., Efremov, D. A., Berkova, G. A., and Perekalin, V. V. *Zh. Obshch. Khim.*, 1981, **51**, 2418.

[296] Efremova, I. F., Berestovitskaya, V. M., and Perekalin, V. V. *Zh. Org. Khim.*, 1984, **20**, 890.

[297] Tkachev, V. V., Atovmyan, L. O., Berestovitskaya, V. M., and Efremov, D. A. *Zh. Strukt. Khim.*, 1988, **29**, 112.

[298] Atovmyan, L. O., Tkachev, V. V., Zolotoi, A. B., and Berestovitskaya, V. M. *Dokl. Akad. Nauk SSSR*, 1986, **291**, 1389.

[299] Atovmyan, L. O., Tkachev, V. V., Atovmyan, E. G., Berestovitskaya, V. M., and Titova, M. V. *Izv. Akad. Nauk SSSR, Ser. Khim.*, 1989, 2312.

[300] Titova, M. V., Berestovitskaya, V. M., and Perekalin, V. V. *Zh. Org. Khim.*, 1979, **15**, 877.

[301] Berestovitskaya, V. M., Titova, M. V., Paperno, T. Ya., and Perekalin, V. V. *Zh. Org. Khim.*, 1984, **30**, 2383.

[302] Berestovitskaya, V. M., Titova, M. V., and Perekalin, V. V. *Zh. Org. Khim.*, 1977, **13**, 2454; 1982, **18**, 1783.

[303] Berestovitskaya, V. M., Speranskii, E. M., and Perekalin, V. V. *Zh. Org. Khim.*, 1979, **15**, 185.

[304] Trukhin, E. V., Berestovitskaya, V. M., and Perekalin, V. V. *Mezhvuz. Sborn. Sint. Str. Khim. Prevraschen Soedin. Azota*, 1993, 3.

[305] Berestovitskaya, V. M., Efremova, I. E., and Berkova, G. A. *Zh. Org. Khim.*, 1993, **29**, in press.

[306] Titova, M. V. and Berestovitskaya, V. M. *Mezhvuz. Sborn. Sint. Str. Khim. Prevraschen Soedin. Azota*, 1993, 10.

[307] Berestovitskaya, V. M., Efremova, I. E., Titova, M. V., Vasilieva, M. V., and Perekalin, V. V. *Tez. Dokl. Vses. Soveshch. Khim. Primenen. Org. Soedin. Ser. Kazan*, 1987, 42.

[308] Berestovitskaya, V. M., Bundule, M. F., Bleidelis, Ya. Ya., and Efremova, I. E. *Zh. Obshch. Khim.*, 1984, **54**, 1182.

[309] Berestovitskaya, V. M., Bundule, M. F., Bleidelis, Ya. Ya., and Efremova, I. E. *Zh. Obshch. Khim.*, 1986, **56**, 375.

[310] Berestovitskaya, V. M., Speranskii, E. M., Perekalin, V. V., and Trukhin, E. V. *Zh. Org. Khim.*, 1974, **10**, 1783.

[311] Efremova, I. E. and Berestovitskaya, V. M. *Zh. Org. Khim.*, 1985, **21**, 1140.

[312] Pozdnyakov, V. P., Ratovskii, G. V., Berestovitskaya, V. M., Chuvashev, D. D., and Efremova, I. E. *Zh. Obshch. Khim.*, 1989, **59**, 175.

[313] Berestovitskaya, V. M., Titova, M. V., Efremova, I. E., and Perekalin, V. V. *Zh. Org. Khim.*, 1986, **21**, 2463.

[314] Speranskii, E. M., Berestovitskaya, V. M., Sulimov, I. G., and Trukhin, E. V. *Zh. Org. Khim.*, 1972, **8**, 1763.

[315] Berestovitskaya, V. M., Titova, M. V., and Perekalin, V. V. *Zh. Org. Khim.*, 1980, **16**, 891.

[316] Pivovarov, A. B., Titova, M. V., Berestovitskaya, V. M., and Perekalin, V. V. *Zh. Org. Khim.*, 1989, **24**, 215.

[317] Vasilieva, M. V., Berestovitskaya, V. M., Berkova, G. A., and Pozdnyakov, V. P. *Zh. Org. Khim.*, 1986, **22**, 428.

[318] Berestovitskaya, V. M., Yakovleva, O. G., Latypova, V. Z., Pozdnyakov, V. P., Vasilieva, M. V., Zolotoi, A. B., and Kargina, N. M. *Zh. Obshch. Khim.*, 1991, 61.

[319] Vasilieva, M. V., Berestovitskaya, V. M., and Perekalin, V. V. *Zh. Org. Khim.*, 1985, **21**, 1580.

[320] Vasilieva, M. V., Berkova, G. A., and Perekalin, V. V. *Zh. Org. Khim.*, 1988, **24**, 436.

[321] Berestovitskaya, V. M., Titova, M. V., Vasilieva, M. V., Lyamina, T. N., and Perekalin, V. V. *Tez. Dokl. Vses. Konf. Khim. Sredstvam Zashchity Rastenii, Sekts 1, Ufa*, 1982, 160.

[322] Berestovitskaya, V. M., Trukhin, E. V., Efremova, I. E., Pivovarov, A. B., and Perekalin, V. V. *Tez. Vses. Konf. Fiz. Aktiv. Veshch., Chernogolovka*, 1989, 37.

SUBJECT INDEX